# Lecture Notes in Control and Information Sciences

Edited by M. Thoma and A. Wyner

For information about Vols. 1-96 please contact your bookseller or Springer-Verlag

# Lecture Notes in Control and Information Sciences

Edited by M. Thoma and A. Wyner

162

C. Canudas de Wit (Ed.)

## Advanced Robot Control

Proceedings of the International Workshop
on Nonlinear and Adaptive Control: Issues
in Robotics, Grenoble, France, Nov. 21-23, 1990

Springer-Verlag Berlin Heidelberg GmbH

**Editor**

Carlos Canudas de Wit
Laboratoire d'Automatique de Grenoble
B.P. 46, ENSIEG-INPG (CNRS u. a. 228)
38402, St. Martin d'Hères
France

ISBN 978-3-540-54169-1    ISBN 978-3-540-47479-1 (eBook)
DOI 10.1007/978-3-540-47479-1

# About the Editor

**Carlos Canudas de Wit** was born in Villahermosa, Tabasco, Mexico, in 1958. He received his B.Sc. degree in electronics and communications from the Technologic of Monterey, Mexico in 1980. From 1981 to 1982 he worked as a research engineer at the Department of Electrical Engineering at the CINVESTAV- del IPN, in Mexico City. In 1984 he received his M.Sc. in the Department of Automatic Control, Grenoble, France. He was visiting researcher in 1985 at The Lund Institute of Technology, Sweden. In 1987 he received his Ph.D. in automatic control from the Polytechnic of Grenoble (Department of Automatic Control), France. Since then he has been working at the same department as a researcher associated with the CNRS, where he teaches and conducts research in the area of adaptive control and robotic control. Dr. Canudas de Wit has written in 1988 a book on Adaptive Control for Partially Known Systems: Theory and Applications (Elsevier Publisher).

# Preface

Fine motion control in robot manipulators became a desired goal in the last decade as the result of new robot morphology and the definition of new tasks involving high velocity motions and end effector tracking precision. Typical examples of the new robot structure conception are the lightweight arms used in space applications. Elasticity of the body is thus explicitly modelled and hence new control design is required. Similarly, the definition of new tasks involving high speed and requiring a high degree of precision render obsolete and inadequate control schemes based on linear and decoupled models.

The need for advanced controllers fulfilling the previous goals ( which can be called model-based controllers) have propitiated the cross-fertilization among the control and robotics communities. Some control techniques, such as nonlinear and adaptive theory, have been applied to solve control problems in the robotic field. Adaptation is, in general, required as a mean to counteract model parameter uncertainties. Nonlinear control theory appears as a natural control design approach due to the nonlinear nature of the mechanical systems under consideration. Although development and research in these two fields are constantly evolving evolution, these techniques ( as well as other related approaches: observer based controllers, learning systems, etc. ) have already be used with success to solve several robot control problems.

The papers included in this book were presented at the "International Workshop on Adaptive and Nonlinear Control: Issues in Robotics", which was held in November 21-23, in Grenoble, France. The aim of the workshop was to present a sample of the most recent contributions in the field of robot control. The worshop was organized by the European Laboratory Network (ELN) in Adaptive and Nonlinear Systems: Application to Robotics", with funds from MNRT-France "Ministere Nationale de la Recherche et la Technologie". This network was founded in September 1989 under the support of the MNRT gathering some of the European laboratories working in adaptive and nonlinear control theory and interested in robotics applications. Member of the first generation of this ENL are : Universita di Roma " La Sapienza"(Italy), Seconta Universita di Roma "Tor Vergata"(Italy), Lund Institute of Technology (Sweden), Laboratoire d'Automatique et Systémes (Belgium), IRIA Sophia-Antipolis (France) and the Laboratoire d'Automatique de Grenoble (France).

The book is organized by subjects. Although the treatment of these subjects is not exhaustive, the contributions presented here give a sample of the state of the art in these themes.

*Adaptation and Learning.* Some control schemes using adaptation and learning are first presented. Adaptation and learning procedures are useful when model parameters are uncertain or unknown. Adaptation laws have been studied exploiting the fact the nonlinear robot model dynamics is linear in the parameters. Learning is an alternative procedure to identify a error sequence in repetitive robot tasks.

*Control of robotic systems with nonholonomic constraints.* A typical example of such a system is a mobile robot having only two degrees of freedom when moving in the *x-y* plane. The main difficulty of controlling this type of system is that, although they are locally controllable, it is not possible to find a smooth state feedback law which asymptotically stabilizes the cart model. Some solutions based on time-varying and nonlinear controllers are presented.

*Control in the task space.* Controlling the robot end-effector in the so-called task space ( in which the task is specified) represents the final goal of the control designer. As such, the mathematical functions describing the mapping between joint coordinates and task space need to be considered. Methods to include additional sensor information in the control loops and the introduction of new concepts of active impedance are here presented.

*Control of flexible robots.* Mechanical flexibility, either due to joint trasmissions ( industrials robots carrying heavy loads involving high-velocity motions are a typical example) or to the link deformations (like the lightweight robots used in space applications) represent one of the most important limiting factors to perform accurate motion. Techniques such as feedback linearization and nonlinear decoupling, are applied to these systems. Robustness of such controllers under the problem of parameters inaccuracies are also analysed.

*Observer based controllers.* The problem of substituting additional sensors, like joint velocity measurements in the case of robot with joint flexibilities, or simply motor velocity in the case of rigid robot, by nonlinear observers is also treated. The main difficulty in this particular setup is that the "separation" principle which consists in designing separately the observer and the controller loops does not apply for nonlinear systems. An additional stability analysis including both control and observer equations is thus required. Model parameter inaccuracies give rise to additional difficulties. An example of a observer-based control design using adaptation is given in the section of Adaptation and Learning.

*Robot control under kinematic singularities.* One major difficulty in formulating the control problem in the Cartesian space is due to kinematic singularities. The correct understanding of the treatment of such singularities is fundamental not only for performing feedback control in the work space but also to realize tasks in the framework of force and position control. In here, the impact of kinematic singularities in the feedback control design and its relation with the concepts of nonlinear controllability are analysed.

The three day workshop was an excellent opportunity to exchange ideas and discus common topics. The nice environment of Grenoble with its beautiful mountains combined with the exoticism of the French cuisine and wines propitiated this exchange. I would like to thank the members of the program committee, M. Di Benedetto, F. Nicolo, L. Nielsen, C. Samson, G. Campion, P. Tomei and L. Dugard for making this event possible. I am also indebted to M. Spong, A. De Luca, B. Espiau, S. Nicosia, Y. Nakamura and S. Arimoto for accepting to participate in this conference and for giving interesting tutorial talks. The local organization was an important support. I would like to thank A. Aubin and S. Seleme for helping in the workshop preparation and M.R. Choisy and M.T. Decotes-Genon for ensuring an efficient organization and a warm reception. The laboratory of Automatic Control of Grenoble (LAG), which belongs to the Polytechnic Institute of Grenoble (INPG) and is associated to the National French Research Foundation ( CRNS), played the roll of host in the workshop organization. Thanks are also due to I.D. Landau, director of the LAG and to M. Garnier vice-director of the INPG for supporting us in this effort. Finally, I would like to thank the society ALEPH-Technologies for kindly accepting to show their robotics activities and developments to the workshop attendees and to the MNRT-France for its financial support.

Grenoble, France, November 1990

Carlos Canudas de Wit
(Laboratory of Automatic Control of Grenoble)

# Table of Contents

## Adaptation and Learning

## Control of robotic systems with nonholonomic constraints.

## Control in the task space.

## Control of flexible robots.

## Observer based control

# Robot control under kinematic singularities

# Robustness of Adaptive Control of Robots: Theory and Experiment *

Fathi Ghorbel

Alan Fitzmorris

Mark W. Spong

Coordinated Science Laboratory

University of Illinois at Urbana-Champaign

1101 W. Springfield Ave.

Urbana, Ill. 61801

### Abstract

It is well known in adaptive control theory that the performance of adaptive controllers can be highly sensitive to the modeling assumptions used to prove convergence. In this paper we discuss the robustness of adaptive control of rigid robots and methods for improving robustness in the face of unmodeled dynamics and external disturbances. Both theoretical and experimental results are presented. Robustness is achieved by modifying the rigid control algorithm in two important ways. First, the rigid robot control law is incorporated into a composite slow/fast control law by adding to it a "fast" control to damp the joint oscillations. Second, so-called $\sigma$-modification is used to ensure boundedness of the estimated parameters.

## 1   Introduction

One of the goals of robotics research is to develop so-called intelligent robots which are capable of adapting their behavior to uncertainties in their environment. Uncertainties arise from many

---

*Research partially supported by the University of Illinois Manufacturing Research Center under Grant No. UFAS 1-5-80405.

sources; unknown loads, grinding forces, part misalignment in assembly, time delays in teleoperation, unknown terrain in mobile robots, etc. Other important sources of uncertainty include uncertainties in the dynamic description of the robot itself, for example, when a "rigid" control algorithm is applied to a flexible robot. For this reason the application of adaptive control techniques in robotics has been an area of intense interest.

At the present time there are a number of "provably correct" adaptive algorithms for motion and force control of rigid robots. By "provably correct" we mean an adaptive control algorithm that uses the full Lagrangian dynamic model (considering the robot as a chain of coupled rigid bodies) and which can be proven to be globally convergent, i.e., the position and velocity tracking errors converge asymptotically to zero with all internal signals (input torque, estimation error, etc.) remaining bounded. A recent tutorial[14] contains details of seven such globally convergent adaptive algorithms. In the meantime, several additional results have been published, including results on exponential stability[25], persistency of excitation[26], improved Lyapunov arguments[27], etc. The result is that rigid robot dynamics are now well understood with respect to the design of adaptive control algorithms.

It is known, however, that the stability of adaptive systems can be highly sensitive to disturbances and unmodeled dynamics. These arise in the robotics context from several sources. External disturbances include many types of interaction with the environment. For example, robotic assembly has been described as a sequence of controlled collisions with the environment. These collision forces can be viewed as disturbances to the controller. A repetitive task, for example, subjects the robot to periodic forcing which, even in non-adaptive control, can excite complex nonlinear dynamic behavior, such as period doubling bifurcations and chaos[29].

Unmodeled dynamics include actuator/sensor dynamics, joint flexibility, link flexibility, and environment dynamics. Of these, joint flexibility is dominant in most manipulator designs. Environment dynamics arise in force and impedance controlled tasks such as assembly and grinding and will become increasingly important in future applications.

Several so-called "instability mechanisms" in adaptive control have been identified[10]. In the robotics context these instability mechanisms may manifest themselves when a rigid model is used as a basis for the design of an adaptive control algorithm. Among the mechanisms leading to instability are:

- 1) Reference trajectories which are "too fast." In other words, if the bandwidth of the reference trajectory is in the same frequency range as the unmodeled dynamics, these dynamics can be excited and drive the system unstable.

- 2) Parameter drift. The estimated parameters do not necessarily converge to their true values even in the ideal case without persistency of excitation conditions on the reference signal. In the presence of unmodeled dynamics, or in the presence of external disturbances, the parameters can drift along an equilibrium manifold until an instability results [15].

- 3) High Gain instability. This type of instability, when the controller gains are too high, is actually due to the loss of passivity from the ideal case and can occur even for non–adaptive algorithms [1].

- 4) Fast adaptation instability. This type of instability occurs when the gains in the parameter update law are too large. Due to the complicated nonlinear structure of robotic systems there are few design rules that can be called upon to design these various gains. At present the choice of such gains in adaptive robot control is largely a trial and error process.

In this paper we restrict our discussion to a treatment of the robustness of adaptive control to joint flexibility and to techniques to enhance robustness. Our design methodology, however, can be used as a basis for designing controllers which are robust to other forms of uncertainty such as actuator dynamics, external disturbances, and other effects. Our approach can be explained intuitively as follows: Using the idea of composite control of singularly perturbed systems a fast feedback control law is first designed to damp the oscillations of the fast variables representing the joint flexibility. Once the fast transients have decayed, the slow part of the system should appear nearly like the dynamics of a rigid robot, which can then be controlled using any number of techniques. Our strategy is then summarized as

$$control_{composite} = control_{slow} + control_{fast} \tag{1}$$

where $control_{slow}$ is designed using a rigid robot model and $control_{fast}$ is designed solely to provide sufficient damping of the fast dynamics. In this paper, we base our design of the slow control on the algorithm of Slotine and Li [18] because it is globally convergent in the absence of joint flexibility, and because its implementation requires only position and velocity measurements.

We will see that the presence of unmodeled dynamics greatly complicates the analysis of the tracking properties of the system. Global convergence is no longer guaranteed for all possible reference trajectories. Using the composite Lyapunov theory for singularly perturbed systems we present sufficient conditions for adaptive trajectory tracking. For point–to–point motion we show that there is always a range of joint stiffness for which convergence is achieved and we quantify the region of convergence. For tracking of (smooth and bounded) reference trajectories we give sufficient conditions for closed loop stability and uniform boundedness of the tracking error. A residual set to which the tracking error converges is quantified. We also show that for special classes of trajectories, which include step responses generated from reference models and certain joint interpolated trajectories we can achieve asymptotic tracking. The actuals details and calculations of the proofs are tedious. For this reason we have omitted most of the calculations from the present paper and the interested reader should consult the thesis [4] for complete proofs.

## 2   Notation and Terminology

In what follows, we use the following standard notation and terminology [3]: $\mathbf{R}_+$ will denote the set of nonnegative real numbers, and $\mathbf{R}^n$ will denote the usual $n$-dimensional vector space over $\mathbf{R}$ endowed with the Euclidean norm $\|\mathbf{x}\| = \{\sum_{i=1}^n x_i^2\}^{\frac{1}{2}}$. $\mathbf{R}^{n \times n}$ denotes the set of all $n \times n$ matrices with real elements. For each matrix $A \in \mathbf{R}^{n \times n}$, we define the induced matrix norm of $A$ corresponding to the Euclidean vector norm $\|A\| = \{\lambda_{\max}(A^T A)\}^{\frac{1}{2}}$, where $\lambda_{\max}(A^T A)$ is the maximum eigenvalue of $A^T A$. We define the standard Lebesgue spaces $\mathbf{L}_\infty$ and $\mathbf{L}_2$ as

$$\mathbf{L}_\infty^n(\mathbf{R}_+) = \{f : \mathbf{R}_+ \to \mathbf{R}^n \text{ such that } f \text{ is Lebesgue measurable and } \|f\|_\infty < \infty\} \qquad (2)$$

where $\|f\|_\infty = \text{ess sup}_{t \in [0,\infty)} \|f(t)\|$,

$$\mathbf{L}_2^n(\mathbf{R}_+) = \{f : \mathbf{R}_+ \to \mathbf{R}^n \text{ such that } f \text{ is Lebesgue measurable and } \|f\|_2 < \infty\} \qquad (3)$$

where $\|f\|_2 = \{\int_0^\infty \|f(t)\|^2 \, dt\}^{\frac{1}{2}}$. Denote by $\mathbf{B_x} \subset \mathbf{R}^{2n}$, $\mathbf{B_\theta} \subset \mathbf{R}^r$, $\mathbf{B_y} \subset \mathbf{R}^{2n}$ the closed balls centered at $\mathbf{x} = 0$, $\tilde{\theta} = 0$, and $\mathbf{y} = 0$ respectively, and let $\mathbf{B} = \mathbf{B_x} \times \mathbf{B_\theta} \times \mathbf{B_y} \subset \mathbf{R}^{2n} \times \mathbf{R}^r \times \mathbf{R}^{2n}$. Also define $\mathcal{B} = \{(\|\mathbf{x}\|, \|\tilde{\theta}\|, \|\mathbf{y}\|) : (\mathbf{x}, \tilde{\theta}, \mathbf{y}) \in \mathbf{B}\} \subset \mathbf{R}_+^3$.

# 3 Singular Perturbation Model

The dynamic equations of a flexible joint manipulator are given by [19]

$$D(q_1)\ddot{q}_1 + C(q_1, \dot{q}_1)\dot{q}_1 + g(q_1) + K(q_1 - q_2) = 0 \tag{4}$$

$$J\ddot{q}_2 - K(q_1 - q_2) = u, \tag{5}$$

where the vectors $q_1 \in \mathbf{R}^n$ and $q_2 \in \mathbf{R}^n$ represent the link angles and motor angles, respectively, $D(q_1)$ is the $n \times n$ inertia matrix for the rigid links, $J$ is a diagonal matrix of actuator inertias reflected to the link side of the gears, $C(q_1, \dot{q}_1)\dot{q}_1$ represents the Coriolis and centrifugal terms, $g(q_1)$ represents the gravitational terms, and $K$ is a diagonal matrix representing the joint stiffness. For notational simplicity we will assume that all joint stiffness constants are the same in which case $K$ may be taken as a scalar. The composite control law $u$ that we consider is given by [20] $u = u_s(q_1, \dot{q}_1, t) + u_f(\dot{q}_1, \dot{q}_2)$, where, $u_f = K_v(\dot{q}_1 - \dot{q}_2)$, $K_v$ is a constant diagonal matrix, and $u_s$ is designed using the following rigid model, obtained by letting the joint stiffness $K$ tend to infinity, [19]

$$(D(q_1) + J)\ddot{q}_1 + C(q_1, \dot{q}_1)\dot{q}_1 + g(q_1) = u_s. \tag{6}$$

We define the variable $z := K(q_2 - q_1)$, and we assume that $K$ is $O(1/\epsilon^2)$, and $K_v$ is $O(1/\epsilon)$, so that we may write $K = K_1/\epsilon^2$, $K_v = K_2/\epsilon$, where $K_1$, $K_2$ are $O(1)$. By substituting the control law $u$ into (4)-(5), and using the definition of $z$, we obtain the singularly perturbed system [20]

$$D(q_1)\ddot{q}_1 + C(q_1, \dot{q}_1)\dot{q}_1 + g(q_1) = z \tag{7}$$

$$\epsilon^2 J\ddot{z} + \epsilon K_2\dot{z} + K_1 z = K_1(u_s - J\ddot{q}_1). \tag{8}$$

We now choose $u_s$ as the adaptive control law of [18] designed for the rigid system (6). The whole adaptive system can therefore be written as

i) **Plant:**

$$D(q_1)\ddot{q}_1 + C(q_1, \dot{q}_1)\dot{q}_1 + g(q_1) = z \tag{9}$$

$$\epsilon^2 J\ddot{z} + \epsilon K_2\dot{z} + K_1 z = K_1(u_s - J\ddot{q}_1). \tag{10}$$

ii) **Controller** (designed for the rigid plant (6)):

$$u_s = (\hat{D}(q_1) + \hat{J})a + \hat{C}(q_1, \dot{q}_1)v + \hat{g}(q_1) - K_D r, \tag{11}$$

where $\hat{D}$, $\hat{J}$, $\hat{C}$ and $\hat{g}$ represent the terms in (6) with estimated values of the parameters, $K_D$ is a diagonal matrix of positive gains,

$$\tilde{q}_1 = q_1 - q^d, \quad v = \dot{q}^d - \Lambda\tilde{q}_1, \quad r = \dot{q}_1 - v = \dot{\tilde{q}}_1 + \Lambda\tilde{q}_1, \quad a = \dot{v}. \tag{12}$$

$\Lambda$ is a constant diagonal matrix, and $q_d(t)$ is the reference trajectory which is at least three times continuously differentiable.

**iii) Parameter Update Law:**

$$\dot{\theta} = -\Gamma^{-1}Y^T(q_1, \dot{q}_1, a, v)r, \tag{13}$$

where $\Gamma$ is a symmetric, positive definite matrix, $\tilde{\theta} = \hat{\theta} - \theta$ is the parameter error, and

$$(D(q_1) + J)a + C(q_1, \dot{q}_1)v + g(q_1) = Y(q_1, \dot{q}_1, a, v)\theta. \tag{14}$$

$Y(q_1, \dot{q}_1, a, v)$ is an $n \times r$ matrix of known functions (regressor), and $\theta$ is an $r$-dimensional vector of parameters.

The plant (9)-(10), the controller (11), and the parameter update law (13) are now transformed into a more suitable singularly perturbed form, details of which can be found in [4]

$$S \quad : \quad \begin{cases} \dot{x} = A_1 x + \Phi\tilde{\theta} + A_3 y \\ \dot{\tilde{\theta}} = -\Gamma\varphi x \\ \epsilon\dot{y} = A_2 y + \epsilon A_2^{-1} B_2 \dot{u}, \end{cases} \tag{15}$$

or equivalently,

$$S \quad : \quad \begin{cases} \dot{p} = f(t, p, y) = \begin{bmatrix} A_1 & \Phi \\ -\Gamma\varphi & 0_{r \times r} \end{bmatrix} p + \begin{bmatrix} A_3 \\ 0_{r \times 2n} \end{bmatrix} y \\ \epsilon\dot{y} = g(t, p, y, \epsilon) = A_2 y + \epsilon A_2^{-1} B_2 \dot{u}, \end{cases} \tag{16}$$

where

$$\bullet \quad x = \begin{bmatrix} \tilde{q}_1 \\ r \end{bmatrix} = T \begin{bmatrix} \tilde{q}_1 \\ \dot{\tilde{q}}_1 \end{bmatrix} \in \mathbf{R}^{2n}, \text{ with the nonsingular linear transformation } T \tag{17}$$

$$T = \begin{bmatrix} I_{n \times n} & 0_{n \times n} \\ \Lambda & I_{n \times n} \end{bmatrix},$$  (18)

- $$p = \begin{bmatrix} \mathbf{x} \\ \tilde{\theta} \end{bmatrix} \in \mathbf{R}^{2n+r},$$  (19)

- $$A_1 = A_1(\mathbf{x}, \mathbf{q}_d, \dot{\mathbf{q}}_d) = \begin{bmatrix} -\Lambda & I_{n \times n} \\ -M(\mathbf{q}_1)^{-1}[C(\mathbf{q}_1, \dot{\mathbf{q}}_1) + K_D] & 0_{n \times n} \end{bmatrix} \in \mathbf{R}^{2n \times 2n},$$  (20)

- $$M(\mathbf{q}_1) = D(\mathbf{q}_1) + J,$$  (21)

- $$\Phi = \Phi(\mathbf{x}, \mathbf{q}_d, \dot{\mathbf{q}}_d, \ddot{\mathbf{q}}_d) = \begin{bmatrix} 0_{n \times r} \\ M(\mathbf{q}_1)^{-1} Y(\mathbf{q}_1, \dot{\mathbf{q}}_1, \mathbf{v}, \mathbf{a}) \end{bmatrix} \in \mathbf{R}^{2n \times r},$$  (22)

- $$A_3 = A_3(\mathbf{x}, \mathbf{q}_d) = \begin{bmatrix} 0_{n \times n} & 0_{n \times n} \\ M(\mathbf{q}_1)^{-1} & 0_{n \times n} \end{bmatrix} \in \mathbf{R}^{2n \times 2n},$$  (23)

- $$\varphi = \varphi(\mathbf{x}, \mathbf{q}_d, \dot{\mathbf{q}}_d) = \begin{bmatrix} 0_{r \times n} & Y^T(\mathbf{q}_1, \dot{\mathbf{q}}_1, \mathbf{a}, \mathbf{v}) \end{bmatrix} \in \mathbf{R}^{r \times 2n},$$  (24)

- $$A_2 = \begin{bmatrix} 0_{n \times n} & I_{n \times n} \\ -J^{-1} K_1 & -J^{-1} K_2 \end{bmatrix} \in \mathbf{R}^{2n \times 2n},$$  (25)

- $$B_2 = \begin{bmatrix} 0_{n \times n} \\ J^{-1} K_1 \end{bmatrix} \in \mathbf{R}^{2n \times n},$$  (26)

- $$\mathbf{u} = \mathbf{u}_s - J \ddot{\mathbf{q}}_1,$$  (27)

- $$\mathbf{y} = \begin{bmatrix} \mathbf{z} \\ \epsilon \dot{\mathbf{z}} \end{bmatrix} + A_2^{-1} B_2 \mathbf{u} \in \mathbf{R}^{2n}.$$  (28)

## 4  Analysis of the Singularly Perturbed System $\mathcal{S}$

System $\mathcal{S}$ is a nonautonomous nonlinear singularly perturbed system in the standard form [12]. $\mathbf{p}$ is the slow variable, and $\mathbf{y}$ is the fast variable. The analysis of system $\mathcal{S}$ follows the techniques of composite Lyapunov functions for nonlinear singularly perturbed systems developed in [17]; see also [12].

The boundary layer system, denoted $S_b$, is defined as

$$S_b \quad : \quad \frac{dy}{d\tau} = g(t, \mathbf{p}, \mathbf{y}(\tau), \epsilon = 0) = A_2 \mathbf{y}, \tag{29}$$

where $\tau = t/\epsilon$ is a stretching time scale. Let $P$ be the symmetric positive definite matrix that satisfies the Lyapunov Equation $A_2^T P + P A_2 = -Q$, where $Q$ is a positive definite matrix. We choose, for the boundary layer system, the Lyapunov Function Candidate $W(\mathbf{y}) = \mathbf{y}^T P \mathbf{y}$.

The reduced system, or quasi–steady state, is defined by setting $\epsilon = 0$ in $S$, that is,

$$\dot{\mathbf{p}} = f(t, \mathbf{p}, \mathbf{y}) = \begin{bmatrix} A_1 & \Phi \\ -\Gamma\varphi & 0_{r \times r} \end{bmatrix} \mathbf{p} + \begin{bmatrix} A_3 \\ 0_{r \times 2n} \end{bmatrix} \mathbf{y} \tag{30}$$

$$0 = g(t, \mathbf{p}, \mathbf{y}, \epsilon = 0) = A_2 \mathbf{y}. \tag{31}$$

Since $A_2$ is invertible, the algebraic equation (31) has the unique root $\mathbf{y} = 0$. The reduced system, denoted $S_r$, is obtained by replacing $\mathbf{y} = 0$ into (30)

$$S_r \quad : \quad \dot{\mathbf{p}} = f(t, \mathbf{p}, \mathbf{y} = 0) = \begin{bmatrix} A_1 & \Phi \\ -\Gamma\varphi & 0_{r \times r} \end{bmatrix} \mathbf{p}, \tag{32}$$

or equivalently,

$$S_r : \begin{cases} \dot{\mathbf{x}} = A_1 \mathbf{x} + \Phi\tilde{\theta} \\ \dot{\tilde{\theta}} = -\Gamma\varphi\mathbf{x}. \end{cases} \tag{33}$$

**Fact 1** *The reduced system $S_r$ is equivalent to the adaptive rigid-joint system* [4].

□

A consequence of Fact 1 is that we can use the same Lyapunov function candidate as that of the adaptive rigid-joint system [14], [22], namely,

$$\begin{aligned} V &= \frac{1}{2}\mathbf{r}^T M(\mathbf{q}_1)\mathbf{r} + \tilde{\mathbf{q}}_1^T \Lambda^T K_D \tilde{\mathbf{q}}_1 + \frac{1}{2}\tilde{\theta}^T \Gamma^{-1}\tilde{\theta} \\ &= V(\mathbf{q}_d, \tilde{\mathbf{q}}_1, \mathbf{r}, \tilde{\theta}) = V(t, \mathbf{x}, \tilde{\theta}) = V(t, \mathbf{p}). \end{aligned} \tag{34}$$

Define

$$F \quad := \quad I_{n\times n} + JD(\mathbf{q}_1)^{-1}, \tag{35}$$

$$\rho(t) \quad := \quad \frac{\partial u}{\partial \mathbf{q}_d}\dot{\mathbf{q}}_d + \frac{\partial u}{\partial \dot{\mathbf{q}}_d}\ddot{\mathbf{q}}_d + \frac{\partial u}{\partial \ddot{\mathbf{q}}_d}\mathbf{q}_d^{(3)}. \tag{36}$$

For $\forall(\mathbf{x},\tilde{\boldsymbol{\theta}},\mathbf{y}) \in \mathbf{B}$, define positive constants $k_1$, $k_2$, $k_3$, and positive functions $k_4(t)$, $k_5(t)$ and $k_g(t)$ such that

$$\bullet(\text{a1}) \;:\; \left\| F\frac{\partial u}{\partial \mathbf{x}}A_3\mathbf{y} + \frac{1}{\epsilon}F\frac{\partial u}{\partial \mathbf{y}}A_2\mathbf{y} \right\| \leq (k_3 + \frac{1}{\epsilon}k_2)\|\mathbf{y}\|, \tag{37}$$

$$\begin{aligned}
\bullet(\text{a2}) \;:\; & \left\| \left\{ F\frac{\partial u}{\partial \mathbf{x}}A_1 - F\frac{\partial u}{\partial \tilde{\boldsymbol{\theta}}}\Gamma\varphi \right\}\mathbf{x} + F\frac{\partial u}{\partial \mathbf{x}}\Phi\tilde{\boldsymbol{\theta}} \right\| \\
& \leq \left\| \left\{ F\frac{\partial u}{\partial \mathbf{x}}A_1 - F\frac{\partial u}{\partial \tilde{\boldsymbol{\theta}}}\Gamma\varphi \right\} \right\| \|\mathbf{x}\| \\
& + \left\| F\frac{\partial u}{\partial \mathbf{x}} \begin{bmatrix} 0 \\ M^{-1}\left( -\tilde{M}(\mathbf{q}_1)\Lambda\dot{\tilde{\mathbf{q}}}_1 - \tilde{C}(\mathbf{q}_1,\dot{\mathbf{q}}_1)\Lambda\tilde{\mathbf{q}} \right) \end{bmatrix} \right\| \\
& + \left\| F\frac{\partial u}{\partial \mathbf{x}} \begin{bmatrix} 0 \\ M^{-1}\left( \tilde{M}(\mathbf{q}_1)\ddot{\mathbf{q}}_d + \tilde{C}(\mathbf{q}_1,\dot{\mathbf{q}}_1)\dot{\mathbf{q}}_d \right) \end{bmatrix} \right\| \\
& + \left\| F\frac{\partial u}{\partial \mathbf{x}} \begin{bmatrix} 0 \\ M^{-1}Y_g(\mathbf{q}_1)\tilde{\boldsymbol{\theta}}_g \end{bmatrix} \right\| \\
& \leq k_1\|\mathbf{x}\| + k_5(t) + k_g(t). 
\end{aligned} \tag{38}$$

$$\bullet(\text{a3}) \;:\; \|F\rho(t)\| \leq k_4(t). \tag{39}$$

Note that the existence of the various constants $k_i$ in the above estimates requires only continuity of the functions involved since the set $\mathbf{B}$ is compact. Note also that $k_5(t)$ is proportional to the norms of the $\dot{\mathbf{q}}_d$ and $\ddot{\mathbf{q}}_d$, and that $k_g(t)$ is proportional to the norm of $\tilde{g} = Y_g(\mathbf{q}_1)\tilde{\boldsymbol{\theta}}_g$.

Consider the following composite Lyapunov function candidate for the singularly perturbed system $\mathcal{S}$

$$\mathcal{V}_1(t,\mathbf{x},\tilde{\boldsymbol{\theta}},\mathbf{y}) = (1-d)\,V(t,\mathbf{x},\tilde{\boldsymbol{\theta}}) + d\,W(\mathbf{y}) \quad,\quad 0 < d < 1, \tag{40}$$

which represents a weighted sum of $V(t,\mathbf{x},\tilde{\boldsymbol{\theta}})$, the Lyapunov function of the reduced system $\mathcal{S}_r$, and $W(\mathbf{y})$, the Lyapunov function of the boundary layer system $\mathcal{S}_b$. Taking into account (a1),(a2), and (a3), a rather lengthy calculation shows, (see [4]), that there exist constants $\alpha_i$, $\beta_i$, $\gamma_i$, $i = 1, 2$ such that the derivative of $V_1$ along the solution trajectories of $S$ satisfies

$$\dot{V}_1(t, \mathbf{x}, \mathbf{y}) \leq - \left[ \begin{array}{cc} \|\mathbf{x}\| & \|\mathbf{y}\| \end{array} \right] P_d \left[ \begin{array}{c} \|\mathbf{x}\| \\ \|\mathbf{y}\| \end{array} \right] + d\mu^2(t), \tag{41}$$

where

$$P_d = \left[ \begin{array}{cc} (1-d)\alpha_1 & -\frac{(1-d)\beta_1 + d\beta_2}{2} \\ -\frac{(1-d)\beta_1 + d\beta_2}{2} & \frac{d}{\epsilon}(\alpha_2 - \gamma_2) - d\gamma_1 \end{array} \right]. \tag{42}$$

and

$$\mu^2(t) = \mu_d^2(t) + \mu_g^2(t); \quad \mu_d(t) = 2\|P\|[k_4(t) + k_5(t)]; \quad \mu_g(t) = 2\|P\|k_g(t). \tag{43}$$

From the expression of $\dot{V}_1$, we observe that the right hand side of (41) consists of a quadratic expression and the term $d\mu^2(t)$. First of all note that the quadratic term does not include the state $\tilde{\theta}$. Also, $P_d$ can be made positive definite for some range of $\epsilon$. In the next section, we consider arbitrary trajectories and give sufficient conditions guaranteeing stability and show that the tracking error converges to a residual set, which we quantify. To achieve this, however, the parameter update law of [18] must be modified by the $\sigma$-modification scheme of [10] and [11].

Recall (a3) and (a2) and note that $F$ is a bounded function since $D(\mathbf{q}_1)$ and $D(\mathbf{q}_1)^{-1}$ are bounded matrices for all $\mathbf{q}_1$. Three important cases on the nature of $\mu_d(t)$ are of special interest:

1. **Case 1:** In the regulation problem, the desired trajectory $\mathbf{q}_d$ is a constant vector ($\mathbf{q}(t) = \bar{\mathbf{q}}_d$) and hence all higher derivatives of $\bar{\mathbf{q}}_d$ are zero. Thus, $\mu_d^2(t)$ is zero and $\mu^2(t) = \mu_g^2(t)$ in (41).

2. **Case 2:** If the desired trajectory is three times continuously differentiable with bounded derivatives, so that $\dot{\mathbf{q}}_d$, $\ddot{\mathbf{q}}_d$, $\mathbf{q}_d^{(3)} \in L_\infty^n$, then $\forall (\mathbf{x}, \tilde{\theta}, \mathbf{y}) \in B$, $\mu_d(t)$ is a bounded function of time ($\mu_d(t) \in L_\infty$). So $\exists \bar{\mu}_d$ a positive real constant such that $\mu_d(t) \leq \bar{\mu}_d$ $\forall t \in \mathbf{R}_+$.

3. **Case 3:** If $\dot{\mathbf{q}}_d$, $\ddot{\mathbf{q}}_d$, $\mathbf{q}_d^{(3)} \in L_1^n \cap L_\infty^n$ (i.e. $\in L_2^n \cap L_\infty^n$), then $\forall (\mathbf{x}, \tilde{\theta}, \mathbf{y}) \in B$, $\mu_d(t) \in L_2^n \cap L_\infty^n$, and furthermore $\lim_{t \to \infty} \dot{\mathbf{q}}_d = 0$, and $\lim_{t \to \infty} \ddot{\mathbf{q}}_d = 0$. For example, the class of bounded desired trajectories which are eventually constant fits into this category. For such class of desired trajectories, let's further assume that $(\mathbf{q}_d(t) - \bar{\mathbf{q}}_d) \in L_1^n \cap L_\infty^n$. Therefore, this class of desired trajectories consists of bounded trajectories that converge to a steady state value $\bar{\mathbf{q}}_d$ as fast as an $L_1^n$ function.

# 5  Tracking Analysis

The tracking (of time varying signals) results presented below can be thought of as extending the results of [24] from the regulation problem to the tracking problem. The extension is nontrivial and exploits the particular nature of robot dynamics and robot tracking problem. We first note that the quadratic term in the Lyapunov derivative expression (41) is only negative semi–definite since it does not contain the parameter estimate error $\tilde{\theta}$. As is well–known in adaptive control circles parameter drift instability may occur as a result. In order to halt this parameter drift several approaches may be used, such as dead–zones, projections, or $\sigma$–modification. In the next section we derive analytical results on robust tracking using $\sigma$–modification to guarantee boundedness of the estimated parameters.

## 5.1  Robustness Via the fixed $\sigma$-modification

In the tracking analysis that follows, we modify the parameter update law in (15), using the fixed $\sigma$-modification scheme [10]. The singularly perturbed system $S$ becomes

$$
S_\sigma \quad : \quad \begin{cases} \dot{\mathbf{x}} = A_1\mathbf{x} + \Phi\tilde{\theta} + A_3\mathbf{y} \\ \dot{\tilde{\theta}} = -\Gamma\varphi\mathbf{x} - \sigma\Gamma\hat{\theta} \\ \epsilon\dot{\mathbf{y}} = A_2\mathbf{y} + \epsilon A_2^{-1}B_2\dot{\mathbf{u}}, \end{cases}
\tag{44}
$$

where $\sigma > 0$ is a scalar. The reduced system now becomes

$$
S_r^\sigma : \begin{cases} \dot{\mathbf{x}} = A_1\mathbf{x} + \Phi\tilde{\theta} \\ \dot{\tilde{\theta}} = -\Gamma\varphi\mathbf{x} - \sigma\Gamma\hat{\theta}. \end{cases}
\tag{45}
$$

The boundary layer system $S_b$ is still defined by equation (29).

The analysis of system $S_\sigma$ is very similar to that of the original singularly perturbed system $S$. In fact we use the same Lyapunov function candidates $V$ (for the reduced system $S_r^\sigma$) and $W$ (for the boundary layer system $S_b$.) Consequently, the composite Lyapunov function candidate $V_1$ given in (40) is also used for the singularly perturbed system $S_\sigma$. $\forall(\mathbf{x}, \tilde{\theta}, \mathbf{y}) \in \mathbf{B}$, define positive constant $k_8$ such that

$$
\bullet(a2)' : \left\| \left\{ F\frac{\partial\mathbf{u}}{\partial\mathbf{x}}A_1 - F\frac{\partial\mathbf{u}}{\partial\tilde{\theta}}\Gamma\varphi \right\}\mathbf{x} - \sigma F\frac{\partial\mathbf{u}}{\partial\tilde{\theta}}\Gamma\hat{\theta} + F\frac{\partial\mathbf{u}}{\partial\mathbf{x}}\Phi\tilde{\theta} \right\| \le k_1\|\mathbf{x}\| + k_5(t) + k_g(t) + \sigma k_6. \tag{46}
$$

Define

$$\mathcal{I} := \left\{ (\|\mathbf{x}\|, \|\tilde{\theta}\|, \|\mathbf{y}\|) \in \mathcal{B} : \; \mathcal{V}_1(t, \mathbf{x}, \tilde{\theta}, \mathbf{y}) \le c_1 \right\}, \tag{47}$$

where $c$ is the largest positive real number such that $\mathcal{I} \subset \mathcal{B}$.

We have the following result.

**Theorem 1** (Fixed $\sigma$-modification, Boundedness of Tracking Errors) *Assume*

1. $\dot{\mathbf{q}}_d$, $\ddot{\mathbf{q}}_d$, $\mathbf{q}_d^{(3)} \in \mathbf{L}_\infty^n$, *so that* $\exists \; \bar{\mu}_d$ *a positive real constant such that* $\mu_d(t) \le \bar{\mu}_d \quad \forall t \in \mathbf{R}_+$ (Case 2).

2. (a1), (a2)$'$, *and* (a3) *are satisfied* $\forall (\mathbf{x}, \tilde{\theta}, \mathbf{y}) \in \mathbf{B}$.

3. $\alpha_2 - \gamma_2 > 0$.

*Define the sets* $\mathcal{D}_{\bar{\mu},\sigma}$ *and* $\mathcal{R}_{\bar{\mu},\sigma}$ *as follows:*

$$\mathcal{D}_{\bar{\mu},\sigma} := \left\{ (\|\mathbf{x}\|, \|\tilde{\theta}\|, \|\mathbf{y}\|) \in \mathcal{B} : \; \begin{bmatrix} \|\mathbf{x}\| & \|\mathbf{y}\| \end{bmatrix} P_d \begin{bmatrix} \|\mathbf{x}\| \\ \|\mathbf{y}\| \end{bmatrix} \right.$$
$$\left. + \frac{1}{2}(1-d)\sigma \|\tilde{\theta}\|^2 \le \frac{1}{2}(1-d)\sigma \|\theta\|^2 + d\sigma\beta_3 + d\bar{\mu}^2 \right\}, \tag{48}$$

*where*

$$\bar{\mu}^2 = \bar{\mu}_g^2 + \bar{\mu}_d^2; \quad \bar{\mu}_g := \sup_{\mathcal{I}} \mu_g(t); \quad \beta_3 > 0. \tag{49}$$

*and*

$$\mathcal{R}_{\bar{\mu},\sigma} := \left\{ (\|\mathbf{x}\|, \|\tilde{\theta}\|, \|\mathbf{y}\|) \in \mathcal{B} : \; \mathcal{V}_1(t, \mathbf{x}, \tilde{\theta}, \mathbf{y}) \le c_{\bar{\mu},\sigma} \right\}, \tag{50}$$

*where* $c_{\bar{\mu},\sigma}$ *is the smallest positive real number such that* $\mathcal{D}_{\bar{\mu},\sigma} \subseteq \mathcal{R}_{\bar{\mu},\sigma}$.

*If* $\bar{\mu}$ *and* $\sigma$ *are such that* $\mathcal{R}_{\bar{\mu},\sigma} \subset \mathcal{I}$, *then* $\exists$ *an upper bound of* $\epsilon$, *namely,*

$$\epsilon_d = \frac{\alpha_1(\alpha_2 - \gamma_2)}{\alpha_1\gamma_1 + \frac{1}{4d(1-d)}[(1-d)\beta_1 + d\beta_2]^2}, \tag{51}$$

*such that all the solution trajectories of the singularly perturbed system* $S_\sigma$ *starting in* $\mathcal{I}$ *converge to the residual set* $\mathcal{R}_{\bar{\mu},\sigma} \; \forall \epsilon \in (0, \epsilon_d)$.

**Proof of Theorem 1:** The results are obtained by manipulating the expression of $\dot{V}_1$ which now includes the state $\tilde{\theta}$ brought by the fixed $\sigma$-modification introduced in the parameter update law. (see [4].)

**Remark :** It can be shown that the maximum value of $\epsilon_d$ occurs at

$$d^* = \frac{\beta_1}{\beta_1 + \beta_2} \tag{52}$$

and is given by

$$\epsilon^* = \epsilon_{d=d^*} = \frac{\alpha_1(\alpha_1 - \gamma_2)}{\alpha_1\gamma_1 + \beta_1\beta_2}. \tag{53}$$

Choosing $d = d^*$ fixes the size of $\mathcal{D}_{\bar{\mu},\sigma}$, and hence that of the residual set $\mathcal{R}_{\bar{\mu},\sigma}$. If the size of $\mathcal{R}_{\bar{\mu},\sigma}$ is made smaller by choosing another $d$, then a smaller upper bound $\epsilon_d$ (i.e. a larger stiffness $K$) results (see [4].)

**Remark :** The advantage of introducing the fixed $\sigma$-modification is that the tracking errors and the parameter errors are ensured to converge to a residual set under the conditions of Theorem 1. As far as the desired trajectory is concerned, it is only required that the latter is bounded and is three times continuously differentiable with bounded derivatives. The price paid by introducing the fixed $\sigma$-modification is that no conclusion about the convergence to zero of the tracking errors can be made even under further restrictive conditions on the desired trajectory such as those of Case 3 above. Moreover, using the switching $\sigma$-modification of [11] for the class of desired trajectories of Case 2, Theorem 1 still applies, and no conclusion about the convergence to zero of the tracking errors can be made. After considering the regulation problem in the next section, we show that for the class of desired trajectories of Case 3, the tracking errors converge to zero if the switching $\sigma$-modification of [11] is used.

## 6 Regulation Analysis

Since in the regulation case $q(t) = \bar{q}_d$ (Case 1 discussed above) and $\mu_d(t) = 0$, the time derivative of the composite Lyapunov function $V_1$ along the solution trajectories of $\mathcal{S}$ is given by (see (41))

$$\dot{V}_1(t,\mathbf{x},\mathbf{y}) \leq - \begin{bmatrix} \|\mathbf{x}\| & \|\mathbf{y}\| \end{bmatrix} P_d \begin{bmatrix} \|\mathbf{x}\| \\ \|\mathbf{y}\| \end{bmatrix} + d\mu_g^2(t). \tag{54}$$

Recall that $\mu_g(t)$ is proportional to $\left\|Y_g \tilde{\theta}_g\right\|$. Therefore, using the above composite Lyapunov function, our only hope to prove stability is when gravity is abscent (i.e. $\mu_g(t) = 0$) or if the parameter update law is modified to insure boundedness of the parameter estimates. A rather lengthy reformulation of the equations of motion shows that such apparent difficulty can be eliminated, and a new composite Lyapunov function $V_2(\mathbf{x}, \tilde{\theta}, \mathbf{y})$ can be derived to show the following result (see [4] for details and proof)

**Theorem 2** (Regulation) *Under similar conditions as in Theorem 1, there exist $\epsilon^* > 0$ such that the equilibrium $\mathbf{x} = 0$, $\dot{\tilde{\theta}} = \hat{\theta} - \theta = 0$, and $\mathbf{y} = 0$ of system $S$ is stable for all $\epsilon \in (0, \epsilon^*)$, and an estimate of the domain of attraction is given by*

$$\Omega^* = \left\{(\mathbf{x}, \tilde{\theta}, \mathbf{y}) \in B : V_2(\mathbf{x}, \tilde{\theta}, \mathbf{y}) \le c_2\right\} \tag{55}$$

*where $c_2$ is a positive constant and $\Omega^* \subset B$. Moreover, $\forall \left(\mathbf{x}(0), \tilde{\theta}(0), \mathbf{y}(0)\right) \in \Omega^*$ we get $\lim_{t \to \infty} \mathbf{x}(t) = 0$, $\lim_{t \to \infty} \mathbf{y}(t) = 0$, $\lim_{t \to \infty} Y_g \tilde{\theta}(t) = 0$.*

□

## 7 Asymptotic Tracking with the switching $\sigma$−modification

The singularly perturbed system with this modification becomes

$$S_{\sigma_s} : \begin{cases} \dot{\mathbf{x}} = A_1 \mathbf{x} + \Phi \tilde{\theta} + A_3 \mathbf{y} \\ \dot{\tilde{\theta}} = -\Gamma \varphi \mathbf{x} - \sigma_s \Gamma \hat{\theta} \\ \epsilon \dot{\mathbf{y}} = A_2 \mathbf{y} + \epsilon A_2^{-1} B_2 \dot{\mathbf{u}}, \end{cases} \tag{56}$$

where $\sigma_s$ is now given by

$$\sigma_s(t) = \begin{cases} 0 & \text{if } \left\|\hat{\theta}(t)\right\| < \theta_0 \\ \sigma_0 \left(\frac{\left\|\hat{\theta}(t)\right\|}{\theta_0} - 1\right) & \text{if } \theta_0 \le \left\|\hat{\theta}(t)\right\| \le 2\theta_0 \\ \sigma_0 & \text{if } \left\|\hat{\theta}(t)\right\| > 2\theta_0. \end{cases} \tag{57}$$

$\sigma_0$ is a positive scalar design parameter. $\theta_0$ is chosen such that $\|\theta\| < \theta_0$, and hence, it reflects our knowledge of the true parameter vector $\theta$. Determining $\theta_0$ is possible since in general the true parameters have known upper and lower bounds.

Since desired trajectories of Case 3 above eventually converge to a fixed value $\bar{q}_d$, it is possible to rewrite the equations of motion of the flexible joint robot where the tracking error is defined as the difference between the actual tracking value and the steady state desired value $\bar{q}_d$. The advantage of this formulation is to transform the problem of tracking the above class of trajectories into a regulation problem with $L_1 \cap L_\infty$ (hence $L_2$) time varying disturbances. Hence, it can be shown that a composite Lyapunov function $V_3(x, \tilde{\theta}, y)$ can be derived to show the following result (see [4] for details and proof)

**Theorem 3** (Switching $\sigma$-modification, Convergence of Tracking Errors) *Define the set*

$$\mathcal{I}^* := \left\{ (\|x\|, \|\tilde{\theta}\|, \|y\|) \in \mathcal{B} : \ V_3(x, \tilde{\theta}, y) \leq c_3 \right\} \tag{58}$$

*where $c_3$ is a positive constant. Under similar conditions as in Theorem 1, there exist $\epsilon^* > 0$ such that all solution trajectories starting in $\mathcal{I}^*$ converge to a residual set $\mathcal{R}_{\mu, \sigma_0} \subset \mathcal{I}^* \ \forall \ \epsilon \in (0, \epsilon^*)$. Furthermore, $\lim_{t \to \infty} x(t) = 0$, $\lim_{t \to \infty} y(t) = 0$, and $\lim_{t \to \infty} Y_g \tilde{\theta}(t) = 0$.*

□

# 8 Experimental Results

In this section, we illustrate the theoretical results discussed above using an experimental single–link flexible joint robot manipulator. We first demonstrate the effectiveness of the composite control strategy by comparing the tracking performance of the arm using only a rigid–based adaptive control law (unstable) and then using a composite control strategy (good tracking). Second, the tracking properties of the composite control method as discussed in Theorem 1 and Theorem 3, are experimentally illustrated. Third, instability mechanisms due to fast desired trajectory and parameter drift, and the use of $\sigma$–modification are also experimentally demonstrated.

The specially constructed single–link flexible joint arm, used in the experiments is shown in Figure 1. The flexible joint consists of two aluminum plates joined by extension springs. The actuator is a large DC motor connected directly to one plate. A hollow aluminum tube (1.5 inch diameter) about 18 inches long is connected to the second plate. Two incremental encoders provide feedback of the motor and link positions to within $1.534 \times 10^{-3}$ radians while velocity information is obtained by filtering the position feedback data. The computer controller is built around a Motorola 68000 microprocessor and includes a floating point math processor. The control algorithms are implemented in PASCAL with an average sampling period of about $5 \times 10^{-3}$ second ($200\ Hz$). Parameter uncertainty is introduced by clamping payloads to the end of the arm. A payload (small wrench) that is approximately 40% of the nominal gravitational load of the arm is used in all experiments.

We model the dynamics of this system as (see Figure 2)

$$I\ddot{q}_1 + Mglsin(q_1) + k(q_1 - q_2) = 0 \tag{59}$$
$$J\ddot{q}_2 + B_2\dot{q}_2 - k(q_1 - q_2) = u \tag{60}$$

Nominal values for the arm parameters without a payload are shown in Table 1. This system is of the form (4)-(5), except for the non-zero damping at the joint. However, as we will see, this damping is not sufficient to stabilize the elastic oscillation of the joint. The related rigid model, obtained in the limit as $k \to \infty$, is

$$(I + J)\ddot{q}_1 + B_2\dot{q}_1 + Mglsin(q_1) = u_s. \tag{61}$$

The coefficient $B_2$ is known with sufficient precision that we simply cancel it in the control law to follow. Only the inertia parameters are affected by varying payloads. We therefore parametrize the rigid system as

$$(I + J)\ddot{q}_1 + B_2\dot{q}_1 + Mglsin(q_1) = Y\theta + B_2\dot{q}_1 = u \tag{62}$$

$$\theta = \begin{bmatrix} \theta_1 \\ \theta_2 \end{bmatrix} = \begin{bmatrix} I + J \\ Mgl \end{bmatrix}, Y = \begin{bmatrix} \ddot{q}_1 & sin(q_1) \end{bmatrix}. \tag{63}$$

The complete description of the control system is now shown in Table 2.

Recall that

$$\tilde{q}_1 = q_1 - q_d, \ v = \dot{q}_d - \lambda\tilde{q}_1, \ r = \dot{\tilde{q}}_1 + \lambda\tilde{q}_1, \ a = \ddot{q}_d - \lambda\dot{\tilde{q}}_1, \tag{64}$$

where $q_d(t)$ is the desired trajectory.

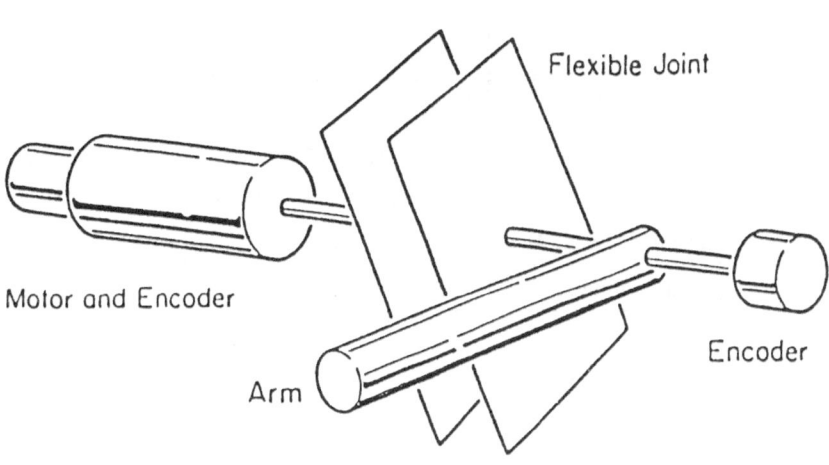

Figure 1: Sketch of Experimental Single–Link Flexible Joint Arm.

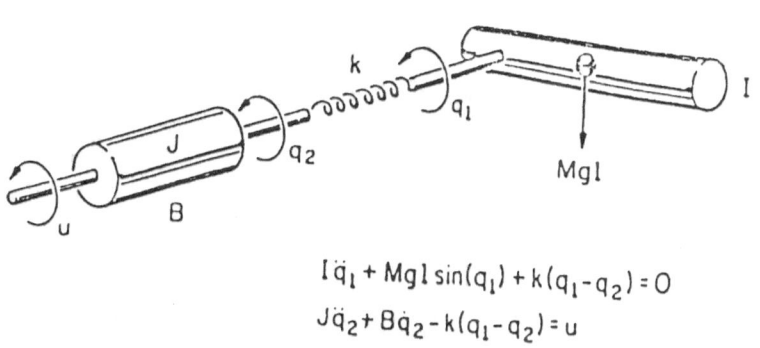

$$I\ddot{q}_1 + Mgl\sin(q_1) + k(q_1-q_2) = 0$$
$$J\ddot{q}_2 + B\dot{q}_2 - k(q_1-q_2) = u$$

Figure 2: Model of Experimental Single–Link Flexible Joint Arm.

| Parameter | Value |
|---|---|
| Link Inertia, $I$ | 0.031 $(kg - m^2)$ |
| Rotor Inertia, $J$ | 0.004 $(kg - m^2)$ |
| Rotor Friction, $B_2$ | 0.007 $(N - m - sec/rad)$ |
| Nominal Load, $Mgl$ | 0.8 $(N - m)$ |
| Joint Stiffness, $k$ | 7.13 $(N - m/rad)$ |

Table 1: Nominal Values of the Arm Parameters

| Plant | $I\ddot{q}_1 + Mgl \, sin(q_1) + k(q_1 - q_2) = 0$ |
|---|---|
| | $J\ddot{q}_2 + B_2\dot{q}_2 - k(q_1 - q_2) = u$ |
| Control | $u = u_s + u_f$ |
| Law | $u_f = K_v(\dot{q}_1 - \dot{q}_2)$ |
| | $u_s = \hat{\theta}_1 a + \hat{\theta}_2 sin(q_1) + B_2\dot{q}_1 - K_d r$ |
| Parameter | $\dot{\hat{\theta}}_1 = -\gamma_1 a r$ |
| Update Law | $\dot{\hat{\theta}}_2 = -\gamma_2 sin(q_1) r$ |

Table 2: Adaptive Composite Control System

## 8.1 Experiment # 1 : Effectiveness of the Adaptive Composite Control Strategy

In this experiment, we demonstrate the effectiveness of the composite control. A desired trajectory $q_d(t)$ consisting of a smooth 90–degree rotation with the arm initially pointing straight down is considered (see Table 3.) First, we neglect joint flexibility and use a rigid joint based adaptive control law (i.e. $u = u_s$, $u_f = 0$) with link variables for feedback. The result is an unstable system. Figure 3 displays the unsatisfactory response of the link. The response of the flexible joint system is bounded only because the joint deflection is limited by mechanical stops. The effectiveness of the adaptive composite control is shown in Figure 4 where the response is stable and the tracking is satisfactory. The gains used in the above two runs of the experiment are shown in Table 4. Note that the desired trajectory is eventually constant, and hence it is of the category of desired

| $q_d(t)$ | $A$(rad) | $\alpha$ (rad/sec) |
|---|---|---|
| $A - Ae^{-\alpha t}(1 + \alpha t)$ | $\frac{\pi}{2}$ | 5 |

Table 3: Desired Trajectory for Experiment # 1

| Gain | $\lambda$ | $K_d$ | $K_v$ | $\gamma_1$ | $\gamma_2$ |
|---|---|---|---|---|---|
| **Adaptive Rigid Control Only** | 10 | 1 | 0 | 0.001 | 5 |
| **Adaptive Composite Control** | 10 | 1 | 0.34 | 0.001 | 5 |

Table 4: Gain Values for Experiment # 1

trajectories of Case 3 discussed above. Consequently, according to Theorem 3, the link position and velocity steady state errors converge to zero. This fact is illustrated in Figure 4 where the position steady state error, $\tilde{q}_1$, is less than the minimum resolution of the encoder after approximately 5 seconds. Note that $\sigma$–modification (required by Theorem 3 in general) is not used in Experiment # 1 since the parameter estimates are bounded without this modification.

## 8.2   Experiment # 2 : Instability Caused by Fast Reference Trajectory

The desired trajectory in this experiment is a smooth 90–degree rotation with the arm initially pointing straight down (see Table 5.) In this experiment, we demonstrate an instability mechanism due to the interaction of the high frequencies in the desired trajectory with joint flexibility and the adaptation law. As shown in Figure 5, the link variable $q_1(t)$ is not tracking the fast desired trajectory. Recall that the response of the flexible joint system is bounded only because the joint deflection is limited by mechanical stops. When the adaptation was switched off, instability disappeared and the response was smooth (not shown here.) This observation shows that this type of instability is characteristic of adaptive systems. The gains used in this experiment are shown in Table 6.

| $q_d(t)$ | $A$(rad) | $\alpha$ (rad/sec) |
|---|---|---|
| $A - Ae^{-\alpha t}(1 + \alpha t)$ | $\frac{\pi}{2}$ | 0.6 |

Table 5: Desired Trajectory for Experiment # 2

| Gain | $\lambda$ | $K_d$ | $K_v$ | $\gamma_1$ | $\gamma_2$ |
|---|---|---|---|---|---|
| Value | 10 | 1 | 0.34 | 0.001 | 15 |

Table 6: Gain Values for Experiment # 2

## 8.3 Experiment # 3 : Parameter Drift Instability and $\sigma$–modification

In this experiment we illustrate the following:

- Without modification of the parameter update law, parameter drift instability mechanism is possible.

- Adding $\sigma$-modification to the parameter update law, we get boundedness of all signals, but nonzero tracking errors (Theorem 1.)

The desired trajectory in this experiment consists of two components. The first is a smooth 90–degree rotation with the arm initially pointing straight down. Once the steady state is reached, a second component consisting of a pure sinusoid with small amplitude and high frequency is added (see Table 7.) Note that the desired trajectory is differentiable everywhere except at the time when the second sinusoidal component is introduced. This can be thought of as disturbance introduced to the system. First, without modifying the parameter update law (recall the description of the complete system in Table 2), we observe from Figure 6 that the link and rotor variables remain bounded for sometime, then the parameter estimates rapidly diverge and all the states become unbounded.

We now introduce the switching $\sigma$–modification in the parameter update law. The description of the complete system becomes as shown in Table 8 (compare with Table 2). As seen in the plots of Figure 7 all signals are now bounded as predicted by Theorem 1. Note that the tracking error

| $q_d(t)$ | | $A$(rad) | $\alpha$ (rad/sec) |
|---|---|---|---|
| $\begin{cases} q_{d1}(t) & \text{if } q_{d1}(t) < 0.98 * A \\ q_{d1}(t) - 0.1sin(30\alpha t) & \text{if } q_{d1}(t) \geq 0.98 * A \end{cases}$ <br> $q_{d1}(t) = A - Ae^{-\alpha t}(1 + \alpha t)$ | | $\frac{\pi}{2}$ | 0.6 |

Table 7: Desired Trajectory for Experiment # 3

| Plant | $I\ddot{q}_1 + Mgl\ sin(q_1) + k(q_1 - q_2) = 0$ <br> $J\ddot{q}_2 + B\dot{q}_2 - k(q_1 - q_2) = u$ |
|---|---|
| Control <br><br> Law | $u = u_s + u_f$ <br> $u_f = K_v(\dot{q}_1 - \dot{q}_2)$ <br> $u_s = \hat{\theta}_1 a + \hat{\theta}_2 sin(q_1) + B_2\dot{q}_1 - K_d r$ |
| Parameter <br><br> Update Law | $\dot{\hat{\theta}}_1 = -\gamma_1 a r - \sigma_{r1}(t)\gamma_1\hat{\theta}_1$ <br> $\dot{\hat{\theta}}_2 = -\gamma_2 sin(q_1)r - \sigma_{r2}(t)\gamma_2\hat{\theta}_2$ <br> $\sigma_{ri}(t) = \begin{cases} 0 & \text{if } \mid\hat{\theta}_i(t)\mid < \theta_{0i} \\ \sigma_{0i}\left(\frac{\mid\hat{\theta}_i(t)\mid}{\theta_{0i}} - 1\right) & \text{if } \theta_{0i} \leq \mid\hat{\theta}_i(t)\mid \leq 2\theta_{0i} \quad\quad i = 1,2. \\ \sigma_{0i} & \text{if } \mid\hat{\theta}_i(t)\mid > 2\theta_{0i} \end{cases}$ |

Table 8: Adaptive Composite Control System with $\sigma$−modification

does not converge to zero but remains bounded since the desired trajectory is bounded but not eventually constant. The gains used in the control law and the parameter update law are shown in Table 9.

# 9    Conclusions

In this paper we have given stability results for a composite adaptive control law for flexible joint robot manipulators. The proofs of all results are contained in [4]. The complexity of the analysis points out the difficulty of the control problem for this class of systems. Although our results

| Gain | $\lambda$ | $K_D$ | $K_v$ | $\gamma_1$ | $\gamma_2$ | $\sigma_{01}$ | $\sigma_{02}$ | $\theta_{01}$ | $\theta_{02}$ |
|------|-----------|-------|-------|-----------|-----------|---------------|---------------|---------------|---------------|
| Value | 10 | 1 | 0.34 | 0.001 | 7 | 100 | 10 | 0.05 | 1.2 |

Table 9: Gain Values for Experiment # 3

give only sufficient conditions for local stability it can be argued, based on what is known about the behavior of adaptive control systems, that this is the best one can do without additional compensation. One promising approach to extend these results would be to incorporate the integral manifold based corrective control idea[19]. See [4] for some results on adaptive integral manifold based corrective control.

# References

[1] Albert, M., and Spong, M.W., "Compensator Design for Robot Manipulators with Flexible Joints," *Proc. Int. Conf. on Systems Engineering*, Dayton, OH, 1987.

[2] Craig, J.J., Hsu, P., and Sastry, S., "Adaptive Control of Mechanical Manipulators," *Proc. IEEE Int. Conf. Robotics and Automation*, San Francisco, CA, Mar. 1986.

[3] Desoer, C., and Vidyasagar, M., *Feedback Systems: Input-Output Properties*, Academic Press, New York, 1975.

[4] Ghorbel, F., "Adaptive Control of Flexible Joint Robot Manipulators : A Singular Perturbation Approach," *Ph.D. Thesis*, Department of Mechanical Engineering, University of Illinois at Urbana–Champaign, 1990.

[5] Ghorbel, F., Hung, J.Y., and Spong, M.W., "Adaptive Control of Flexible Joint Manipulators," *Proc. 1989 IEEE Int. Conf. Robotics and Automation*, pp. 1188–1193, Phoenix, AZ, 1989.

[6] Ghorbel, F., Hung, J.Y., and Spong, M.W., "Adaptive Control of Flexible Joint Manipulators," *IEEE Control Systems Magazine*, Vol. 9, No. 7, pp. 9–13, December, 1989.

[7] Ghorbel, F., and Spong, M.W., "Stability Analysis of Adaptively Controlled Flexible Joint Manipulators," *Coordinated Science Laboratory Report, UILU–ENG–90–2246*, University of Illinois at Urbana-Champaign, September, 1990.

[8] Hung, J.Y., "Robust Control of Flexible Joint Robot Manipulators," *Ph.D. Thesis*, Department of Electrical and Computer Engineering, University of Illinois at Urbana–Champaign, 1989.

[9] Hung, J.Y., Bortoff, S., Ghorbel, F., and Spong, M.W., "A Comparison of Feedback Linearization and Singular Perturbation Techniques for the Control of Flexible Joint Robots," *Proc. 1989 ACC*, pp. 25-30, Pittsburgh, PA, 1989.

[10] Ioannou, P.A., and Kokotović, P.V., "Instability Analysis and Improvement of Robustness of Adaptive Control," *Automatica*, Vol. 20, No. 5, pp. 583–594, 1984.

[11] Ioannou, P.A., and Tsakalis, K., "A Robust Direct Adaptive Controller," *IEEE Trans. Automatic Control,*, AC-31, pp. 1033–1043, November, 1986.

[12] Kokotović, P.V., Khalil, H.,K., and O'Reilly, J., *Singular Perturbation Methods in Control: Analysis and Design*, Academic Press, Inc., London, 1986.

[13] Middleton, R.H., and Goodwin, G.C., "Adaptive Computed Torque Control for Rigid Link Manipulators," *Systems and Control Letters*, Vol. 10, 9–16, (1988).

[14] Ortega, R., and Spong, M.W., "Adaptive Control of Rigid Robots: A Tutorial," *Proc. IEEE Conf. on Decision and Control*, pp. 1575–1584, Austin, TX, 1988.

[15] Riedle, B.D., and Kokotović, P.K., "Integral Manifolds of Slow Adaptation," *IEEE Transactions on Automatic Control*, Vol. AC-31, no. 4, pp. 316–324, 1986.

[16] Rohrs, C.E., Valavani, L., Athans, M., and Stein, G., "Robustness of Adaptive Control Algorithms in the Presence of Unmodeled Dynamics," *Proc. IEEE Conf. on Decision and Control*, pp. 3–11, Orlando, FL, 1982.

[17] Saberi, A., and Khalil, H.,K., *Quadratic-Type Lyapunov Functions for Singularly Perturbed Systems*, IEEE Trans. Autom. Con., Vol AC-29, No. 6, June 1984.

[18] Slotine, J.-J. E., and Li, W., "On the Adaptive Control of Robot Manipulators", *Int. J. of Robotics Research*, Vol. 6, No. 3, pp. 49-59, 1987.

[19] Spong, M.W., "Modeling and Control of Elastic Joint Manipulators", *J. Dyn. Sys., Meas. and Control*, Vol. 109, 310-319, 1987.

[20] Spong, M.W., "Adaptive Control of Flexible Joint Manipulators", *Systems and Controls Letters*, Vol. 13, pp. 15-21, 1989.

[21] Spong, M.W., "The Control of Flexible Joint Robots: A Survey," in *New Trends and Applications of Distributed Parameter Control Systems*, Lecture Notes in Pure and Applied Mathematics, G. Chen, E.B. Lee, W. Littman, and L. Markus, Eds., Chapter 12, pp. 355–383, Marcel Dekker Publishers, New York, 1990.

[22] Spong, M. W., Ortega, R., and Kelly, R., "Comments on 'Adaptive Manipulator Control'," *IEEE Transactions on Automatic Control*, Vol. AC–35, no. 6, pp. 761–762, 1990.

[23] Spong, M.W., and Vidyasagar, M., *Robot Dynamics and Control*, John Wiley & Sons, Inc., New York, 1989.

[24] Taylor, D.G., Kokotović, P.V., Marino, R., and Kanellakopoulos, I., "Adaptive Regulation of Nonlinear Systems with Unmodeled Dynamics," *IEEE Transactions on Automatic Control*, Vol. AC–34, no. 4, pp. 405–412, 1989.

[25] Sadegh, N., and Horowitz, R., "An Exponentially Stable Adaptive Control Law for Robotic Manipulators," *Proc. American Control Conf.*, San Diego, May, 1990.

[26] Schwartz, H.M., and Warshaw, G., "On the Richness Condition for Robot Adaptive Control," *ASME Winter Annual Meeting*, DSC-Vol. 14, pp 43-49, Dec. 1989.

[27] Bayard, D.S., and Wen, J.T., "A New Class of Control Laws for Robotic Manipulators-Part 2. Adaptive Case," *Int. J. of Control*, 47(5):1387-1406, 1988.

[28] Middleton, R.H., and Kokotovic, P.V., "Boundedness Properties of Simple Indirect Adaptive Control Systems," submitted to 1991 ACC and *IEEE Trans. on Automatic Cont.*, September, 1990.

[29] Streit, D.A., Krousgrill, C.M., and Bajaj, A.K., "Nonlinear Response of Flexible Robotic Manipulators Performing Repetitive Tasks," *ASME J. Dyn. Sys., Meas., and Cont.*, Vol. 111, pp. 470-480, Sept., 1989.

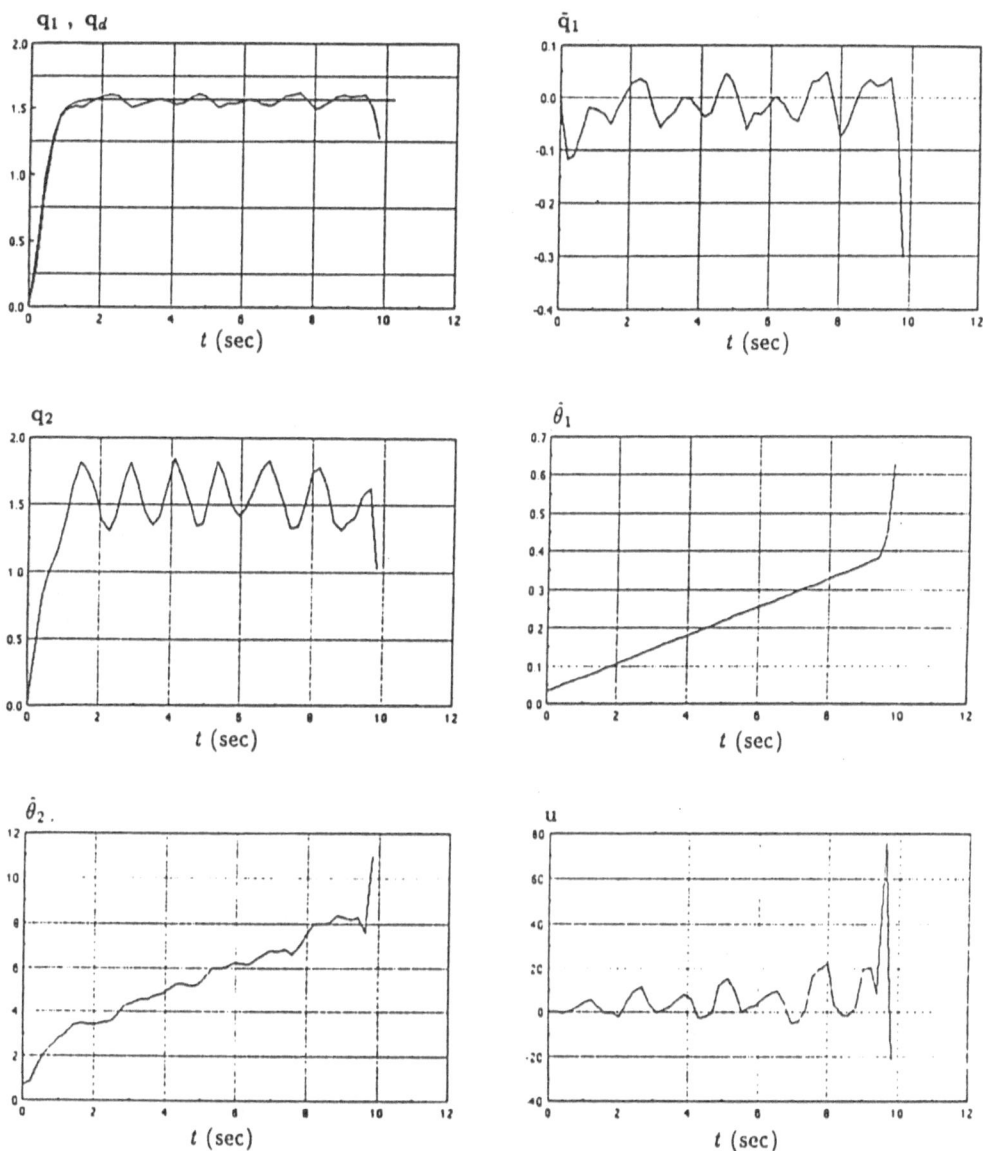

Figure 3. Response of Flexible Joint System with only Adaptive Rigid Control

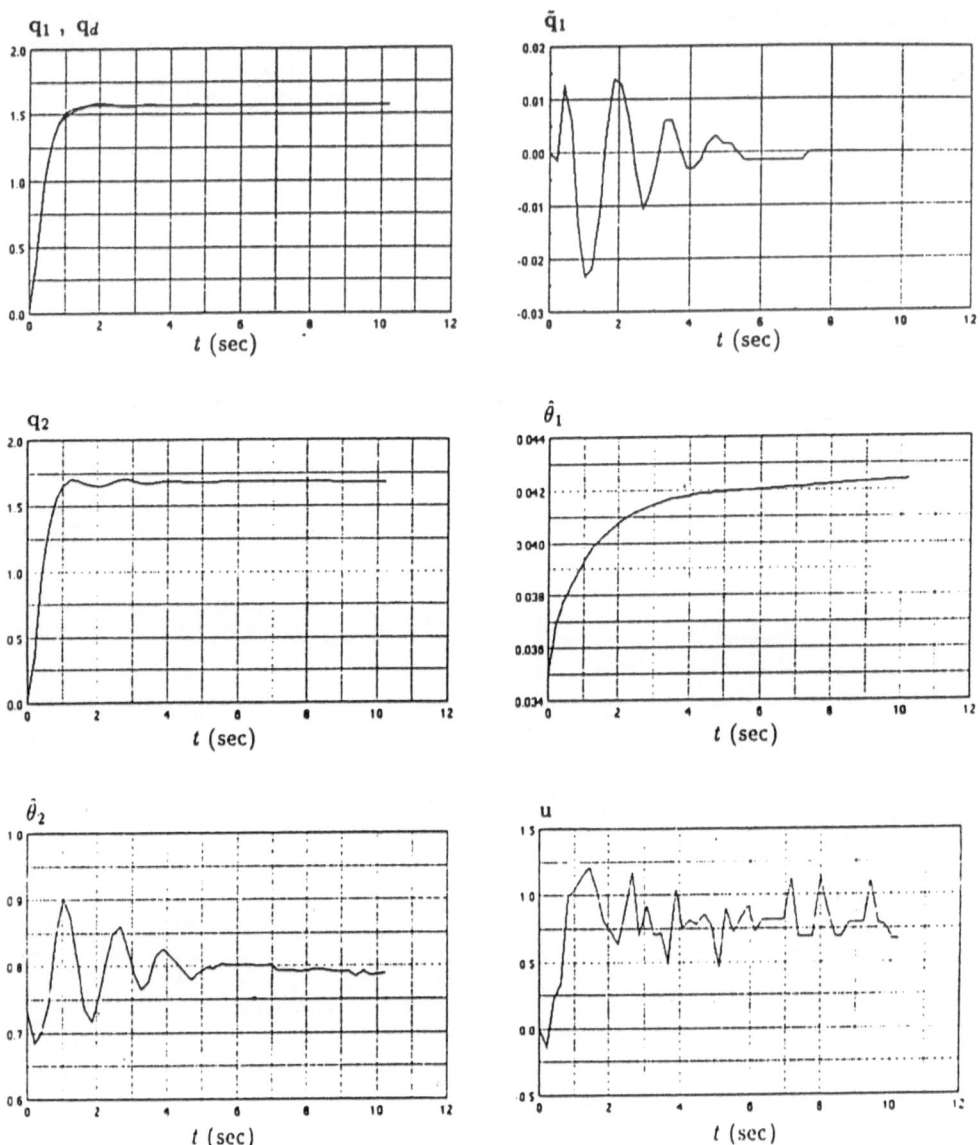

Figure 4. Response of Flexible Joint System with Adaptive Composite Control

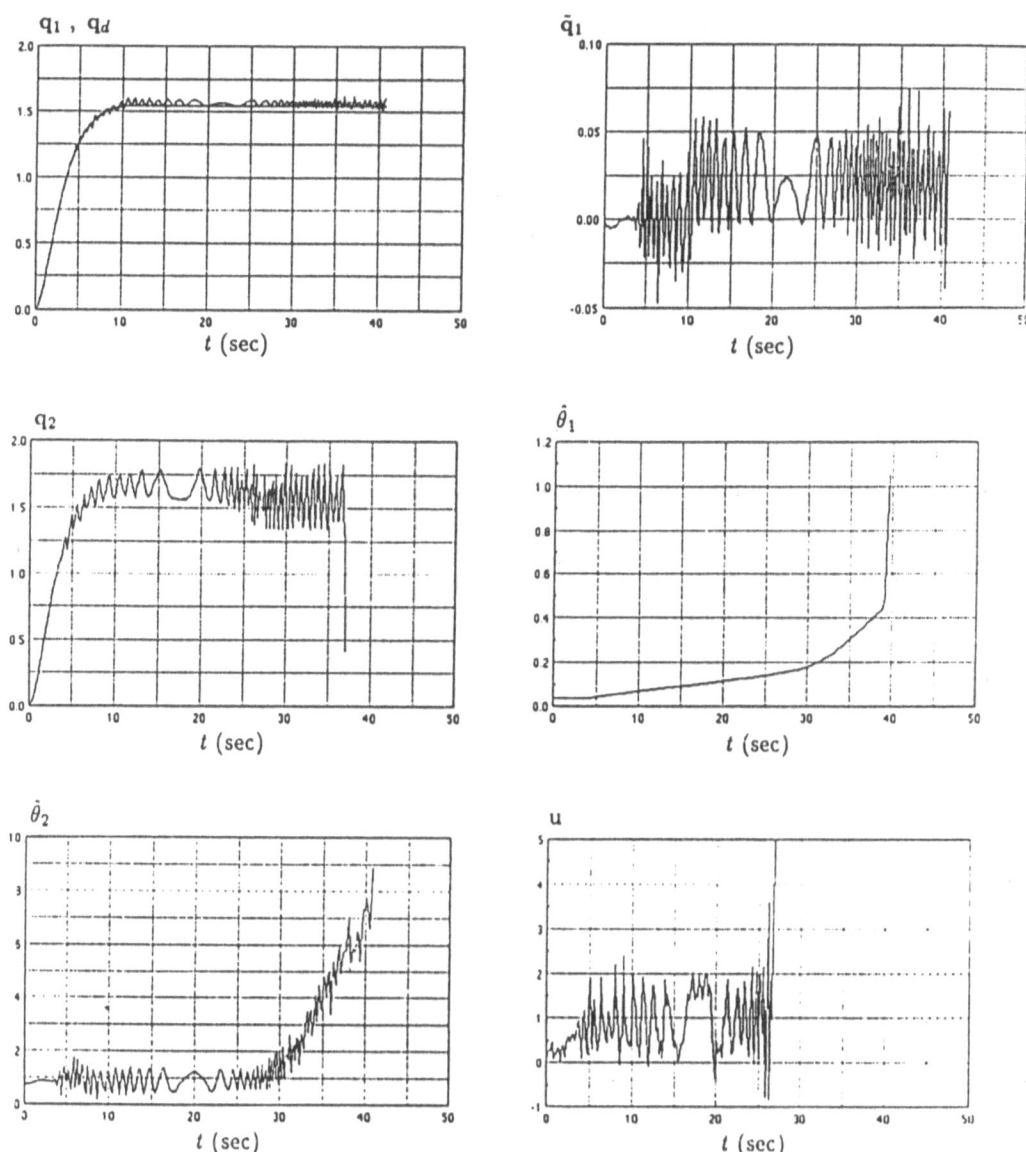

Figure 5. Instability Caused by Fast Reference Trajectory

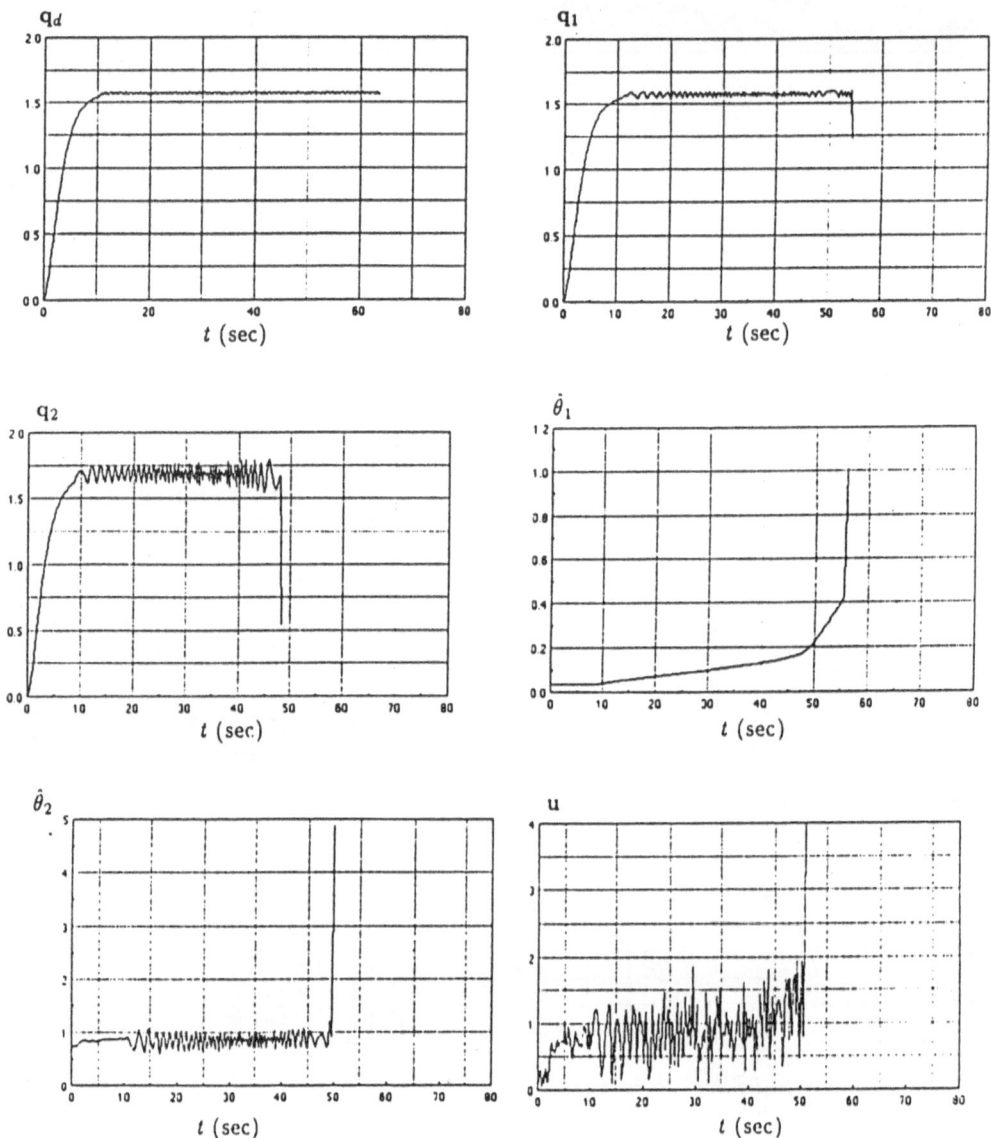

Figure 6. Parameter Drift Instability

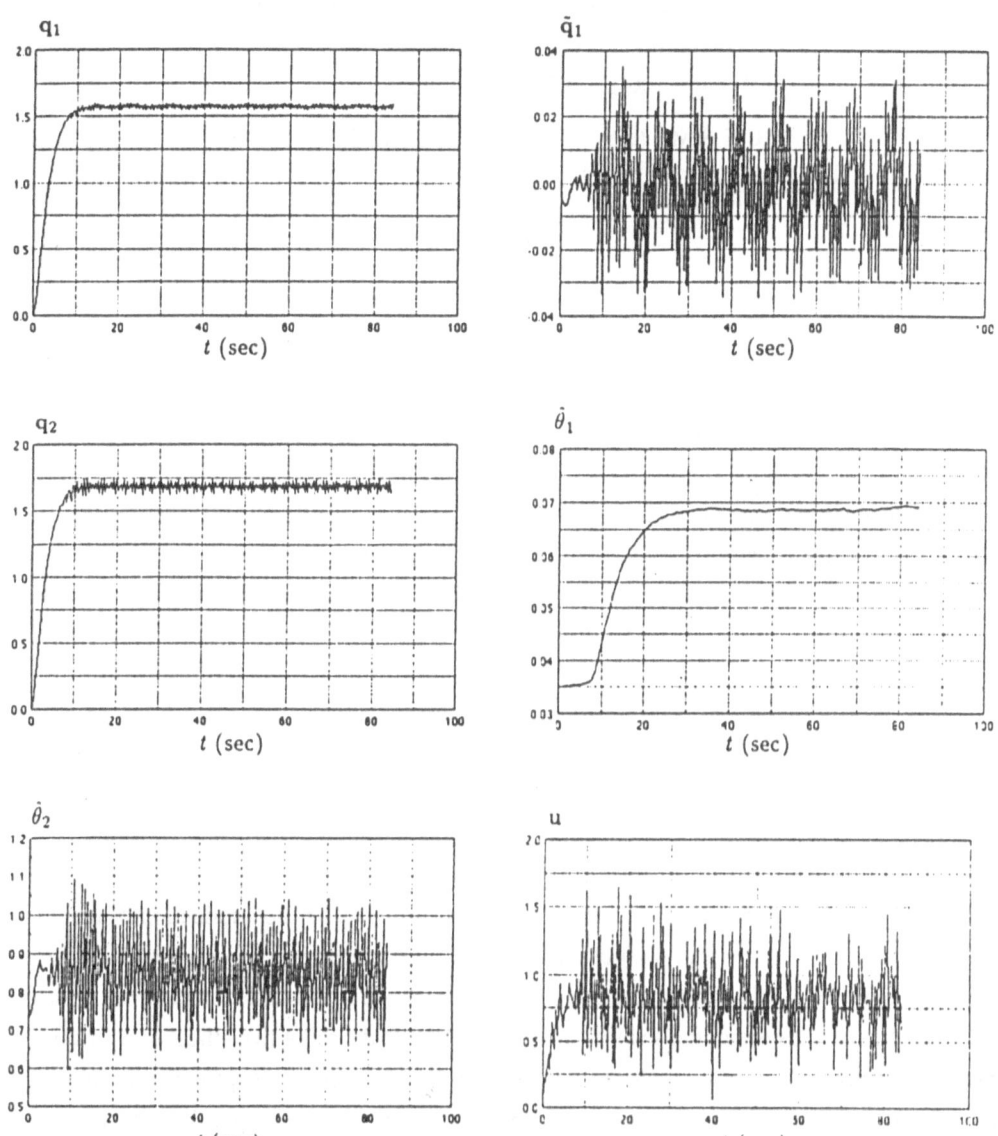

Figure 7. System Response with $\sigma$-modification

# ENERGY BASED ADAPTIVE ROBOTS CONTROLLER

K. EL SERAFI, W. KHALIL

E.N.S.M , Lab. d'Automatique UAR C.N.R.S 823
1 Rue de la Noë , 44072 Nantes cedex
FRANCE

## Abstract

This paper presents a new indirect adaptive controller for robots. The control law is a computed torque type, which can be calculated directly using fast recursive Newton-Euler algorithm. The inertial parameters of the robot are updated through the minimization of an energy prediction error, function of the joint positions and velocities. The calculation cost of the complete adaptive control law for the 6 degree of freedom PUMA robot is given.

## 1–Introduction

In order to increase the performance of robots, in speed and accuracy, controllers based on dynamic model are usually proposed. If the robot inertial parameters are well known, computed torque controllers guarantee the exact tracking of the desired trajectories. To deal with the poor knowledge of the values of the inertial parameters, the development of adaptive controllers is required.

The first works on adaptive control of robots are based on the application of methods developed for linear systems [1, 2], but such methods do not lead to an overall adaptive scheme that can be proved stable. Horowitz and Tomizuka [3] are the first to take into account some of the dynamics of the system, the stability proof assumes that the robot dynamics is much slower than the adaptation law which leads to limit the robot speed. Craig et al.[4] formulate a globally convergent nonlinear decoupling adaptive controller avoiding this approximation but their method requires to estimate or measure the joint accelerations, and assumes that the inverse of the estimated inertia matrix remains bounded. Slotine and Li [5] and Sadegh and Horowitz [6] prove the global asymptotic stability of a direct adaptive control law which compensates the nonlinearities of the robot equations and estimates on-line the inertial parameters of the links. They utilize the Lyapunov function approach and the passivity property of the robot to prove the stability of the control scheme, but the on-line implementation of the proposed control law is difficult. Landau and Horowitz [7] demonstrate the stability of Slotine adaptive controller using hyper stability theorem. At the same time Kelly and Ortega [8] use the input–output approach to prove the stability of the same control law.

Middleton and Goodwin [9] propose an indirect nonlinear decoupling adaptive controller ensuring global tracking convergence and avoiding acceleration measurements, but its implementation is difficult because it needs the explicit calculation of the time derivative of the inverse of the estimated inertia matrix and requires that the inverse of this matrix remains bounded. Li and Slotine [10, 11] develop a computed torque indirect adaptive controller which guarantees global asymptotic stability and avoids the inversion of the estimated inertia matrix, but the need to calculate the time derivative of this matrix increases the computational cost of the control law. The common disadvantage of the indirect controllers [9, 10, 11], which are based on a model obtained by filtering the dynamic model of the robot, is that there is no general method to calculate the regressor matrix of the adaptation law, and that the estimation of the inertial parameters needs to measure or calculate filtered values of the joint torques.

In this paper we propose an indirect computed torque adaptive control scheme. The control law is a computed torque one, while the adaptation mechanism is based on robot energy model. It can be considered as an extension of the off-line identification method proposed by Gautier and Khalil [12]. The proposed control law is simple to calculate on-line. It avoids the acceleration measurements and have no constraint on the estimated inertia matrix.

## 2–The dynamic model of robot

The robot is composed of n-rigid links in serial chain. The equations of motion of the system can be written as :

$$\Gamma = A(q)\ddot{q} + B(q, \dot{q})\dot{q} + Q(q) \tag{1}$$

where:

$A(q)$: is the (nxn) symmetric and positive definite robot inertia matrix.

$B(q, \dot{q})\dot{q}$ : is the (nx1) vector of the centrifugal and Coriolis forces.

$q, \dot{q}, \ddot{q}$: the (nx1) vectors of the joint positions, velocities, and accelerations.

$\Gamma$: (nx1) vector of joint torques or forces.

$Q$ : (nx1) vector of gravity torques.

The dynamic model (1) is linear in terms of a suitably selected set of links and load inertial parameters [13] and can be written in the form:

$$\Gamma = W(q, \dot{q}, \ddot{q})\, x \tag{2}$$

Where, $W$ is an (nx10n) matrix function of $q, \dot{q}, \ddot{q}$ and the constant geometric parameters of the robot, while $x$ is the (10nx1) vector of the links inertial parameters. We suppose that the constant geometric parameters are accurately known.

The vector x is given as:

$$x = [\ x_1^T\ x_2^T\ \quad \ldots\ \quad x_n^T]^T \tag{3}$$

Where $x_j$ is the (10x1) vector of the inertial parameters of link j. It is defined as:

$$x_j \quad = [\ XX_j\ XY_j\ XZ_j\ YY_j\ YZ_j\ ZZ_j\ MX_j\ MY_j\ MZ_j\ M_j\ ]^T$$

where :

$XX_j$ , ...., $ZZ_j$ are the elements of the inertia tensor of link j about the origin of its frame,

$MX_j$, $MY_j$, $MZ_j$ are the elements of the first moments of link j,

$M_j$ the mass of link j

Thus there are theoretically 10n inertial parameters for the robot. But some of these parameters have no effect on the dynamic model and some others can be regrouped. The minimum inertial parameters affecting the dynamic model constitute also the parameters to be adapted in an adaptive control strategy [14,15]. Using these parameters equation (2) will be:

$$\Gamma = D\ (\ q,\ \dot{q},\ \ddot{q})\ X \tag{4}$$

where D is (nxm) matrix, and X is the (mx1) minimum inertial parameters vector. The columns of D constitute a set of m independent columns of W [15].

## 3–Robot energy model

The robot total energy is linear in terms of the elements of the inertial parameters vector X, it can be written as:

$$H(t) \ = \ \sum_{i=1}^{m} h_i\ X_i \tag{5}$$

Where:

$H(t) = E(t) + U(t)$,

E: The kinetic energy of the robot,

U: The potential energy of the robot,

$h_i = \dfrac{\partial H}{\partial X_i}$ is function of q and $\dot{q}$.

In matrix form (5) can be written as :

$$H(t) = h\ (t)\ X \tag{6}$$

where h is (1xm) row vector

From the energy theorem , we obtain [12]:

$$\int_{t0}^{t} \dot{q}^T(\tau)\ \Gamma(\tau)\ d\tau = [\ h(t)\ -\ h(t0)]\ X \tag{7}$$

which can be written as:

$$\int_{t0}^{t} \dot{q}^T(\tau)\, \Gamma(\tau)\, d\tau \;=\; \Delta h\,(\,q(t),\; \dot{q}(t))\; X \tag{8}$$

where

$$\Delta h\,(q,\dot{q}) = h(t) - h(t0)$$

Denoting the $\dfrac{\partial H}{\partial X_i}$ functions corresponding to the inertial parameters of link j by the row vector:

$$h_j \;=\; [\,h_{XXj}\;\; h_{XYj}\;\; h_{XZj}\;\; h_{YYj}\;\; h_{YZj}\;\; h_{ZZj}\;\; h_{MXj}\;\; h_{MYj}\;\; h_{MZj}\;\; h_{Mj}\,],$$

the elements of $h_j$ can be calculated as follows [12]:

$$
\begin{aligned}
h_{XXj} &= \tfrac{1}{2}\,\omega_{1j}\,\omega_{1j}\\
h_{XYj} &= \omega_{1j}\,\omega_{2j}\\
h_{XZj} &= \omega_{1j}\,\omega_{3j}\\
h_{YYj} &= \tfrac{1}{2}\,\omega_{2j}\,\omega_{2j}\\
h_{YZj} &= \omega_{2j}\,\omega_{3j}\\
h_{ZZj} &= \tfrac{1}{2}\,\omega_{3j}\,\omega_{3j}\\
h_{MXj} &= \omega_{3j}\,V_{2j} - \omega_{2j}\,V_{3j} - g^T\;{}^0A_j\,a_1\\
h_{MYj} &= \omega_{1j}\,V_{3j} - \omega_{3j}\,V_{1j} - g^T\;{}^0A_j\,a_2\\
h_{MZj} &= \omega_{2j}\,V_{1j} - \omega_{1j}\,V_{2j} - g^T\;{}^0A_j\,a_3\\
h_{Mj} &= \tfrac{1}{2}\,{}^jV_j^T\,{}^jV_j - g^T\;{}^0P_j
\end{aligned}
\tag{9}
$$

where:

* $\omega_{kj}$ is the $k^{th}$ component of the rotational velocity of link j referred to coordinates frame j,
* $V_{kj}$ is the $k^{th}$ component of the linear velocity of link j referred to coordinates frame j,
* ${}^j\omega_j = [\,\omega_{1,j}\;\; \omega_{2,j}\;\; \omega_{3,j}\,]^T$
* ${}^jV_j = [\,V_{1,j}\;\; V_{2,j}\;\; V_{3,j}\,]^T$
* ${}^0A_j$ is the 3x3 matrix defining the orientation of the coordinates frame of link j with respect to frame 0.
* ${}^0P_j$ defines the position of the origin of the coordinates frame of link j with respect to frame 0.
* g is the acceleration of gravity.
* $a_1$, $a_2$ and $a_3$ are the following unit vectors:

$$a_1 = [1 \quad 0 \quad 0]^T, \qquad a_2 = [0 \quad 1 \quad 0]^T, \; a_3 = [0 \quad 0 \quad 1]^T$$

## 4–Adaptive computed torque control

The adaptive control scheme is composed of a computed torque control law, combined with an adaptation mechanism to update on-line the estimated inertial parameters.

### 4–1 Control law

We use a computed torque feed-forward dynamics compensation combined with a local PD feedback controller. The control law is given as:

$$\Gamma = \hat{\Gamma} - K_p\, e - K_v\, \dot{e} \tag{10}$$

where:

$\hat{\Gamma}$ is the feed-forward dynamics compensation vector which is calculated from the robot inverse dynamics as a function of the desired trajectory ($q_d$, $\dot{q}_d$, $\ddot{q}_d$) and the estimated inertial parameters $\hat{X}$

$K_p$ and $K_v$ represent the feedback gain matrices, which are constant, diagonal and positive definite.

$e$ and $\dot{e}$ are the tracking position and velocity errors, defined as:

$$e = q - q_d \tag{11}$$

$$\dot{e} = \dot{q} - \dot{q}_d \tag{12}$$

Using (4) the control law (10) can also be written as :

$$\Gamma = D(q_d, \dot{q}_d, \ddot{q}_d)\hat{X} - K_p\, e - K_v\, \dot{e} \tag{13}$$

The main advantage of this controller is that the computation of this feed-forward signal is approximately free from noise. Also in the case of adaptation of the model parameters there is no constraint on the estimated inertia matrix, which is the main source of instability in some previously proposed algorithms [4, 10, 11].

### 4–2 On-line estimation of the inertial parameters

Inertial parameters adaptation will be based on the robot energy model given by equation (8). It is to be noted that the $\Delta h$ vector is function of $q$ and $\dot{q}$ and not of $\ddot{q}$.

Using this model, we can define the energy prediction error $\delta(\hat{X})$ corresponding to an estimate $\hat{X}$ of the inertial parameter vector $X$ as:

$$\delta(\hat{X}) = \Delta h(q, \dot{q})\, (\hat{X} - X) \tag{14}$$

or

$$\delta(\hat{X}) = \Delta h(q, \dot{q}) \, \tilde{X} \tag{15}$$

where

$$\tilde{X} = \hat{X} - X$$

Using (8), we obtain that the energy prediction error $\delta(\hat{X})$ can be calculated on-line from the following relation:

$$\delta(\hat{X}(t)) = \Delta h(q, \dot{q}) \, \hat{X}(t) - \int_{t0}^{t} \dot{q}(\tau)^T \Gamma(\tau) \, d\tau \tag{16}$$

Equation (14) can be used to update the estimated inertial parameters. We propose to use a gradient algorithm defined as:

$$\dot{\hat{X}} = - K \, \Delta h^T(q, \dot{q}) \, \delta(t) \tag{17}$$

where $K$ is symmetric positive definite matrix. The adaptation law (17) is asymptotically convergent, and have the following properties :

1) $\tilde{X}$ is bounded          2) $\delta(t)$ belongs to $L_2$

These properties can be demonstrated using the following Lyapunov function:

$$V(t) = 1/2 \, \tilde{X}^T K^{-1} \, \tilde{X} \tag{18}$$

Thus:

$$\dot{V}(t) = \tilde{X}^T K^{-1} \dot{\tilde{X}} \tag{19}$$

and using equation (17)

$$\dot{V}(t) = - \tilde{X}^T \Delta h^T (q, \dot{q}) \, \delta \tag{20}$$

$$\dot{V}(t) = - \delta^2 \tag{21}$$

Thus the estimated parameters will converge exponentially to the real values if the trajectories are sufficiently exciting.

Figure (1) represents the block diagram of the proposed adaptive controller.


## 5–Stability of the adaptive controller

From (1) and (10) the error dynamic equation of the robot is given as :

$$\ddot{e} + \kappa_v \, A^{-1}(q) \, (\mu \, e + \dot{e}) + \zeta(q, \dot{q}) = 0 \tag{22}$$

with

$$\zeta(q, \dot{q}) = - A^{-1}(q) \, [\hat{A}(q_d) \ddot{q}_d + (\hat{B}(q_d, \dot{q}_d) - B(q, \dot{q})) + (\hat{Q}(q_d) - Q(q))] + \ddot{q}_d \tag{23}$$

Where the feedback gain matrices are given as :

$$K_v = \kappa_v \, I \qquad \qquad K_p = \mu \, K_v$$

with $\kappa_v$ and $\mu$ are positive scalars.

Using the general stability theory developed by Samson in [16], it can be seen that the error dynamics equation (22) satisfies the stability conditions of samson's theory. Then there exist a minimum value of $\kappa_v$ above which the stability of the system is ensured. The minimum value of $\kappa_v$ is proportional to the modelling errors, and inversly proportional to the upper bound of the tracking errors.

As a conclusion the proposed adaptive control law is stable if appropriate feedback gains are used. Also since the parameters error vector $\widetilde{X}$ is proved to be decreasing, the position and velocity error bounds will decrease as the parameters converge to the real values.

## 6–Computation complexity

From equations (13) and (17), we need to calculate $\hat{\Gamma}$ and $\Delta h$. The calculation of $\hat{\Gamma}$ can be carried out using the customized Newton-Euler algorithm [13]. The calculation of $\Delta h$ is easy and can be performed using equation (9). The elements of the $h$ vector corresponding to the minimum inertial parameters of the 6 d.o.f. PUMA robot are given in appendix 1, they have been obtained using the software package SYMORO [17]. The minimum set of inertial parameters of this robot are given in table 1 [14]. The computation cost of the proposed control and adaptation laws, in the case of the 6 d.o.f. PUMA , are given in tables 2 and 3. The inertial parameters of all the links are assumed unknown, and will be updated on-line. The integration is carried out by trapezoidal rule. The total number of operations is equal to 404 additions and 521 multiplications.

| Link | XX | XY | XZ | YY | YZ | ZZ | MX | MY | MZ | M |
|------|------|------|------|----|------|-------|--------|--------|----|---|
| 1 | 0 | 0 | 0 | 0 | 0 | $ZZR1$ | 0 | 0 | 0 | 0 |
| 2 | $XXR_2$ | $XY_2$ | $XZR_2$ | 0 | $YZ_2$ | $ZZR_2$ | $MXR_2$ | $MY_2$ | 0 | 0 |
| 3 | $XXR_3$ | $XYR_3$ | $XZ_3$ | 0 | $YZ_3$ | $ZZR_3$ | $MXR_3$ | $MYR_3$ | 0 | 0 |
| 4 | $XXR_4$ | $XY_4$ | $XZ_4$ | 0 | $YZ_4$ | $ZZR_4$ | $MX_4$ | $MYR_4$ | 0 | 0 |
| 5 | $XXR_5$ | $XY_5$ | $XZ_5$ | 0 | $YZ_5$ | $ZZR_5$ | $MX_5$ | $MYR_5$ | 0 | 0 |
| 6 | $XXR_6$ | $XY_6$ | $XZ_6$ | 0 | $YZ_6$ | $ZZ_6$ | $MX_6$ | $MY6$ | 0 | 0 |

**Table 1**: The minimum set of inertial parameters of PUMA

| Computation of | Number of operations | |
|---|---|---|
| | Additions | Multiplications |
| Inverse dynamic calculation | 249 | 265 |
| $e$ , $\dot{e}$ | 12 | 0 |
| $k_p e$ | 0 | 6 |
| $k_v \dot{e}$ | 0 | 6 |
| $\Gamma$ | 12 | 0 |
| Total number | 273 | 277 |

**Table 2**: Control law computation cost when adapting all the inertial parameters.

| Computation of | Number of operations | |
|---|---|---|
| | Additions | Multiplications |
| $h(q, \dot{q})$ | 53 | 130 |
| $\Delta h(q, \dot{q})$ | 36 | 0 |
| $\delta$ | 42 | 42 |
| $\dot{\hat{x}}$ | 0 | 72 |
| Total number | 131 | 244 |

**Table 3**: Adaptation law computation cost when adapting all the inertial parameters.

## 7–Simulation results

Simulation studies are conducted for the PUMA 560 robot. The values of the inertial parameters used in the simulations are those given by Armstrong [18]. The simulation is limited to the first three links by blocking the wrist. The nominal values of the minimum inertial parameters of the robot are given in tables 4 and 5. The inertial parameters of the resulting third link are supposed unknown and will be adapted on-line. The parameters XX3, YY3, ZZ3, MY3, M3 will be only concerned by the adaptation procedure, the nominal values of the other parameters are equal to zero. The following feedback and adaptation gains are used:

$$K_v = \text{diag} (30.0, 40.0, 30.0), \qquad K_p = \text{diag} (225.0, 400.0, 225.0),$$

$$K = \text{diag} (5.0, \ 0.07, 0.08, 0.03, 0.08).$$

| Link Nº | XX | XY | XZ | YY | YZ | ZZ |
|---------|--------|---------|--------|--------|-------|--------|
| 1 | 0.0 | 0.0 | 0.0 | 0.0 | 0.0 | 2.995 |
| 2 | −0.4738 | −0.0071 | 0.0189 | 0.0 | .0017 | 5.3301 |
| 3 | 0.3373 | 0.0 | 0.0 | 0.0168 | 0.0 | 0.3560 |

**Table 4:** Nominal values of the minimum parameters of the inertia matrices.

| Link Nº | MX | MY | MZ | M |
|---------|-----|---------|--------|------|
| 1 | 0.0 | 1.1832 | 0.1044 | 0.0 |
| 2 | 0.0 | 1.1832 | 0.1044 | 0.0 |
| 3 | 0.0 | −1.0067 | 0.0 | 7.04 |

**Table 5:** Nominal values of the minimum first moments and mass parameters.

The maximum joint velocities are [1.43, 0.935, 2.432] rad/sec.

The maximum joint accelerations are [8.512, 5.565, 12.691] rad/sec$^2$.

The desired trajectory, figure 2, is chosen such that the motion of each joint will be not in phase with the others. The trajectory of link 1 is a polynomial of 5$^{th}$ degree, generated between 0.0 and 4.18 rad, the desired trajectory of the second joint is a sinusoidal function starting at 0.0 and ends at 0.0 with maximum displacement 6.28 rad, and the third joint is a sinusoidal multi period trajectory generated between 0.0 and 6.28 rad. Initial and final velocities and accelerations are supposed to be zero. The simulation is extended beyond the final position for testing the static behavior of the system. Simulations are carried out for the following two cases:
1- Zero initial values of the estimated inertial parameters, the initial estimation error equal to − 100 %

2- The initial values of the estimated inertial parameters equal to 200 % its real values, the initial estimation error equal to + 100 %.

Each case is compared with the case without adaptation $K = 0$.

The results can be summarized as follows:

For the first case:
- Figure (3) represents the norm of the position error of the three joints.
- Figure (4) represents the norm of the velocity error of the three joints.
- Figures (5) give the evolution of the estimation of the inertial parameters.
- Figure (6) shows the evolution of the energy error.

For the second case:
- The corresponding results are given in figures (7), (8), (9), (10) respectively.

We remark the reduction in the tracking errors in the presence of the adaptation. Also we find that most of the inertial parameters converge towards their real values.

## 8–Conclusion

In this paper we present an adaptive computed torque controller. The inertial parameters are updated through the evaluation of an energy error function depending on the joint positions and velocities. The calculation of the adaptive law for the 6 degree of freedom PUMA robot illustrates that it can be calculated on-line. Simulation results on the first three links of the PUMA robot show that most of the estimated parameters converge towards the real values, and that the error in position and velocity decrease when using adaptation.

## References

[1] S. Dubowsky, D. T. Forges  " The application of model-referenced adaptive control to robotic manipulators ", Trans. of ASME, J. of Dynamic Systems, Measurements, and Control, Vol. 101, 1979, pp. 193 - 200.

[2] T.C.Hsia, " Adaptive control of robot manipulators–a review" Proc. IEEE  Conf. on Robotics and Automation San Francisco, April 1986, pp.183 - 189

[3] R.Horowitz, M.Tomizuka,"An adaptive control scheme for mechanical manipulators. Compensation of nonlinearity and decoupling control", Presentation at the Winter Meeting of ASME, Dynamic Systems and Control Division, Chicago, 1980

[4] J.J Craig, P. Hsu, S.S. Sastry, " Adaptive control of mechanical manipulators " Proc.1986 IEEE  Conf. on Robotics and Automation, San Francisco, April 1986, pp. 190 - 195.

[5] J.J.Slotine , W.Li , " Adaptive manipulator control. A case study " Proc. IEEE Conf. on Robotics and Automation, Raleigh, March 1987, pp. 1392 - 1400.

[6] N.Sadegh, R.Horowitz," Stability analysis of an adaptive controller for robotic manipulators " Proc. IEEE Conf. on Robotics and Automation, Raleigh, March 1987, pp. 1223–1229.

[7]I.D.Landau,R.Horowitz" Synthesis of adaptive controllers for robot  manipulators using a passive feedback System Approach." Proc.IEEE  Int. Conf. on Robotics and Automation Philadelphia,  April 1988, pp. 1028 - 1033.

[8] R.Kelly,R.Ortega " Adaptive control of robot manipulators : An  input–output approach " Proc.IEEE Conf. on Robotics and Automation  Philadelphia, April 1988, pp. 699 - 703

[9] R. H. Middleton, G. C. Goodwin, " Adaptive computed torque control for rigid link manipulators ", Systems & Control Letters, Vol.10 , 1988, pp. 9 – 16.

[10] W.Li, J.J.Slotine " Indirect adaptive robot control "  Proc.IEEE Conf. on Robotics and Automation Philadelphia, April 1988, pp. 704 - 709

[11] W.Li, J.J.Slotine  " An indirect adaptive robot controller ", Systems & Control Letters, Vol.12 , 1989, pp. 259 – 266.

[12] M. Gautier, W. Khalil, "On the identification of the Inertial parameters of robots" 27th IEEE-CDC Conf., Austin, Dec. 1988, pp.2264 - 2269

[13] W.Khalil, J.F.Kleinfinger,"Minimum operations and minimum parameters of the dynamic models of tree structure robots", IEEE J. Robotics and Automation, Vol. RA-3(6), Dec. 1987, pp.517-526.

[14] M.Gautier and W.Khalil, "Direct calculation of minimum set of inertial parameters of serial robots", IEEE Transactions on Robotics and Automation, Vol. (6), June 1990, pp.386-373.

[15] M.Gautier, "Numerical calculation of the base inertial parameters. Proc. IEEE Conf. on Robotics and Automation, Cincinatti 1990, pp. 1020-1025.

[16]C.Samson, "Robust control of a class of non-linear systems and applications to robotics.Int.Journal of adaptive control and signal processing, Vol.1,1987, pp.49-68.

[17]W.Khalil, "SYMORO: Symbolic modelling of robots, user guide", LAN, ENSM, 1990.

[18] B. Armstrong, O. Khatib, J. Burdick " The explicit dynamic model and inertial parameters of the PUMA 560 arm ", Proc.IEEE Conf. on Robotics and Automation, San Francisco, California, April 1986. pp. 510 - 518.

## Appendix 1

Notations: $\dot{q}_j = QPj$ , $\cos\theta_j = Cj$ , $\sin\theta_j = Sj$

gravity acceleration : $g = [\,0 \qquad 0 \qquad G3\,]^T$

Equations:

```
hZZ1   = 0.5*QP1*QP1
DU91   = - G3
hXX2   = 0.5*S2*QP1*S2*QP1
hXY2   = S2*QP1*C2*QP1
hXZ2   = -S2*QP1*QP2
hYZ2   = - C2*QP1*QP2
hZZ2   = 0.5*QP2*QP2
DU72   = -S2*DU91
DU82   = -C2*DU91
W13    = -C3*S2*QP1-S3*C2*QP1
V13    = -C3*C2*QP1*R3+S3*(QP2*D3+S2*QP1*R3)
W23    = S3*S2*QP1-C3*C2*QP1
V23    = S3*C2*QP1*R3+C3*(QP2*D3+S2*QP1*R3)
W33    = QP2+QP3
V33    = C2*QP1*D3
hXX3   = 0.5*W13*W13
hXY3   = W13*W23
hXZ3   = W13*W33
hYZ3   = W23*W33
hZZ3   = 0.5*W33*W33
DE73   = V23*W33-V33*W23
DE83   = V33*W13-V13*W33
DU73   = S3*DU82+C3*DU72
DU83   = C3*DU82-S3*DU72
```

```
W14     = C4*W13-S4*W33
V14     = C4*(V13-W33*R4)-S4*(V33+W13*R4-W23*D4)
W24     = -S4*W13-C4*W33
V24     = -S4*(V13-W33*R4)-C4*(V33+W13*R4-W23*D4)
W34     = W23+QP4
V34     = V23+W33*D4
hXX4    = 0.5*W14*W14
hXY4    = W14*W24
hXZ4    = W14*W34
hYZ4    = W24*W34
hZZ4    = 0.5*W34*W34
DE74    = V24*W34-V34*W24
DE84    = V34*W14-V14*W34
DU74    = C4*DU73
DU84    = -S4*DU73
DU94    = DU83
W15     = C5*W14+S5*W34
V15     = C5*V14+S5*V34
W25     = -S5*W14+C5*W34
V25     = -S5*V14+C5*V34
W35     = -W24+QP5
hXX5    = 0.5*W15*W15 h
hXY5    = W15*W25
hXZ5    = W15*W35
hYZ5    = W25*W35
hZZ5    = 0.5*W35*W35
DE75    = V25*W35+V24*W25
DE85    = -V24*W15-V15*W35
DU75    = S5*DU94+C5*DU74
DU85    = C5*DU94-S5*DU74
DU95    = -DU84
W16     = C6*W15-S6*W35
V16     = C6*V15+S6*V24
W26     = -S6*W15-C6*W35
V26     = -S6*V15+C6*V24
W36     = W25+QP6
hXX6    = 0.5*W16*W16
hXY6    = W16*W26
hXZ6    = W16*W36
hYZ6    = W26*W36
hZZ6    = 0.5*W36*W36
DE76    = V26*W36-V25*W26
DE86    = V25*W16-V16*W36
DU76    = -S6*DU95+C6*DU75
DU86    = -C6*DU95-S6*DU75

hMX2     = DE72 + DU72
hMY2     = DE82 + DU82
hMX3     = DE73 + DU73
hMY3     = DE83 + DU83
hMX4     = DE74 + DU74
hMY4     = DE84 + DU84
hMX5     = DE75 + DU75
hMY5     = DE85 + DU85
hMX6     = DE76 + DU76
hMY6     = DE86 + DU86
```

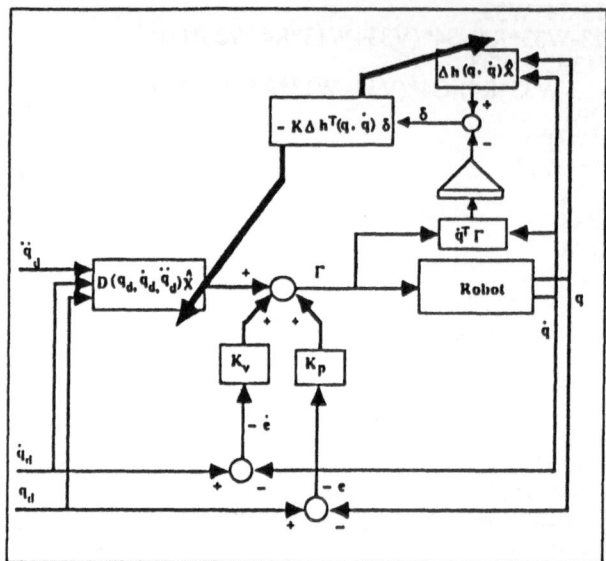

Figure 1 Energy based indirect adaptive controller

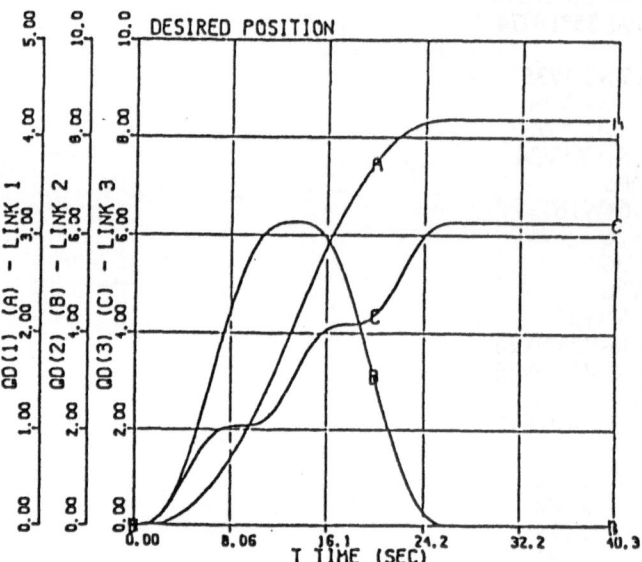

Figure 2 The desired trajectory (position in rd).
A - Axis 1, B - Axis 2, C - Axis 3

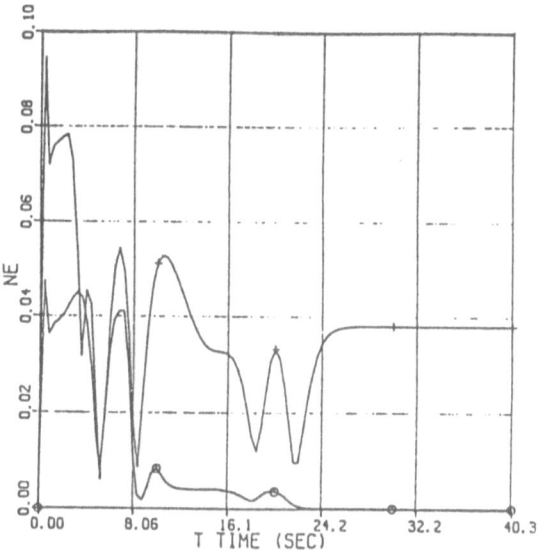

**Figure 3** Norm of position tracking error in rd.
Initial parameters error = – 100 %, (⊕ with adaptation, + without adaptation).

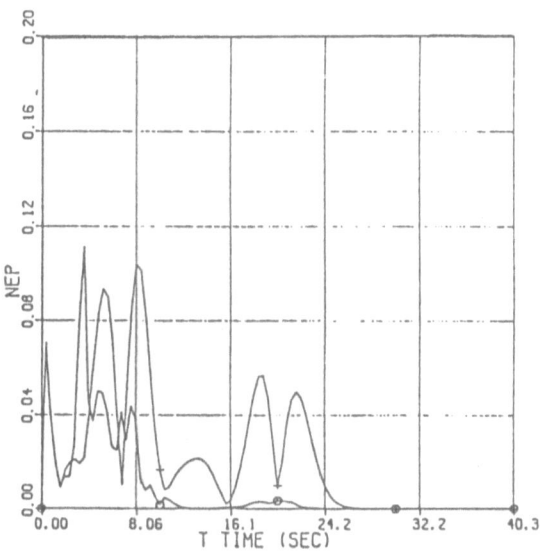

**Figure 4** Norm of velocity error in rd/sec.
Initial parameters error = – 100 %, (⊕ with adaptation, + without adaptation).

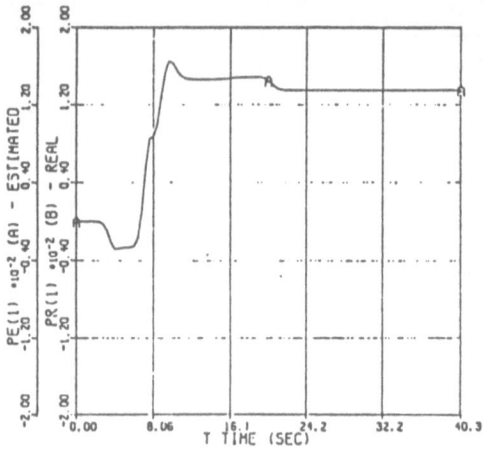

**Figure 5-a** Evolution of the estimation
of XX3 in Kg.m2

**Figure 5-b** Evolution of the estimation
of YY3 in Kg.m2

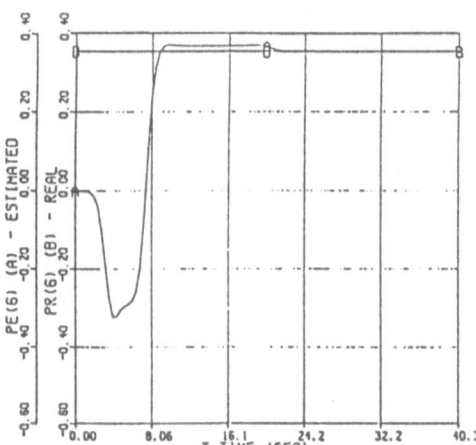

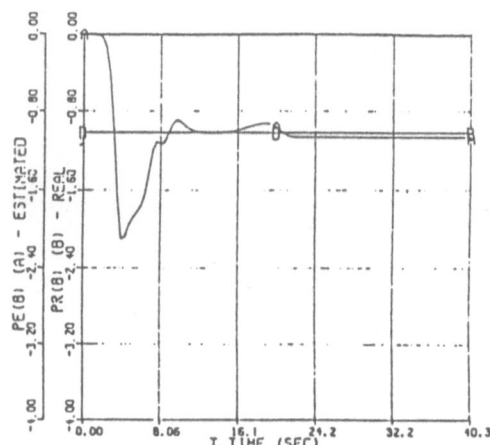

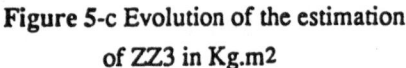

**Figure 5-c** Evolution of the estimation
of ZZ3 in Kg.m2

**Figure 5-d** Evolution of the estimation
of MY3 in Kg.m

45

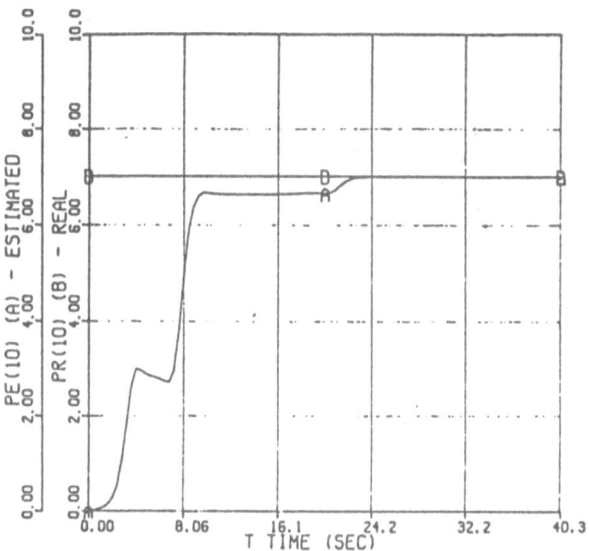

**Figure 5-e** Evolution of the estimation of M3 in Kg.

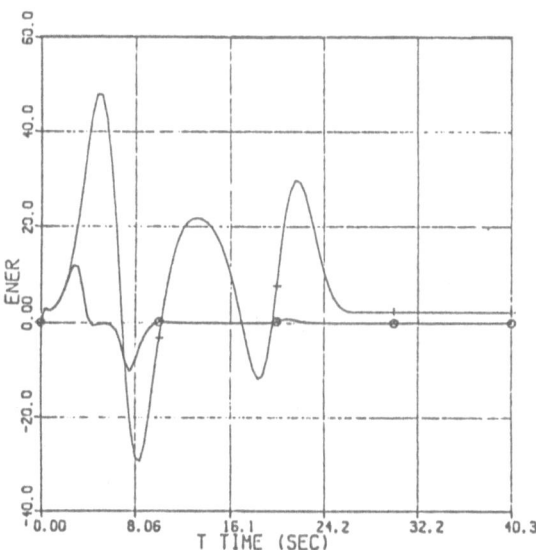

**Figure 6** The robot total energy error, (⊕ with adaptation, + without adaptation).

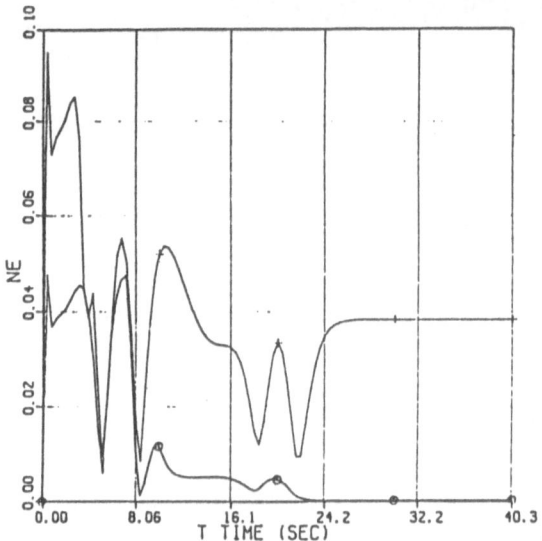

**Figure 7** Norm of position tracking error in rd.
Initial parameters error = + 100 %, (⊕ with adaptation, + without adaptation).

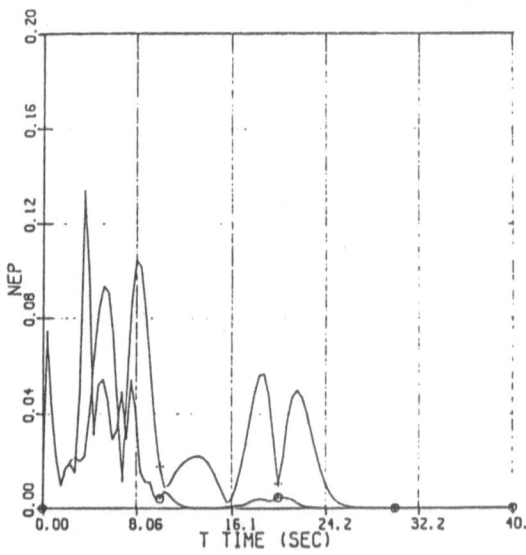

**Figure 8** Norm of velocity error rd/sec.
Initial parameters error = + 100 %, (⊕ with adaptation, + without adaptation).

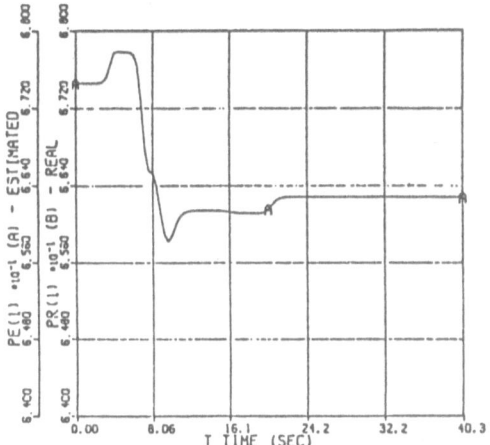

**Figure 9-a** Evolution of the estimation
of XX3 in Kg.m2

**Figure 9-b** Evolution of the estimation
of YY3 in Kg.m2

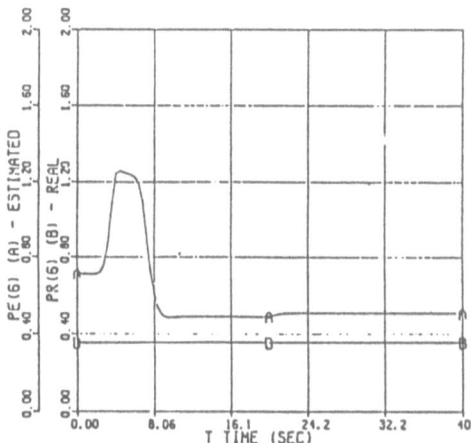
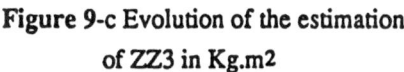

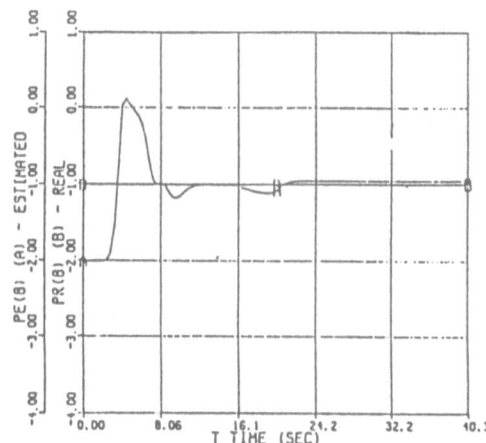

**Figure 9-c** Evolution of the estimation
of ZZ3 in Kg.m2

**Figure 9-d** Evolution of the estimation
of MY3 in Kg.m

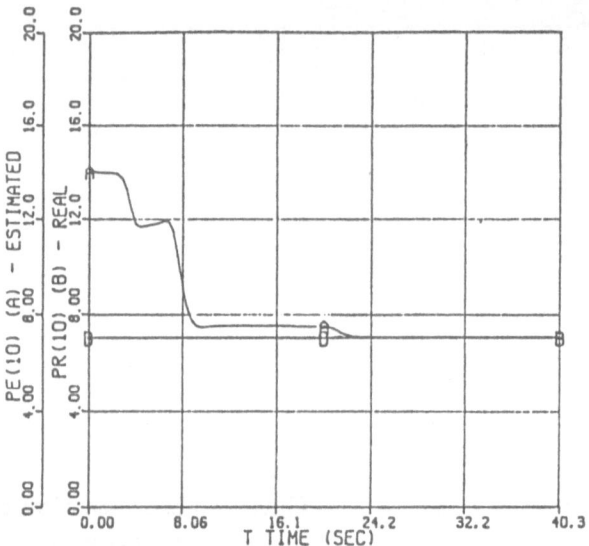

**Figure 9-e** Evolution of the estimation of M3 in Kg.

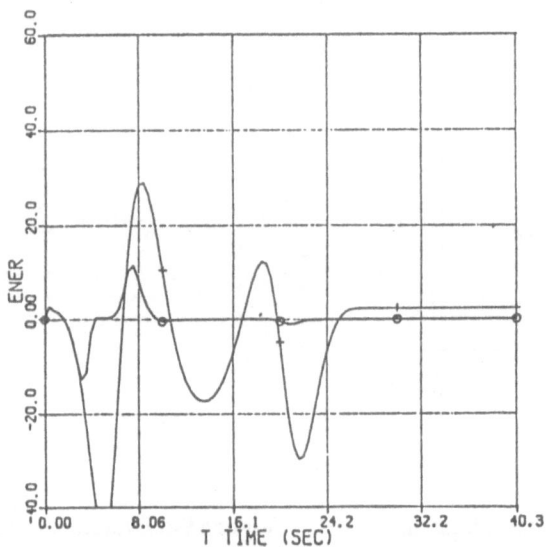

**Figure 10** The robot total energy error, (⊕ with adaptation, + without adaptation).

# Passivity of Robot Dynamics Implies Capability of Motor Program Learning

Suguru Arimoto

Faculty of Engineering, University of Tokyo

Bunkyo-ku, Tokyo, 113 Japan

## Abstract

Learning control is a new approach to the problem of skill refinement for robot arms by iterative training. It is considered to be a mathematical model of motor program learning for skilled motions in the central nervous system.

This paper proposes a class of learning control algorithms with a forgetting factor $1 >> \alpha > 0$ and without differentiation of velocity signals, which updates the input command by $u_{k+1} = (1 - \alpha)u_k + \alpha u_0 + \Phi e_k$, where $u_k$ and $e_k$ stand for command input and velocity error at $k$-th exercise respectively. The robustness of this learning control with respect to reinitialization errors, fluctuations of robot dynamics, and measurement noises is studied in detail. It is shown that not only the passivity of robot dynamics but also the exponential passivity of displacement dynamics on errors and difference dynamics between consecutive trials play a crucial role in proving the uniform boundedness of transient behaviors and the convergence in the progress of learning. Furthermore, two methods of learning called "interval training" and "selective learning" are proposed, which updates $u_0$ in the long-term memory every after several trials by the current command input $u_k$ or by the best command input among the past trials. The effectiveness of these methods in acceleration of the speed of convergence is also discussed.

# 1 Introduction

The culture of mankind began with use and creation of tools. Amongst all creation man is the most skillful with his fingers and hands. We humans owe such skilled motions to inherent abilities of learning. Our babies are so clumsy with their hands that they are unable to manipulate a knife and fork. But they are able to improve their motions from repeated exercise.

The above observation led recently to the proposal of learning control theory for improvement of robot motions. The concept of learning control differs from that of conventional control methodology. It is a discipline for a class of mechanical robots and mechatronics systems, which is based on autonomous self-training. It can be regarded as a mathematical model of motor program learning in the process of skill refinement.

Given a desired trajectory $y_d(t)$ on $t \in [0, T]$ to the robot, learning control generates the command input $u_k(t)$ for the robot at trial $k$ so that the output $y_k(t)$ is as close as possible to the desired one $y_d(t)$. A new input $u_{k+1}$ at the next time is generated by a simple recursive learning law $u_{k+1} = F(u_k(t), e_k(t))$ where $e_k(t) = y_d(t) - y_k(t)$. The author and his colleagues [1]– [6] proposed so far two types of learning law (see Fig.1):

$$u_{k+1} = u_k(t) + \Gamma \frac{d}{dt}(y_d(t) - y_k(t)), \tag{1}$$

$$u_{k+1} = u_k(t) + \Phi(y_d(t) - y_k(t)). \tag{2}$$

In both cases both measured output $y_k$ and desired output $y_d$ stand for velocity signals of joint coordinates, and $\Gamma$ and $\Phi$ are constant gain matrices. It has been shown that the D-type learning control defined by eq.(1) with an appropriate gain matrix $\Gamma$ is convergent in a sense that the output trajectory approaches the desired one with the repetition of operation, namely, $y_k(t) \rightarrow y_d(t)$ uniformly in $t \in [0, T]$ as $k \rightarrow \infty$. A similar but weaker result on convergence of the P-type learning control described by eq.(2) has also been obtained by the author [4], but the argument in its proof is based on a linearized dynamics model of the robot manipulator around the desired motion trajectory and ignorance of higher terms. In addition, it was implicitly assumed in both cases that the manipulator must take the same initial position and velocity at every operation trial and none of fluctuations of dynamics and measurement noise arises throughout the repetition of

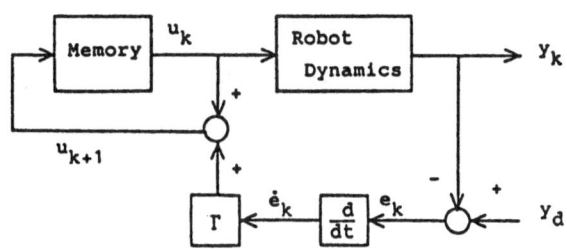

Fig.1 (a) D-type learning control

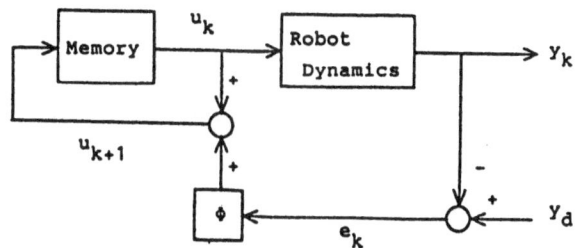

Fig.1 (b) P-type learning control

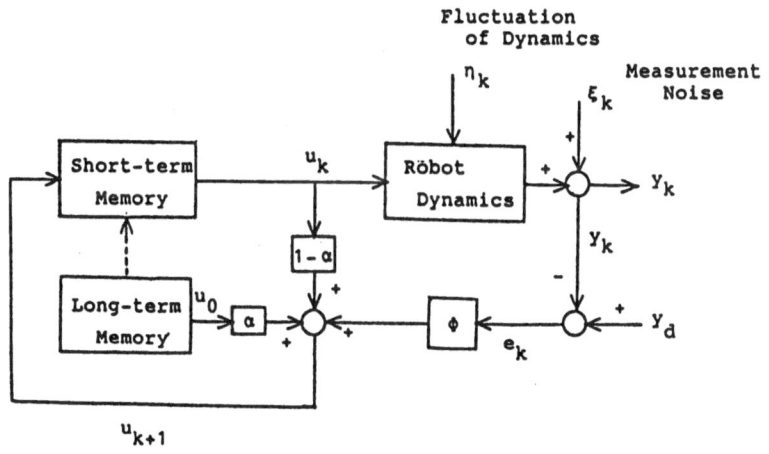

Fig. 2 Learning control with
a forgetting factor $\alpha$

exercise. Although most of present industrial robots may satisfy well approximately these conditions because of their superiority of repeatability precision, it is important to assure the technical soundness and robustness of the learning control with respect to small but persistent errors of initialization, fluctuations of dynamics, and measurement noise during operation.

In the present paper we study in detail robustness problems of P-type learning control with respect to initialization errors, fluctuations of dynamics, and measurement noise. The reason of avoiding the D-type learning is that numerical differentiation of velocity signals may cause more noises and even amplify them. By introducing a forgetting factor $0 < \alpha << 1$ in such a way (see Fig.2) that

$$u_{k+1}(t) = (1 - \alpha)u_k(t) + \alpha u_0(t) + \Phi(y_d(t) - y_k(t)) \tag{3}$$

we prove the uniform boundedness of motion trajectories on the basis of a variety of passivity properties of robot dynamics. These are 1) the passivity of direct robot dynamics, 2) the exponential passivity of direct robot dynamics, 3) the exponential passivity of displacement dynamics in terms of $r_k(t)(= q_k(t) - q_d(t))$, and 4) the exponential passivity of difference dynamics in terms of $d_k(t)(= q_{k+1}(t) - q_k(t))$. With the aid of these properties inherent to robot dynamics and introduction of a new concept of learning called "interval training", we finally present a rough sketch of the proof for the convergence that motion trajectories approach an $\alpha$-neighborhood of the desired one with repetition of self-trainings.

## 2 Passivity of Robot Dynamics

We consider a class of articulated robot arms, namely, serial-link manipulators whose all joints are of revolute-type. The dynamics of such a robot arm can be described in terms of joint coordinates vector $q = (q^1, \ldots, q^n)^T$ in the following way:

$$(H_0 + H(q))\ddot{q} + (B_0 + \dot{H}(q))\dot{q} - \frac{\partial K}{\partial q} + g(q) = u \tag{4}$$

where $H$ denotes an inertia matrix, $H_0$ a positive diagonal matrix representing inertial terms of internal load distribution of actuators, $K = \dot{q}^T(H_0 + H(q))\dot{q}/2$ the kinetic energy,

$g(q)$ a vector of gravity terms, $u$ a vector of input torques generated at servo actuators, $B_0$ a positive definite matrix representing damping factors and coefficients of electro-motive forces. It is well known that the inertia matrix $H(q)$ is symmetric and positive definite and, moreover, each entry of $H$ is constant or a trigonometric function in components of joint vector $q$. Hence, $H(q)$ and any of partial derivatives of $H(q)$ with respect to $q^i$ are uniformly Lipschitz continuous in $q$. Next observe that eq.(4) can be written in the form

$$(H_0 + H(q))\ddot{q} + (B_0 + \frac{1}{2}\dot{H}(q))\dot{q} + S(q, \dot{q}) + g(q) = u \tag{5}$$

where

$$S(q, \dot{q})\dot{q} = \frac{1}{2}\dot{H}(q)\dot{q} - \frac{\partial}{\partial q}\left\{\frac{1}{2}\dot{q}^T H(q)\dot{q}\right\}. \tag{6}$$

As was pointed out first by Arimoto and Miyazaki [7][8] and later but independently by Koditschek [9], $S(q, \dot{q})$ is skew symmetric, in other words,

$$\dot{q}^T S(q, \dot{q})\dot{q} = 0 \quad \text{and more generally} \quad r^T S(q, \dot{q})r = 0. \tag{7}$$

Now we show the passivity of robot dynamics (5). More precisely, the following property follows directly from the skew-symmetry of matrix $S$ in eq.(5).

**Proposition 1** For eq.(5) the passivity of velocity vector $\dot{q}$ with respect to torque input $u$ is valid, i.e., it follows that

$$\int_0^t \dot{q}^T(\tau)u(\tau)d\tau \geq \gamma \tag{8}$$

for any $t \geq 0$ and a fixed constant $\gamma$ depending only on the initial state $x(0)(=(q(0), \dot{q}(0)))$.

In fact, we see that

$$\int_0^t \dot{q}^T(\tau)u(\tau)d\tau$$
$$= \int_0^t \dot{q}^T\left[(H_0 + H(q))\ddot{q} + (B_0 + \frac{1}{2}\dot{H}(q))\dot{q} + S(q, \dot{q})\dot{q} + g(q)\right] d\tau$$
$$= \int_0^t \left[\frac{d}{d\tau}\left\{\frac{1}{2}\dot{q}^T(H_0 + H(q))\dot{q} + G(q)\right\} + \dot{q}^T B_0\dot{q}\right] d\tau$$
$$= \int_0^t \dot{q}^T B_0\dot{q}d\tau + V(t) - V(0), \tag{9}$$

where $V(t)$ is defined by

$$V(t) = \frac{1}{2}\dot{q}^T(t)\left\{H_0 + H(q(t))\right\}\dot{q}(t) + G(q(t)) \tag{10}$$

and $G(q)$ denotes the potential function induced by the gravity force, i.e.,

$$g(q) = (\partial G/\partial q^1, \ldots, \partial G/\partial q^n)^T. \tag{11}$$

Since the constant term of potential is arbitrary, it is reasonable to assume that

$$\min_q G(q) = 0. \tag{12}$$

Then, $V(t) \geq 0$ and therefore eq.(9) implies

$$\int_0^t \dot{q}^T(\tau)u(\tau)d\tau \geq -V(0) = \gamma \tag{13}$$

which proves Proposition 1.

The passivity of robot dynamics is quite natural as well as the passivity of electrical lumped-parameter circuits (see Fig.3). In fact, the left-hand side of eq.(8) implies the total work done by generated torques at actuators during the time interval $[0, T]$ (see Fig.4). This quantity is reasonably equivalent to the increase or decrease of total energy, $V(t) - V(0)$, plus the energy consumption during the time interval $[0, T]$ as shown in eq.(9).

Next consider the case that the PD feedback is used as an inner loop in actuator servos, which is typical in industrial manipulators. This is described by

$$u = v + A(q_d - q) - B_1\dot{q} \tag{14}$$

where $q_d$ is a given reference position input and $v$ is an external feedforward input. Substitution of eq.(14) into eq.(5) yields

$$(H_0 + H(q))\ddot{q} + (B + \frac{1}{2}\dot{H}(q))\dot{q} + S(q,\dot{q})\dot{q} + g(q) + A(q - q_d) = v, \tag{15}$$

where $B = B_0 + B_1$. Let us denote by $v_d$ the ideal input that realizes the given desired output $y_d(t)(= \dot{q}_d(t))$. If $y_d(t)$ is differentiable and $\dot{y}_d(t)$ is piecewise continuous on $t \in [0, T]$, $v_d(t)$ must satisfy

$$(H_0 + H(q_d))\ddot{q}_d + (B + \frac{1}{2}\dot{H}(q_d))\dot{q}_d + S(q_d, \dot{q}_d)\dot{q}_d + g(q_d) = v_d. \tag{16}$$

Next define the displacement vector by $r = q - q_d$, which satisfies

$$(H_0 + H(q_d + r))\ddot{r} + (B + \frac{1}{2}\dot{H}(q_d + r))\dot{q} + S(q_d + r, \dot{q}_d + \dot{r})\dot{r} + Ar + f = \Delta v \tag{17}$$

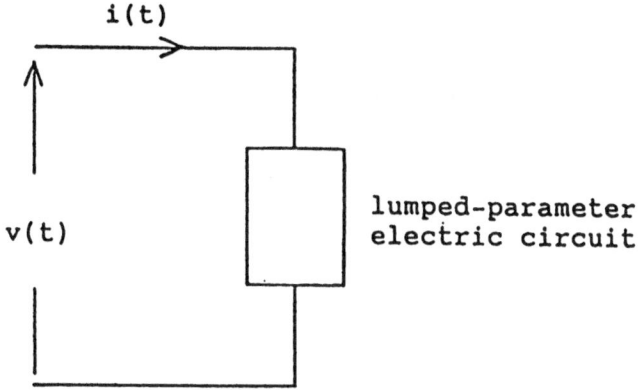

Fig.3 A lumped-parameter electrical circuit satisfies
the passivity condition:
$$\int_0^t v(\tau)i(\tau)d\tau \geq \gamma.$$

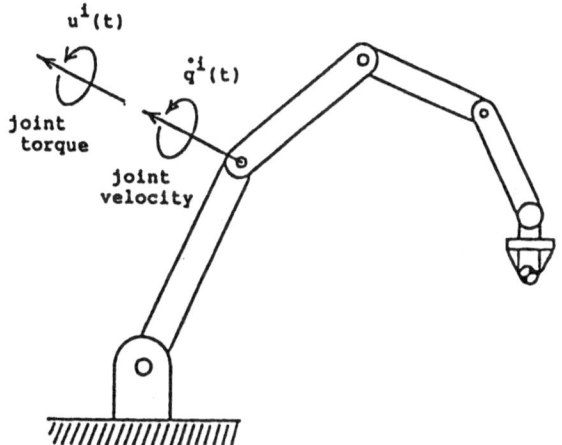

Fig.4 The total work done by torques generated at joint
actuators is irreversible. This implies the passivity
of robot dynamics:
$$\int_0^t \dot{q}^T(\tau)u(\tau)d\tau \geq \gamma.$$

where

$$\Delta v \;=\; v - v_d, \qquad\qquad r = q - q_d,$$

$$
\begin{aligned}
f \;&=\; f(r,\dot r)\\
&=\; \{H(q_d + r) - H(q_d)\}\,\ddot q_d + \frac{1}{2}\left\{\dot H(q_d + r) - \dot H(q_d)\right\}\dot q_d\\
&\quad + \{S(q_d + r, \dot q_d + \dot r) - S(q_d, \dot q_d)\}\,\dot q_d + g(q_d + r) - g(q_d)
\end{aligned}
\tag{18}
$$

Note that every entry of $H(q)$ is constant or sinusoidal function of components of $q$ and every entry of $S(q,\dot q)$ is linear in $\dot q$. Therefore, $f$ is linear in $\dot r$ and hence can be rewritten into the following forms:

$$f = E(q_d, \dot q_d, \ddot q_d)r + F(q_d, \dot q_d, r)\dot r + h(q_d, \dot q_d, \ddot q_d, r) \tag{19}$$

where all linear terms of $\dot r$ in $f$ in eq.(18) are firstly recast into the second term of the right hand side and hence the remaining terms in the right hand side become irrelevant to $\dot r$. In detail,

$$F(q_d, \dot q_d, r)\dot r = \frac{1}{2}\left\{\sum_{i=1}^{n} H^i(q_d + r)\dot r^i\right\}\dot q_d + \{S(q_d + r, \dot q_d + \dot r) - S(q_d + r, \dot q_d)\}\,\dot q_d \tag{20}$$

where $H^i = \partial H/\partial r^i$, and

$$
\begin{aligned}
&E(q_d, \dot q_d, \ddot q_d)r + h(q_d, \dot q_d, \ddot q_d, r)\\
&= g(q_d + r) - g(q_d) + \frac{1}{2}\left[\sum_{i=1}^{n}\left\{H^i(q_d + r) - H^i(q_d)\right\}\dot q_d^i\right]\dot q_d\\
&\quad + \{H(q_d + r) - H(q_d)\}\,\ddot q_d + \{S(q_d + r, \dot q_d) - S(q_d, \dot q_d)\}\,\dot q_d.
\end{aligned}
\tag{21}
$$

Note again that $F$ and $h$ are periodic in $r$ and thereby all entries and components of $F$ and $h$ are bounded provided all components of $\ddot q_d$ are piecewise continuous and hence bounded. According to these observations, we see that there exist constants $\rho_0 > 0$ and $\rho_1 > 0$ such that

$$\left|\dot r^T f\right| \le \left|\dot r^T Er\right| + \left|\dot r^T F\dot r\right| + \left|\dot r^T h\right| \le \rho_0 r^T r + \rho_1 \dot r^T \dot r \tag{22}$$

for any $r$ and $\dot r$.

Now we show the exponential passivity of the displacement dynamics described by eq.(17).

**Proposition 2**  Assume that a given desired output $y_d(t) = (\dot{q}_d(t))$ is differentiable and its derivative is piecewise continuous. Then, for the displacement dynamics described by eq.(17) there is a positive constant $\lambda$ such that

$$\int_0^t e^{-\lambda \tau} \dot{r}^T(\tau) \Delta v(\tau) d\tau \geq \gamma \tag{23}$$

for any $t \geq 0$ and a fixed constant $\gamma$ depending on only the initial state $(r(0), \dot{r}(0))$.

To prove this, we observe that

$$\int_0^t e^{-\lambda \tau} \dot{r}^T \Delta v(\tau) d\tau$$

$$= \int_0^t e^{-\lambda \tau} \dot{r}^T(\tau) \Big[ (H_0 + H(q_d + r)) \ddot{r} + (B + \frac{q}{2} \dot{H}(q_d + r)) \dot{r}$$

$$\qquad + S(q_d + r, \dot{q}_d + \dot{r}) \dot{r} + Ar + f \Big] d\tau$$

$$= \frac{1}{2} \int_0^t \frac{d}{d\tau} \left[ e^{-\lambda \tau} \left\{ \dot{r}^T (H_0 + H(q_d + r)) \dot{r} + r^T Ar \right\} \right] d\tau$$

$$\qquad + \frac{1}{2} \int_0^t \lambda e^{-\lambda \tau} \left\{ \dot{r}^T (H_0 + H(q_d + r)) \dot{r} + r^T Ar \right\} d\tau$$

$$\qquad + \int_0^t e^{-\lambda \tau} \left\{ \dot{r}^T B \dot{r} + \dot{r}^T f(r, \dot{r}) \right\} d\tau$$

$$\geq e^{-\lambda t} U(r(t), \dot{r}(t)) - U(r(0), \dot{r}(0))$$

$$\qquad + \int_0^t e^{-\lambda \tau} \left[ \lambda U(r(\tau), \dot{r}(\tau)) + \dot{r}^T(\tau) B \dot{r}(\tau) \right] d\tau$$

$$\qquad - \int_0^t e^{-\lambda \tau} \left\{ \rho_0 r^T(\tau) r(\tau) + \rho_q \dot{r}^T(\tau) \dot{r}(\tau) \right\} d\tau \tag{24}$$

where

$$U(r, \dot{r}) = \frac{1}{2} \left\{ \dot{r}^T (H_0 + H(q_d + r)) \dot{r} + r^T Ar \right\}. \tag{25}$$

Next we define a scalar function

$$W(\lambda; r, \dot{r}) = \lambda U(r, \dot{r}) + \dot{r}^T B \dot{r} - \rho_0 r^T r - \rho_1 \dot{r}^T \dot{r}$$

$$= \frac{1}{2} r^T (\lambda A - 2\rho_0 I) r + \frac{1}{2} [\lambda \{H_0 + H(q_d + r)\} + 2B - 2\rho_1 I] \dot{r} \tag{26}$$

which becomes positive definite in $r$ and $\dot{r}$ with an appropriate choice for positive $\lambda$. From this it follows that

$$\int_0^t e^{-\lambda \tau} \dot{r}^T \Delta v d\tau \geq e^{-\lambda t} U(r(t), \dot{r}(t)) - U(r(0), \dot{r}(0)) + \int_0^t W(\lambda; r(\tau), \dot{r}(\tau)) d\tau$$

$$\geq -U(r(0), \dot{r}(0)) = \gamma \tag{27}$$

which proves Proposition 2.

As a special case of Proposition 2 by putting $q_d = 0$ and $u_d = 0$ we obtain the exponential passivity for the direct robot dynamics described by eq.(15). This means that the exponential passivity is a more generalized and hence "weaker" concept than regular passivity.

# 3   Uniform Boundedness of Trajectories

By introducing a forgetting factor $\alpha > 0$ into P-type learning scheme we show the uniform boundedness of all trajectories during repetitive trainings under the existence of reinitialization errors, fluctuations of dynamics, and measurement noises. Suppose that at every trial of operation the velocity vector of joint coordinates is measured in the following way

$$y_k(t) = \dot{q}_k(t) + \xi_k(t) \tag{28}$$

where $\xi_k(t)$ is the measurement error satisfying

$$\|\xi_k\|_\infty \overset{\Delta}{=} \sup_{t \in [0,T]} \|\xi_k(t)\| \leq \varepsilon_1 \tag{29}$$

and $\|\xi\|$ denotes the euclidean norm of vector $\xi$, i.e.,

$$\|\xi\| = \left\{ \sum_{i=1}^n \left| \xi^i \right|^2 \right\}^{1/2}. \tag{30}$$

Then the update law of learning with a forgetting factor is described by

$$v_{k+1}(t) = (1 - \alpha)v_k(t) + \alpha v_0(t) + \Phi e_k(t) \tag{31}$$

where

$$e_k(t) = y_d(t) - y_k(t) = \dot{q}_d(t) - (\dot{q}_k(t) + \xi_k(t)). \tag{32}$$

In practice we need not know the details of description of manipulator dynamics. We only assume that an adequate gain matrix $\Phi$ is chosen to be real symmetric and positive definite in advance. When a robot arm exercises repeatedly by updating its actuator inputs in accordance with the learning law described by eq.(31), it arises firstly a problem whether whole trajectories $(q_k(t), \dot{q}_k(t))$ can stay in a bounded domain in $R^{2n}$ for all $k$. We are going to prove this uniform boundedness of motion trajectories during repetitive exercises under the assumption that the motion of the arm is subject to

$$(H_0 + H(q_k))\ddot{q}_k + (B + \frac{1}{2}\dot{H}(q_k))\dot{q}_k + S(q_k, \dot{q}_k)\dot{q}_k + g(q_k) + A(q_k - q_d) = v_k + \eta_k \tag{33}$$

where $\eta_k$ denotes the fluctuation of dynamics at the $k$-th exercise and it satisfies

$$\|\eta_k\|_\infty \le \varepsilon_2 \tag{34}$$

for a small constant $\varepsilon_2 > 0$. Note that we need the full description of robot dynamics, i.e., the knowledge of eq.(33), only for presenting a mathematical rigorous proof of the uniform boundedness and convergence of trajectories when the robot arm is subject to repetitive trainings. Finally we assume that, at every $k$-th trial, the arm is reinitialized within a prescribed precision, that is,

$$\max\{\|q_k(0) - q_d(0)\|, \|\dot{q}_k(0) - \dot{q}_k(0)\|\} \le \varepsilon_3. \tag{35}$$

Now we show one of main results of the paper.

**Theorem 1**  Assume that all $\varepsilon_1 > 0$, $\varepsilon_2 > 0$, and $\varepsilon_3 > 0$ are small in comparison with $\alpha > 0$. Then, for a given piecewise continuous initial input $v_0(t)$, all trajectories of solutions generated by the set of recursive equations (28), (31), (32), and (33) with initial conditions $(q_k(0), \dot{q}_k(0))$ satisfying eq.(35) are bounded uniformly in $k$.

**proof**  From eq.(31) it follows that

$$\Delta v_{k+1} - \alpha \Delta v_0 = (1 - \alpha)\Delta v_k + \Phi e_k \tag{36}$$

which gives rise to

$$(\Delta v_{k+1} - \alpha \Delta v_0)^T \Phi^{-1}(\Delta v_{k+1} - \alpha \Delta v_0)$$
$$= (1 - \alpha)^2 \Delta v_k^T \Phi^{-1} \Delta v_k + 2(1 - \alpha)\Delta v_k^T e_k + e_k^T e_k. \tag{37}$$

The left hand side can be written in the form

$$(\Delta v_{k+1} - \alpha \Delta v_0)^T \Phi^{-1}(\Delta v_{k+1} - \alpha \Delta v_0)$$
$$= (1 - \alpha)\Delta v_{k+1}^T \Phi^{-1} \Delta v_{k+1} - \alpha(1 - \alpha)\Delta v_0^T \Phi^{-1} \Delta v_0$$
$$+ \alpha(\Delta v_{k+1} - \Delta v_0)^T \Phi^{-1}(\Delta v_{k+1} - \Delta v_0). \tag{38}$$

Substitution of this into eq.(37) yields

$$(1 - \alpha)\Delta v_{k+1}^T \Phi^{-1} \Delta v_{k+1}$$
$$= (1 - \alpha)^2 \Delta v_k^T \Phi^{-1} \Delta v_k + \alpha(1 - \alpha)\Delta v_0^T \Phi^{-1} \Delta v_0$$
$$-\alpha(\Delta v_{k+1} - \Delta v_0)^T \Phi^{-1}(\Delta v_{k+1} - \Delta v_0) + 2(1 - \alpha)\Delta v_k^T e_k + e_k^T e_k. \tag{39}$$

By substituting eq.(36) into the right hand side of eq.(39), we obtain

$$(1-\alpha)\Delta v_{k+1}^T \Phi^{-1}\Delta v_{k+1}$$

$$= (1-\alpha)^2\Delta v_k^T \Phi^{-1}\Delta v_k + \alpha(1-\alpha)\Delta v_0^T \Phi^{-1}\Delta v_0$$

$$-\alpha\left\{(1-\alpha)(\Delta v_k - \Delta v_0) + \Phi e_k\right\}\Phi^{-1}\left\{(1-\alpha)(\Delta v_k - \Delta v_0) + \Phi e_k\right\}$$

$$+2(1-\alpha)\Delta v_k^T e_k + e_k^T e_k$$

$$= (1-\alpha)^2\Delta v_k^T \Phi^{-1}\Delta v_k + \alpha(1-\alpha)\Delta v_0^T \Phi^{-1}\Delta v_0$$

$$-\alpha(1-\alpha)^2(\Delta v_k - \Delta v_0)^T\Phi^{-1}(\Delta v_k - \Delta v_0)$$

$$+2(1-\alpha)^2\Delta v_k^T e_k + 2\alpha(1-\alpha)\Delta v_0^T e_k + (1-\alpha)e_k^T e_k. \tag{40}$$

Note that

$$2(1-\alpha)^2\Delta v_k^T e_k + 2(1-\alpha))\Delta v_0^T e_k$$

$$= -2(1-\alpha)^2\Delta v_k^T(\dot{r}_k + \xi_k) - 2\alpha(1-\alpha)\Delta v_0^T(\dot{r}_k + \xi_k)$$

$$= -2(1-\alpha)^2\Delta v_k^T \dot{r}_k - 2(1-\alpha)^2(\Delta v_k - \Delta v_0)^T\xi_k - 2(1-\alpha)\Delta v_0^T(\alpha\dot{r}_k + \xi_k)$$

$$\leq -2(1-\alpha)^2\Delta v_k^T \dot{r}_k + \alpha(1-\alpha)^2(\Delta v_k - \Delta v_0)^T\Phi^{-1}(\Delta v_k - \Delta v_0)$$

$$+\alpha^{-1}(1-\alpha)^2\xi_k^T\Phi\xi_k + \alpha(1-\alpha)\Delta v_0^T \Phi^{-1}\Delta v_0$$

$$+\alpha^{-1}(1-\alpha)(\alpha\dot{r}_k + \xi_k)^T\Phi(\alpha\dot{r}_k + \xi_k), \tag{41}$$

in which we used the inequality

$$2x^T y \leq \alpha^{-1}x^T\Phi x + \alpha y^T\Phi^{-1}y. \tag{42}$$

Substituting eq.(41) into eq.(40), we have

$$(1-\alpha)\Delta v_{k+1}^T\Phi^{-1}\Delta v_{k+1}$$

$$\leq (1-\alpha)^2\Delta v_k^T\Phi^{-1}\Delta v_k + 2\alpha(1-\alpha)\Delta v_0^T\Phi^{-1}\Delta v_0 - 2(1-\alpha)^2\Delta v_k^T\dot{r}_k$$

$$+(1-\alpha)\left[(\dot{r}_k + \xi_k)^T\Phi(\dot{r}_k + \xi_k) + \alpha^{-1}(1-\alpha)\xi_k^T\Phi\xi_k + \alpha^{-1}(\alpha\dot{r}_k + \xi_k)^T\Phi(\alpha\dot{r}_k + \xi_k)\right]$$

$$\leq (1-\alpha)^2\Delta v_k^T\Phi^{-1}\Delta v_k + 2\alpha(1-\alpha)\Delta v_0^T\Phi^{-1}\Delta v_0$$

$$-2(1-\alpha)^2\Delta v_k^T\dot{r}_k + 2\dot{r}_k^T\Phi\dot{r}_k + 2\alpha^{-1}\xi_k^T\Phi\xi_k. \tag{43}$$

Now we notice that, similarly to eq.(17), the displacement of $q_k$ from $q_d$ is subject to

$$(H_0 + H(q_d + r_k))\ddot{r}_k + (B + \frac{1}{2}\dot{H}(q_d + r_k))\dot{r}_k$$

$$+S(q_d + r_k, \dot{q}_d + \dot{r}_k)\dot{r}_k + Ar_k + f_k = \Delta v_k + \eta_k \tag{44}$$

where

$$f_k = E(q_d, \dot{q}_d, \ddot{q}_d)r_k + F(q_d, \dot{q}_d, r_k)\dot{r}_k + h(q_d, \dot{q}_d, \ddot{q}_d, r_k). \tag{45}$$

By the same reason as mentioned in derivation of eq.(22), we have

$$\left| \dot{r}_k^T f_k \right| \le \rho_0 r_k^T r_k + \rho_1 \dot{r}_k^T \dot{r}_k. \tag{46}$$

Hence, similarly to the argument given in the proof of Proposition 2, we obtain

$$- \int_0^t e^{-\lambda\tau} \dot{r}_k^T(\tau) \Delta v_k(\tau) d\tau$$

$$\le U(r_k(0), \dot{r}_k(0)) - e^{-\lambda t} U(r_k(t), \dot{r}_k(t))$$

$$- \int_0^t e^{-\lambda\tau} W(\lambda; r_k(\tau), \dot{r}_k(\tau)) d\tau + \int_0^t e^{-\lambda\tau} \dot{r}_k^T(\tau) \eta_k(\tau) d\tau \tag{47}$$

where $U$ and $W$ are defined by eqs.(25) and (26) respectively. Multiplying eq.(43) by $e^{-\lambda\tau}$, taking the integral of it over $[0, t]$, and substituting eq.(47) into the resultant inequality, we obtain

$$(1 - \alpha) \int_0^t e^{-\lambda\tau} \Delta v_{k+1}^T \Phi^{-1} \Delta v_{k+1} d\tau$$

$$\le (1 - \alpha)^2 \int_0^t e^{-\lambda\tau} \Delta v_k^T \Phi^{-1} \Delta v_k d\tau + 2\alpha(1 - \alpha) \int_0^t e^{-\lambda\tau} \Delta v_0^T \Phi^{-1} \Delta v_0 d\tau$$

$$+ 2(1 - \alpha)^2 U(r_k(0), \dot{r}_k(0)) - 2(1 - \alpha)^2 e^{-\lambda t} U(r_k(t), \dot{r}_k(t))$$

$$- \int_0^t e^{-\lambda\tau} \left[ 2(1 - \alpha)^2 W(\lambda; r_k, \dot{r}_k) - 2\dot{r}_k^T \Phi \dot{r}_k - \dot{r}_k^T \Phi \dot{r}_k \right] d\tau$$

$$+ 2\alpha^{-1} \int_0^t e^{-\lambda\tau} \xi_k^T \Phi \xi_k d\tau + \int_0^t e^{-\lambda\tau} \eta_k^T \Phi^{-1} \eta_k d\tau. \tag{48}$$

This inequality can be rewritten in the form

$$s_{k+1}(t) \le (1 - \alpha)s_k(t) + 2\alpha s_0(t) - V_k(t) + c_3\varepsilon_3^2 + \left\{ c_1\varepsilon_1^2/\alpha + c_2\varepsilon_2^2 \right\} \frac{1 - e^{-\lambda t}}{\lambda}, \tag{49}$$

where

$$s_k(t) = (1 - \alpha) \int_0^t e^{-\lambda\tau} \Delta v_k^T(\tau) \Phi^{-1} \Delta v_k(\tau) d\tau, \tag{50}$$

$$V_k(t) = 2(1 - \alpha)^2 e^{-\lambda t} U(r_k(t), \dot{r}_k(t))$$

$$+ \int_0^t e^{-\lambda\tau} \left[ 2(1 - \alpha)^2 W(\lambda; r_k(\tau), \dot{r}_k(\tau)) - 3\dot{r}_k^T \Phi \dot{r}_k \right] d\tau, \tag{51}$$

and $c_1$, $c_2$, and $c_3$ are positive constants such that

$$2(1 - \alpha)^2 U(r_k(0), \dot{r}_k(0)) \le c_3\varepsilon_3,$$

$$c_1 = 2 \| \Phi \|, \qquad c_2 = \left\| \Phi^{-1} \right\|. \tag{52}$$

Since the content of [ ] in eq.(51) can be positive definite in $r_k$ and $\dot{r}_k$ with a sufficient large $\lambda > 0$, $V_k(t)$ always is non-negative under an appropriate choice of $\lambda > 0$. Then, it follows from eq.(49) that

$$
\begin{aligned}
s_k(t) &\leq (1-\alpha)^k s_0(t) + \frac{1-(1-\alpha)^k}{\alpha}\{2\alpha s_0(t) + \varepsilon(t)\} \\
&\leq 2s_0(t) + \varepsilon^2(t)/\alpha,
\end{aligned}
\tag{53}
$$

where

$$
\varepsilon^2(t) = c_3\varepsilon_3^2 + \left\{c_1\varepsilon_1^2/\alpha + c_2\varepsilon_2^2\right\}\frac{1-e^{-\lambda t}}{\lambda}.
\tag{54}
$$

Since $\varepsilon_1$ is sufficiently small in comparison with $\alpha$, $s_k(t)$ is bounded from above uniformly in $k$. On the other hand, it follows from eq.(49) that

$$
\begin{aligned}
V_k(t) &\leq (1-\alpha)s_k(t) - s_{k+1}(t) + 2\alpha s_0(t) + \varepsilon^2(t) \\
&\leq 2s_0(t) + \varepsilon^2(t)/\alpha.
\end{aligned}
\tag{55}
$$

In particular, it holds that

$$
2(1-\alpha)^2 e^{-\lambda t} U(r_k(t), \dot{r}_k(t)) \leq V_k(t) \leq 2s_0(t) + \varepsilon^2(t)/\alpha
\tag{56}
$$

which implies the uniform boundedness of $r_k(t)$ and $\dot{r}_k(t)$ in $k$. Thus the theorem has been proved completely.

The problem of uniform boundedness of trajectories generated by iterative learning was first dealt with by Bondi et al [10] for the case of D-type learning law. However, their argument is based on a linearized technique around the desired trajectory and an implicit assumption that the initial command input $v_0$ must be sufficiently close to the command input $v_d$. In contrast, Theorem 1 in this paper assure the uniform boundedness for P-type learning schemes in a global sense.

The robustness problem under the existence of reinitialization errors, fluctuations of dynamics, and measurement noises was first discussed by the author [11] in the case of PID-type learning scheme. However, the argument was based on an assumption that the initial trajectory (and thus all subsequent ones) lies in a neighborhood of the desired one and hence the robot dynamics can be considered to be represented by a linearized model (a linear time-varying mechanical system). Very recently Heinzinger et al [12] attacked the same robustness problem for a class of D-type learning control and proved without use of any linearization that the learning input and corresponding output trajectories converge to neighborhoods of their desired ones respectively. In addition, they made a comment by illustraiting a counter-example that such a robustness property is not valid for a class of P-type learning. Theorem 1 reverses this hypothesis by introducing a forgetting factor into the P-type update law of learning.

The original idea of use of a forgetting factor into learning control schemes is due to Heinzinger et al [12], but it was introduced into only D-type learning. In our previous paper [13] we first attacked with the aid of linearization the robustness problem of P-type learning control by introduction of a forgetting factor.

## 4   Convergence of Trajectories

We now discuss the convergence problem of motion trajectories in an $\alpha$-neighborhood of the desired one in the progress of learning.

Let us consider the sequence of functions $\{r_k(t)\}$ which are uniformly bounded. Since derivatives $\dot{r}_k(t)$ are also bounded uniformly in $k$, the sequence $\{r_k(t)\}$ is equi-continuous. Then, by virtue of Ascoli-Arzela's theorem, it is possible to find a uniformly convergent subsequence in $\{r_k\}$. However, since a bias input $v_0$ exists in the update law in eq.(31), limiting functions may differ each other dependingly on choice of subsequences. This observation motivates a more detailed analysis of convergence discussed below.

To have an insight into the problem, let us consider again eq.(49). By summing up eq.(49) from $k = 0$ to $k = K - 1$, we obtain

$$\sum_{k=0}^{K-1} \left\{ 2(1-\alpha)^2 e^{-\lambda T} U(r_k, \dot{r}_k) + \int_0^T e^{-\lambda t} E(\alpha, \lambda; r_k, \dot{r}_k) dt + \alpha s_k(T) \right\}$$
$$\leq s_0(T) - s_K(T) + 2\alpha K s_0(T) + K\varepsilon^2(T) \leq (1 + 2\alpha K)s_0(T) + K\varepsilon^2(T) \quad (57)$$

where

$$E(\alpha, \lambda; r, \dot{r}) = 2(1-\alpha)^2 W(\lambda; r, \dot{r}) - 3\dot{r}^T \Phi \dot{r}. \quad (58)$$

Let us now suppose that the training is repeated until $k = K$ where $K$ is around $1/\alpha$. Denote by $k^*$ the trial $k$ at which the content of $\{\}$ in the summation of the left hand side of eq.(57) is minimum during these K trials. Then,

$$2(1-\alpha)^2 e^{-\lambda T} U(r_{k^*}(T), \dot{r}_{k^*}(T)) + \int_0^T e^{-\lambda t} E(\alpha, \lambda; r_{k^*}, \dot{r}_{k^*}) dt + \alpha s_{k^*}(T) \le 3\alpha s_0(T) + \varepsilon^2(T) \tag{59}$$

which means

$$U(r_{k^*}(T), \dot{r}_{k^*}(T)) \le O(\alpha), \tag{60}$$

$$\int_0^T e^{-\lambda t} E(\alpha, \lambda; r_{k^*}, \dot{r}_{k^*}) dt \le O(\alpha). \tag{61}$$

Note that $E(\alpha, \lambda; r, \dot{r})$ is a quadratic positive definite function of $r$ and $\dot{r}$. Thus, if $v_0(t)$ in the long-term memory is refreshed by the best input $v_{k^*}(t)$ during the past $K$ trials and the training is restarted (see Fig.5) then it may be expected that the quadratic term of $\Delta v_0$ in eq.(49), that is, $s_0(t)$ itself, may become of order $\alpha$. This implies that if the best input is found during the next set of trials from $i = K$ to $2K - 1$ then it leads to the quadratic error of order $\alpha^2$. Thus, it would be reasonable to expect that after $mK$ trials by this scheme the quadratic error would be of order $\alpha^m$. We call this special class of learning control "selective learning" which is schematically illustrated by Fig.5. In order to accelerate the speed of convergence, the selective learning seems practical and attractive and, by experimental studies [14], we are able to observe this tendency. However, form the theoretical viewpoint it is not so easy to prove the convergence of trajectories in a more rigorous sense that the size of neighborhood attractors is less than $O(\alpha)$.

To be surprising, a little simplified version of learning, which dose not use the selection for the best input but use the refreshment of the long-term memory ever after $K$ trials, is also effective and moreover leads to a simple proof for the convergence with an attractor of order $\alpha$. We call this learning scheme "interval training", which corresponds to a typical training method adopted in athletic and swimming sports (see Fig.6).

To prove the convergence in a strict sense, we need again eq.(49) which can be written in particular as follows:

$$s_{k+1}(T) \le (1-\alpha)s_k(T) + 2\alpha s_0(T) - V_k(T) + \varepsilon^2(T). \tag{62}$$

At an early stage of iterative learning, $V_k(T)$ must be large and hence $s_k(T)$ may be decreasing $k$ as far as $V_k(T)$ is larger than $\alpha(2s_0(T) - s_k(T)) + \varepsilon^2(T)$. Suppose that at the

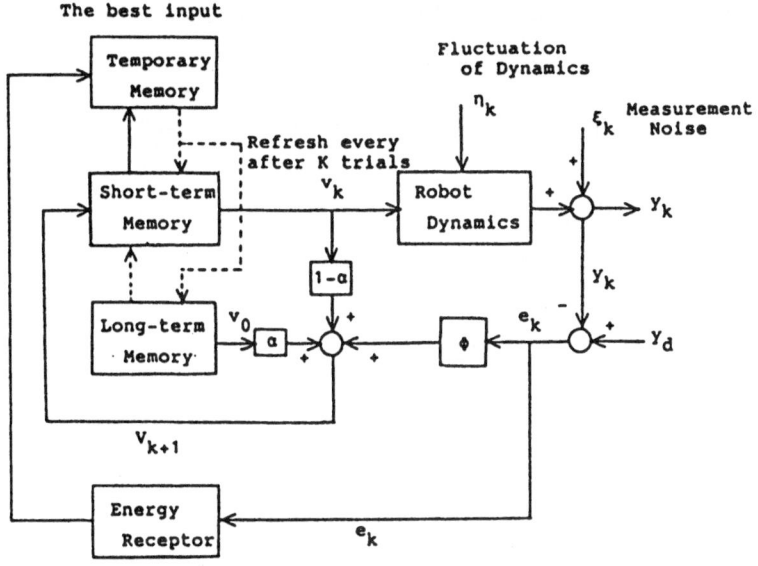

Fig.5 Selective learning. The long-term memory is refreshed
by the best input in the past every after K trials.

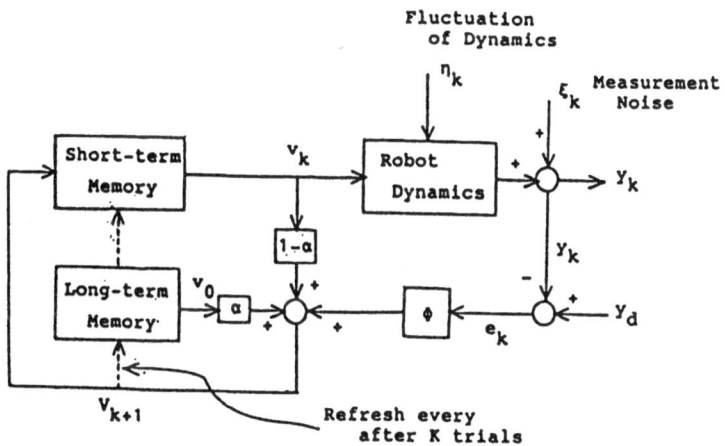

Fig.6 Interval training. The long-term memory is refreshed
every after K trials.

first time $V_k(T)$ becomes less than $3\alpha s_0(T)$ when $k = \bar{k}$. We show that $\bar{k} \leq K$ provided $s_0(T)$ is larger than $5\varepsilon^2(T)/\alpha$ and $K = O(\alpha^{-1})$. From eq.(62) we obtain

$$
\begin{aligned}
s_k(T) &\leq (1-\alpha)^k s_0(T) + \sum_{i=0}^{k-1}(1-\alpha)^{k-i-1}\left[2\alpha s_0(T) + \varepsilon^2(T) - V_i(T)\right] \\
&\leq (1-\alpha)^k s_0(T) - \sum_{i=0}^{k-1}(1-\alpha)^{k-i-1}\left[\alpha s_0(T) - \varepsilon^2(T)\right] \\
&\leq 2(1-\alpha)^k s_0(T) - s_0(T) + \varepsilon^2(T)/\alpha
\end{aligned}
\tag{63}
$$

for any $k(\leq \bar{k})$. Since $log(1-\alpha) \leq -\alpha$, we have

$$
s_k(T) \leq (2e^{-\alpha k} - 1)s_0(T) + \varepsilon^2(T)/\alpha.
\tag{64}
$$

The right hand side of this inequality is decreasing and becomes negative when $k$ is nearly equal to $1/\alpha$ since $(2/e - 1) < -0.2$. From this we conclude that $\bar{k} \leq K$.

Once the value of $V_{\bar{k}}(T)$ becomes of $O(\alpha)$, we then need to assure that $V_k(T)$ will remain to be $O(\alpha)$ after $k > \bar{K}$. In the present paper we do not go further into the details of the proof since it is quite involved and sophisticated, but instead we point out that the proof will be pursued in a modified argument similar to that given in our previous paper [13]. In the argument the exponential passivity of difference dynamics described in terms of $d_k(\overset{\triangle}{=} q_{k+1} - \{(1-\alpha)q_k + \alpha q_0\})$ plays a crucial role.

Finally, we make a comment that, by devising a similar but more sophisticated argument, it is possible to prove the convergence of trajectories to an $\varepsilon$-neighborhood of the desired one under the selective learning scheme.

# 5   Conclusion

A class of P-type learning control algorithms with a forgetting factor and without taking differentiation of velocity signals is proposed and its robustness property with respect to reinitialization errors, fluctuations of dynamics, and measurement noises is analyzed. The uniform boundedness of motion trajectories during the progress of learning is proved by clarifying the exponential passivity of the displacement dynamics. To accelerate the speed of convergence of motion trajectories to an $\alpha$-neighborhood of the desired one, two special strategies called "interval training" and "selective learning" are proposed, by which a rough sketch of the convergence proof is given.

# References

[1] S. Arimoto, S. Kawamura, and F. Miyazaki, "Bettering operation of robots by learning," *Journal of Robotic Systems,* Vol. 1, pp. 123–140, 1984.

[2] S. Arimoto, S. Kawamura, and F. Miyazaki, "Bettering operation of robots by learning; A new control theory for servomechanism and mechatronics systems," *Proc. 23rd IEEE Conf. Decision and Control,* Las Vegas, NV, pp. 1064–1069, 1984.

[3] ibid., "Can mechanical robots learn by themselves?," in *"Robotic Research" The Second International Symposium,* H. Hanafusa & H. Inoue, Eds., MIT Press, Cambridge, Massachusetts, pp. 127–134, 1985.

[4] S. Arimoto, "Mathematical theory of learning with applications to robot control", *Proc. of 4th Yale Workshop on Applications of Adaptive Systems Theory,* Yale University, New Haven, Connecticut, pp. 379–388, 1985.

[5] S. Kawamura, F. Miyazaki, and S. Arimoto, "Hybrid position/force control of manipulators based on learning method," *Proc. '85 Inter. Conf. on Advanced Robotics,* Tokyo, Japan, pp. 235–242, 1985.

[6] S. Kawamura, F. Miyazaki, and S. Arimoto, "Realization of robot motion based on a learning method," *IEEE Trans. on Systems, Man, and Cybernetics,* Vol. SMC-18, No. 1, pp 126–134, 1988.

[7] S. Arimoto and F. Miyazaki, "Stability and robustness of PID feedback control for robot manipulators of sensory capability," in *"Robotic Research" The First International Symposium,* by M. Brady & R.P. Paul, Eds., MIT Press,Cambridge, Massachusetts, pp. 783–799, 1984.

[8] ibid., "Asymptotic stability of feedback control laws for robot manipulators," *Proc. IFAC Symp. on Robot Control '85,* Barcelona, Spain, pp. 447–452, 1985.

[9] D.E. Koditschek, "Natural motion for robot arms," *Proc. of 23rd IEEE Conf. DEcision and Control,* Las Vegas, NV, pp. 733-755, 1987.

[10] P. Bondi, G. Casalino, and L. Gambardella, "On the iterative learning control theory for robotic manipulators," *IEEE J. of Robotics and Automation*, Vol. 4, No. 1, pp. 14–22, 1988.

[11] S. Arimoto, S. Kawamura, F. Miyazaki, "Convergence, stability, and robustness of learning control schemes for robot manipulator," in M.J. Jamshidi, L.Y.S. Luh, and M. Shahinpoor (eds.), *Recent Trends in Robotics: Modeling, Control, and Education*, pp. 307–316, Elsevier Sciences Publishing Co., Inc., New York, 1986.

[12] G. Heinzinger, D. Fenwick, B. Paden, and F. Miyazaki, "Robust learning control," *Proc. 28th IEEE Conf. Decision and Control*, Tampa, Florida, Dec. 13–15, 1989.

[13] S. Arimoto, "Robustness of learning control for robot manipulators," *Proc. of the 1990 IEEE International Conference on Robotics and Automation*, pp. 1523–1528, Cincinnati, Ohio, May 13–18, 1990.

[14] Y. Nanjo and S. Arimoto, "Experimental studies on robustness of a learning method with a forgetting factor for robotic motion control," submitted to '91 ICAR, Pisa, Italy, 1990.

[15] S. Arimoto, "Learning control theory for robotic motion," *International Journal of Adaptive Control and Signal Processing*, Vol. 4, No. 6, pp. 543–564, 1990.

[16] S. Arimoto, T. Naniwa, and H. Suzuki, "Robustness of P-type learning control with a forgetting factor for robotic motions," *Proc. of 29th IEEE Conference on Decision and Control*, Honolulu, Hawaii, Dec. 5–7, 1990.

# Adaptive Control of Robot Manipulators via Velocity Estimated Feedback

C. CANUDAS DE WIT and N. FIXOT

Laboratoire d'Automatique de Grenoble ENSIEG-INPG

B.P. 46, 38402 Saint Martin d'Hères, France

**Abstract.** Adaptive controllers have been proposed as a means of counteracting robot model parameter inaccuracies under the assumption of full state measurements (position and velocities), see Slotine and Li; 1988. Since velocity measurements are often contaminated by high levels of noise coinstraining the system performance, nonlinear controller integrating nonlinear observers have been studied and have proved to be locally exponentially stable ( Canudas, et al; 1989,1990) provided that the model parameters are exactly known. Extensions to the case of model parameter uncertainties have been studied in Canudas and Fixot; 1990, where the same type of local exponentially stability is obtained but additional assumptions on the variation of the inertia matrix eigenvalues are needed. This paper proposes a control scheme which combines an adaptive control law with a sliding observer and needs no additional assumptions on the variation of the inertia matrix eigenvalues A local stable closed-loop system results from this combination.

# 1. Introduction

Recently, some works have been concerned with the problem of controlling robot manipulators by integrating into the control loop nonlinear observers in order to estimate the joint velocities ( Canudas-de-Wit et al, 1989, 1990, Canudas-de-Wit and Fixot, 1990, Nicosia and Tomei, 1990 ). The reason for doing this is due to the closed-loop limitations imposed by the measured velocity noise when joint velocities are obtained by sensors such as tachometers and by the impossibility of obtaining a good velocity estimate by simple techniques such as position interpolation, in particular for low velocities. Besides the practical interest that the idea represents, the problem of controlling nonlinear systems via estimated state feedback has its own relevance and hence merits a certain degree of attention.

The problem of designing nonlinear observers using full nonlinear model dynamics of a revolute robot manipulator was first treated in Canudas-de-Wit and Slotine (1989). They proposed to use the so called "sliding observers", in which some of the physical robot properties are explicitly exploited to show asymptotic convergence of the observation error vector. Sliding observers are a transposition of the switching controllers to the problem of state observation in nonlinear systems (Slotine et al, 1986). Sliding control design consists in defining a switching surface in the phase plane which is rendered attractive by the action of the switching terms. The dynamics on the switching surface is determined by Filippov's solution concept, see Filippov (1960), which indicates that the system dynamic behaviour within the switching surface can be formally described as an average combination of the dynamics of each side of the discontinuous surface. The interest in such controllers lies in the fact that they possess good robustness properties vis-a-vis to model uncertainties, i.e. the trajectories lying on the switching surface are unaffected under process gain variations (see Glad; 1987, for a survey of robustness on nonlinear state feedback) and hence control and observer design can be performed with little knowledge about the system model parameters. Their main drawback is that sliding controllers generate "chattering" motion on the switching surface. "Chattering" is unsuitable because it adds an important amount of high-frequency components to the control law which has discontinuities, see Fig. 1(a). However, sliding control theory applied to the problem of state estimation does not necessarily imply that the control law, which depends on the estimated state vector, contains dominant components of infinite high frequency. Indeed, as shown by Fig.1(b), the observer dynamic equations with discontinuous righthand side act as a low-pass filter reducing the amplitude of the high-frequency components. "Chattering" thus appears at the estimation level rather as numerical problem than as a physical drawback in the control law.

The work developed by (Canudas-de-Wit and Slotine, 1989 and Canudas-de-Wit *et al*, 1989, 1990; Nicosia and Tomei, 1990(a) ) assumes that the robot parameters are exactly known. Extension to the case model parameter inaccuracies is treated in Canudas-de-Wit and Fixot; 1990, which proposes a robust control scheme following the ideas of Leitmann, 1981 on the uncertain linear systems and their application to robotics (Spong and Vidyasagar, 1989). As in these works, local asymptotic stability is obtained provided that the control law contains discontinuities, the idea is sketched by Fig.1(c). Although "Chattering" can be reduced by replacing the discontinuous switching function by a kind of saturation nonlinear control, asymptotic stability is lost and substituted by uniformely ultimate stability (u.u.s), or practical stability ( the tracking error does not tend to zero but to a closed region around it, in finite time). As an other alternative, high-gain smooth control design was suggested by Nicosia and Tornambè (1989). Their approach also yields u.u.s, since perfect tracking is not ensured unless infinite gains are used.

This paper presents a new approach for designing robust controllers via state space feedback. The robot model parameters are assumed to be unknown and velocity measurements are assumed not to be available. State observation and parameter adaptation are performed simultaneously. The adaptation law, the observer gains and the control law, are designed on the reduced order manifold which results from the invariance of the switching surface. With respect to the previous work of Canudas-de-Wit and Fixot, 1990, the asymptotic stability of the

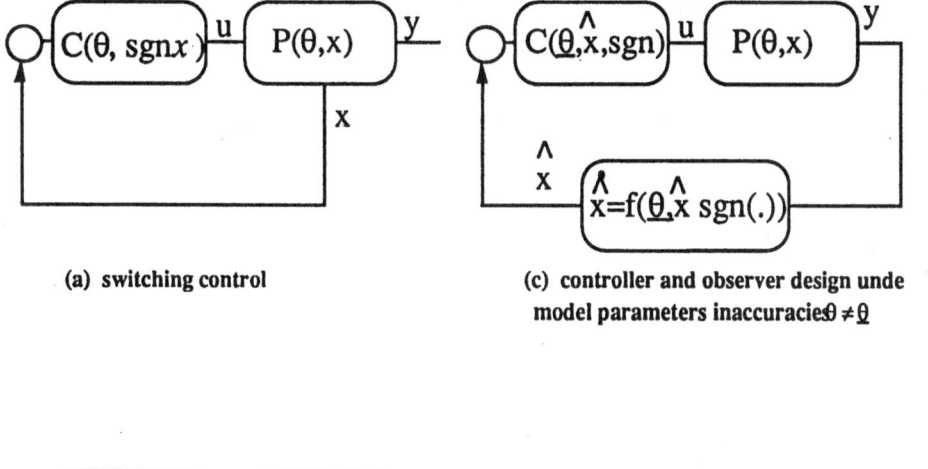

(a) switching control

(c) controller and observer design unde model parameters inaccuracies $\hat{\theta} \neq \underline{\theta}$

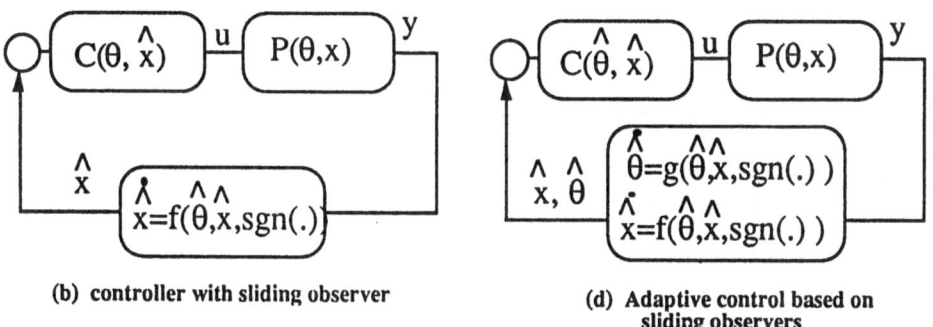

(b) controller with sliding observer

(d) Adaptive control based on sliding observers

# Figure 1.

closed-loop system resulting from the approach presented in this paper is not conditioned by a particular type of inertia matrix variations. On the other hand, the introduction of an adaptation loop may also be motivated by a reduction of "chattering" at control law level since the control law only depends on the estimated state and parameter vectors and hence contains no terms proportional to discontinuities (although the adaptation law and the observer dynamics have discontinuities in the righthand side of the respective differential equations), see Fig 1.(d).

# 2. Problem formulation

The dynamic model of a rigid robot having n revolute joints resulting from the Lagrange equations is expressed as :

$$H(q) \ddot{q} + C(q, \dot{q}) \dot{q} + \tau_g(q) = Y(q, \dot{q}, \ddot{q}) \theta = \tau \tag{2.1}$$

where $q$, $\dot{q}$, $\ddot{q}$, vectors of $\mathcal{R}^n$ represent the link displacements, velocities and accelerations, respectively. $H(q)$ is the $n \times n$ definite positive inertia matrix. $C(q, \dot{q}) \dot{q}$ represents the Coriolis and centripetal

forces. $\tau_g(q)$ is the gravity vector and $\tau$ is the applied motor torques. Friction are neglected in this presentation. The parametrization $\tau = Y\theta$ simply means that the inertial parameter vector $\theta$ of $\mathcal{R}^m$ (where $m \leq 11n$) enters linearly in equation (2.1). The $n \times m$ matrix Y collects state information, in terms of position, velocity and acceleration.

In this paper, the parameter vector $\theta$ is assumed to be unknown, while the structure of the information matrix Y is assumed to be available. The problem is to design a control law using only joint position feedback able to ensure asymptotic tracking of desired trajectory specified in terms of joint positions, velocities and accelerations $q_d$, $\dot{q}_d$, $\ddot{q}_d$.

Introducing $x_1 = q$, $x_2 = \dot{q}$, model (2.1) can be rewritten in the following state space representation.

$$\dot{x}_1 = x_2 \tag{2.2a}$$

$$\dot{x}_2 = \beta (x_1, x_2) + H(x_1)^{-1} \tau \tag{2.2b}$$

where $\beta$ is a nonlinear fonction defined as,

$$\beta(x_1, x_2) = - H(x_1)^{-1} [C(x_1, x_2) x_2 + \tau_g (x_1)] \qquad (2.3)$$

Slotine and Li (1988) have proposed a direct adaptive algorithm for the case of robot manipulators, assuming that the states $x_1$, $x_2$ are assumed to be physically measurable. They have introduced the following auxiliary variables:

$$\dot{z} = \dot{q}_d - \Lambda \tilde{q} \qquad (2.4)$$

$$\ddot{z} = \ddot{q}_d - \Lambda \dot{\tilde{q}} \qquad (2.5)$$

$$s = x_2 - \dot{z} = \dot{\tilde{q}} + \Lambda \tilde{q} \qquad (2.6)$$

where $\tilde{q} = x_1 - q_d$ is the tracking error vector and $\Lambda$ is a constant definite positive matrix.

They have proposed the following control law:

$$\tau_0 = \widehat{H}(x_1) \ddot{z} + \widehat{C}(x_1, x_2) \dot{z} + \widehat{\tau}_g(x_1) - K_D s \qquad (2.7)$$

where $\widehat{H}(x_1)$ and $\widehat{C}(x_1, x_2)$ are the estimates of $H(x_1)$ and $C(x_1, x_2)$ respectively and $K_D$ is a design definite positive constant matrix. In this paper, velocity vector $x_2$ is assumed to be unknown.

Now, with the following definitions:

$$\ddot{z}' = \ddot{q}_d - \Lambda(\widehat{x}_2 - \dot{q}_d) = \ddot{z} - \Lambda \tilde{x}_2 \qquad (2.8)$$

$$s' = \widehat{x}_2 - \dot{z} = s + \tilde{x}_2 \qquad (2.9)$$

, where $\widehat{x}_1$ and $\widehat{x}_2$ are the estimates of $x_1$ and $x_2$, respectively and $\tilde{x}_1 = \widehat{x}_1 - x_1$ , $\tilde{x}_2 = \widehat{x}_2 - x_2$ are observation vectors. And from the substitution of the variables defined by equations (2.4)-(2.6) by the above defined variables (2.8) and (2.9) in the control law (2.7), we get,

$$\tau = \widehat{H}(x_1) \ddot{z}' + \widehat{C}(x_1, \widehat{x}_2) \dot{z} + \widehat{\tau}_g(x_1) - K_D s' \qquad (2.10)$$

$$= \hat{H}(x_1)\, (\ddot{z} - \Lambda\, \tilde{x}_2) + \hat{C}(x_1, x_2 + \tilde{x}_2)\, \dot{z} + \hat{\tau}_g(x_1) - K_D\, (s + \tilde{x}_2)$$

Since the elements of $\hat{C}(x_1, \hat{x}_2)\, \dot{z}$, namely $\hat{x}_2^T \hat{N}_i \dot{z}$, can be expressed as: $\hat{x}_2^T \hat{N}_i \dot{z} = (x_2 + \tilde{x}_2)^T \hat{N}_i \dot{z}$, then

$$\hat{C}(x_1, x_2 + \tilde{x}_2)\, \dot{z} = \hat{C}(x_1, x_2)\, \dot{z} + \hat{C}(x_1, \dot{z})\, \tilde{x}_2$$

and hence $\tau$ can be expressed in terms of $\tau_0$ as :

$$\tau = \tau_0 - \hat{H}(x_1)\, \Lambda\, \tilde{x}_2 + \hat{C}(x_1, \dot{z})\, \tilde{x}_2 - K_D \tilde{x}_2$$

$$\tau = \tau_0 + W(x_1, \dot{z}, \hat{\theta})\, \tilde{x}_2 \qquad (2.11)$$

where W is given as

$$W(x_1, \dot{z}, \hat{\theta}) = - \hat{H}(x_1)\, \Lambda + \hat{C}(x_1, \dot{z}) - K_D \qquad (2.12)$$

and, $\hat{\theta}$ is the estimate of the unknown parameter vector $\theta$.

Introducing this control law in the robot dynamics (2.1) we obtain :

$$H(x_1)\, \dot{x}_2 + C(x_1, x_2)\, x_2 + \tau_g(x_1) = \hat{H}(x_1)\, \ddot{z} + \hat{C}(x_1, x_2)\, \dot{z} + \hat{\tau}_g(x_1) - K_D s + W(x_1, \dot{z}, \hat{\theta})\, \tilde{x}_2$$

by subtracting $H(x_1)\, \ddot{z} + C(x_1, x_2)\, \dot{z}$ in both sides, we obtain

$$H(x_1)\, \dot{s} + C(x_1, x_2)s = \tilde{H}(x_1)\, \ddot{z} + \tilde{C}(x_1, x_2)\, \dot{z} + \tilde{\tau}_g(x_1) - K_D s + W(x_1, \dot{z}, \hat{\theta})\, \tilde{x}_2$$

using the following reparametrization:

$$\tilde{H}(x_1)\, \ddot{z} + \tilde{C}(x_1, x_2)\, \dot{z} + \tilde{\tau}_g(x_1) = Y(x_1, x_2, \dot{z}, \ddot{z})\, \tilde{\theta}$$

the dynamic behaviour of s is given by :

$$\dot{s} = H^{-1}(x_1)[ -(C(x_1, x_2) + K_D)s + Y(x_1, x_2, \dot{z}, \ddot{z})\,\tilde{\theta} + W(x_1, \dot{z}, \hat{\theta})\,\tilde{x}_2 ] \qquad (2.13)$$

The observer proposed to estimate states $x_1$ and $x_2$, is given by the following differential equation with righthand side discontinuities,

$$\dot{\hat{x}}_1 = \hat{x}_2 - \Gamma_1 \tilde{x}_1 - \Lambda_1 \text{ sgn } (\tilde{x}_1) \qquad (2.14a)$$

$$\dot{\hat{x}}_2 = - \Lambda_2 \text{ sgn } (\tilde{x}_1) - W(x_1, \dot{z}, \hat{\theta}) (s' - \Lambda_1 \text{ sgn } (\tilde{x}_1)) + v \qquad (2.14b)$$

where W, defined in (2.12), is introduced to compensate for the Ws' vector included in the control law, and $v$ is introduced to robustify the observer error dynamics which is given as:

$$\dot{\tilde{x}}_1 = \tilde{x}_2 - \Gamma_1 \tilde{x}_1 - \Lambda_1 \text{ sgn } (\tilde{x}_1) \qquad (2.15a)$$

$$\dot{\tilde{x}}_2 = - \Lambda_2 \text{ sgn } (\tilde{x}_1) - W(x_1, \dot{z}, \hat{\theta}) (s' - \Lambda_1 \text{ sgn } (\tilde{x}_1)) + v + \eta \qquad (2.15b)$$

with,

$$\eta = - \beta(x_1, x_2) - H^{-1}(x_1)\,\tau$$

These equations (2.15) together with equation (2.13) describe the complete closed loop system error dynamics. The design vector $v$ and an adaptation law for $\hat{\theta}$ are still to be determined so that s, $\tilde{x}_1$, and $\tilde{x}_2$ tend exponentially to zero. These design vectors will also contain discontinuities in terms of sgn $(\tilde{x}_1)$.

The parameter vector $\theta$ is assumed to be time invariant, so that:

$$\dot{\hat{\theta}} = \dot{\tilde{\theta}} = f(\text{sgn}(\tilde{x}_1), ...) \quad ; \quad v = v(\text{sgn } (\tilde{x}_1), ...)$$

The complete error dynamic equations are thus given by:

$$\dot{s} = -H^{-1}(x_1)\left[(C(x_1, x_2) + K_D)\, s - Y(x_1, x_2, \dot{z}, \ddot{z})\,\tilde{\theta} - W(x_1, \dot{z}, \hat{\theta})\,\tilde{x}_2\right] \qquad (2.16a)$$

$$\tilde{x}_1 = \tilde{x}_2 - \Gamma_1 \tilde{x}_1 - \Lambda_1 \, \text{sgn} \, (\tilde{x}_1) \tag{2.16b}$$

$$\tilde{x}_2 = - W(x_1, \dot{z}, \hat{\theta}) \, (s' - \Lambda_1 \, \text{sgn} \, (\tilde{x}_1)) - \Lambda_2 \, \text{sgn} \, (\tilde{x}_1) + \eta + v \, (\text{sgn}(\tilde{x}_1), \dots) \tag{2.16c}$$

$$\tilde{\theta} = f(\text{sgn} \, (\tilde{x}_1), \dots) \tag{2.16d}$$

# 3. Reduced order manifold dynamics

Systems containing discontinuities in the righthand side have been previously studied by Filippov, 1960 and Aizerman-Piatnitsky, 1978 among other references. Filippov's solution concept indicates that the dynamics on the switching surface is an average of the dynamics on each side of the discontinuity surface.

It is also easy to show, by simple choice of $\Lambda_1$ i.e.; $\Lambda_1 = \lambda_1 I$, that $\tilde{x}_1 = 0$ is invariant as long as

$$\|\tilde{x}_2\| \le \|\Lambda_1\| = \sup_{x \in \mathcal{R}} \frac{\|\Lambda_1 \, x\|}{\|x\|} = \lambda_1 \tag{3.1}$$

This region characterized by $\tilde{x}_1 = 0$, $\|\tilde{x}_2\| \le \lambda_1$ is known as the sliding patch. And the dynamic behaviour in this patch is, according to Filippov's solution concept, given as :

$$\dot{s} = - H^{-1}(x_1) \left[ (C(x_1, x_2) + K_D)s - Y(x_1, x_2, \dot{z}, \ddot{z}) - W(x_1, \dot{z}, \hat{\theta}) \, \tilde{x}_2 \right] \tag{3.2a}$$

$$\tilde{x}_2 = - W(x_1, \dot{z}, \hat{\theta}) \, s - \Lambda_2 \Lambda_1^{-1} \tilde{x}_2 + \eta + v(\Lambda_1^{-1} \tilde{x}_2, \dots) \tag{3.2b}$$

$$\tilde{\theta} = f(\Lambda_1^{-1} \tilde{x}_2, \dots) \tag{3.2c}$$

Note that in the sliding patch :

$$s = s' - \Lambda_1 \, \text{sgn} \, (\tilde{x}_1) \tag{3.3}$$

$$x_2 = \hat{x}_2 - \Lambda_1 \, \text{sgn} \, (\tilde{x}_1) \tag{3.4}$$

$$\ddot{z} = \ddot{z}' + \Lambda \, \Lambda_1^{-1} \, \text{sgn} \, (\tilde{x}_1) \tag{3.5}$$

the set $\mathcal{P}$, defined as,

$$\mathcal{P} = \left\{ (s, \tilde{x}_1, \tilde{x}_2, \tilde{\theta}) : \tilde{x}_1 = 0 \text{ and } \|\tilde{x}_2\| \le \lambda_1 \right\} \tag{3.6}$$

determines the sliding patch in the complete error space within which the dynamics (3.2) is valid.

The following design consists in finding an adaptation law $f(\cdot)$ and an expression for $v$, such that system dynamics (3.1) exponentially tends to zero while the states $(s, \tilde{x}_1, \tilde{x}_2, \tilde{\theta})$ remain inside the set $\mathcal{P}$.

In what follows, we assume that initial conditions can be chosen such that :

$$\tilde{x}_1 (0) = \hat{x}_1(0) - x_1(0) = 0 \tag{3.7}$$

which is always possible since the joint positions are assumed to be measured.

The expression of $\eta$ is :

$$\eta = - \, H^{-1}(x_1) \, C(x_1, x_2) \, x_2 - H^{-1}(x_1) \, \tau_g(x_1) + H^{-1} \, (x_1) \, \tau$$

then, according to the robot model properties given in Canudas and al, 1990 (boundedness of the inertia matrix, of its inverse, of the gravity components, and the square velocity boundedness of the Coriolis and centripetal forces), constants $\sigma_0$, $\sigma_1$ and $\sigma_2$ exist such that :

$$\|\eta\| \le \sigma_0 \, \|x_2\|^2 + \sigma_1 + \sigma_2 \, \|\tau\|$$

and using (3.4),the above inequality can be rewritten as :

$$\|\eta\| \le \sigma_0 \, \|\hat{x}_2 - \Lambda_1 \, \text{sgn}(\tilde{x}_1)\|^2 + \sigma_1 + \sigma_2 \, \|\tau\|$$

$$\le \sigma_0 \|\widehat{x}_2\|^2 + \sigma_0 \lambda_1 + \lambda_1^2 + \sigma_1 + \sigma_2 \|\tau\| = \varphi\,(\widehat{x}_2, \tau) \qquad\qquad (3.8)$$

The scalar and positive function $\varphi$ thus defines a measurable upperbound of $\|\eta\|$ within $\mathcal{P}$.

# 4. Stability in the sliding patch

Introduce the following scalar positive definite function,

$$V = \frac{1}{2}\left[ s^T\, Hs + \widetilde{x}_2^{\,T}\, \widetilde{x}_2 + \widetilde{\theta}^{\,T}\, \Gamma\, \widetilde{\theta} \right] \qquad\qquad (4.1)$$

with $\Gamma$ being a constant definite positive matrix, the time derivative of $V$ is then given as :

$$\dot{V} = s^T\, H\,\dot{s} + s^T\, \frac{\dot{H}}{2}\, s + \widetilde{x}_2^{\,T}\, \dot{\widetilde{x}}_2 + \widetilde{\theta}^{\,T}\, \Gamma\, \dot{\widetilde{\theta}}$$

$$= s^T\left[ Y\,\widetilde{\theta} - K_D\, s - C\, s + W\,\widetilde{x}_2 \right] + s^T\, \frac{\dot{H}}{2}\, s$$

$$+ \widetilde{x}_2^{\,T}\left[ -\Lambda_0\,\widetilde{x}_2 - W^T\, s + \eta + v \right] + \widetilde{\theta}^{\,T}\, \Gamma\, \dot{\widetilde{\theta}}$$

with $\Lambda_0 = \Lambda_2\, \Lambda_1^{-1}$.

Assuming that the chosen parametrization for C corresponds to the Christoffel symbols and hence $(\frac{\dot{H}}{2} - C)$ is skew symmetric, the above expression becomes :

$$\dot{V} = -s^T K_D s - \widetilde{x}_2^T \Lambda_0 \widetilde{x}_2 + \widetilde{\theta}^{\,T}[Y^T s + \Gamma\dot{\widehat{\theta}}] + \widetilde{x}_2^T\,[\eta + v]$$

This suggests defining $v$ and $\hat{\theta}$ as

$$\hat{\theta} = -\Gamma^{-1} Y^T (x_1, x_2, \dot{z}, \ddot{z}) s \qquad (4.2)$$

$$v = \begin{cases} -\varphi(\tilde{x}_2, \tau) \dfrac{\tilde{x}_2}{\|\tilde{x}_2\|} & \text{if} \quad \|\tilde{x}_2\| \neq 0 \\[2mm] 0 & \text{if} \quad \|\tilde{x}_2\| = 0 \end{cases} \qquad (4.3)$$

they are functions of unmeasured quantities, but in $\mathcal{P}$, they can be computed indirectly by using the expressions (3.3)-(3.5) as :

$$\hat{\theta} = f(x_1, \tilde{x}_1, \hat{x}_2, \dot{z}, \dot{z}', s')$$

$$= -\Gamma^{-1} Y(x_1, \hat{x}_2 - \Lambda_1 \, \text{sgn}(\tilde{x}_1), \dot{z}, \dot{z}' + \Lambda \, \Lambda_1 \, \text{sgn}(\tilde{x}_1)) \, (s' - \Lambda_1 \, \text{sgn}(\tilde{x}_1)) \qquad (4.4)$$

and,

$$v = v(\hat{x}_2, \tau, \tilde{x}_1)$$

$$= \begin{cases} -\varphi \dfrac{(\hat{x}_2, \tau)}{\lambda_1} \Lambda_1 \, \text{sgn}(\tilde{x}_1) & \text{if} \, \|\Lambda_1 \, \text{sgn}(\tilde{x}_1)\| \neq 0 \\[3mm] 0 & \text{if} \, \|\Lambda_1 \, \text{sgn}(\tilde{x}_1)\| = 0 \end{cases} \qquad (4.5)$$

with the above definitions of $v$ and $\hat{\theta}$, we obtain (for $\tilde{x}_2 \neq 0$)

$$\dot{V} = -s^T K_D s - \tilde{x}_2^T \Lambda_0 \tilde{x}_2 + \tilde{x}_2^T (\eta + v)$$

$$\leq -k_D \|s\|^2 - \lambda_0 \|\tilde{x}_2\|^2 + \|\tilde{x}_2\| \, \varphi - \frac{\tilde{x}_2^T \tilde{x}_2}{\|\tilde{x}_2\|} \varphi$$

$$\leq -k_D \|s\|^2 - \lambda_0 \|\tilde{x}_2\|^2 \qquad (4.6)$$

where $k_D = \lambda_{min} K_D$, $\lambda_0 = \lambda_{min} \Lambda_0$, and we have used the upperbound (3.8) on $\eta$.

when $\tilde{x}_2 = 0$, we simply obtain :

$$\dot{V} \leq - k_D \, \|s\|^2 \tag{4.7}$$

A closed set in $\mathcal{P}$ has still to be determinate so that all trajectories with initial conditions in this subset do not leave the sliding patch $\mathcal{P}$.

Defining $e^T = (s^T, \tilde{x}_2{}^T, \tilde{\theta}^T)$ and $\beta_{sup}$ and $\beta_{min}$ as

$$\beta_{sup} = \lambda_{sup} \begin{bmatrix} H & 0 & 0 \\ 0 & I & 0 \\ 0 & 0 & \Gamma \end{bmatrix} \quad ; \quad \beta_{min} = \lambda_{min} \begin{bmatrix} H & 0 & 0 \\ 0 & I & 0 \\ 0 & 0 & \Gamma \end{bmatrix}$$

V can be upperbounded as :

$$\beta_{min} \, \|e\|^2 \leq V \leq \beta_{sup} \, \|e\|^2 \tag{4.8}$$

Since $\dot{V}$ is negative, $V(t) \leq V(0)$ , for all $t \geq 0$.

$$\beta_{min} \, \|e(t)\|^2 \leq V(t) \leq V(0) \leq \beta_{sup} \, \|e(0)\|^2$$

then,

$$\|e(t)\|^2 \leq \frac{\beta_{sup}}{\beta_{min}} \, \|e(0)\|^2 \tag{4.9}$$

therefore, $e(0)$ is chosen according to the following inequality:

$$\|e(0)\|^2 \leq \frac{\beta_{min}}{\beta_{sup}} \, \lambda_1 \tag{4.10}$$

so that, $\|e(t)\|^2 \leq \lambda_1$ and hence $e(t)$ remains in the sliding patch $\mathcal{P}$. Indeed, we have proved the following Theorem.

**Theorem.**

Consider the observer equation (2.14) together with the control law (2.10), with $v$ and $\overset{\wedge}{\theta}$ defined as in (4.2) and (4.3). Assume also that $\lambda_1$ verifies the inequality (4.10) and that the initial conditions verify $\widehat{x}_1(0) = x_1(0)$ then,

(i) $\quad \lim_{t \to \infty} \|s\|^2 = 0$

(ii) $\quad \lim_{t \to \infty} \|\widetilde{x}_2\|^2 = 0$

(iii) $\quad \|\widetilde{\theta}\|^2 < \infty \quad$ for all $t$

Furthermore, following Slotine and Li's arguments , we can also conclude that

(iv) $\quad \lim_{t \to \infty} \widetilde{q} = 0$

(v) $\quad \lim_{t \to \infty} \overset{\approx}{q} = 0$

**Remarks :**

The convergence of $s$ and $\widetilde{x}_2$ to zero is exponential, as indicated by the previous analysis. Since $\widetilde{q}$ and $s$ are related by a first order stable filter as :

$$\widetilde{q} = [pI + \Lambda]^{-1} s$$

the tracking error $\widetilde{q}$ will also converge exponentially to zero.

# 5. Conclusions

This work can also be understood as a particular control design of nonlinear systems where adaptation and observation are performed simultaneously while asymptiotically stability is obtained. We believe that these results are possible due to the switching terms introduced in the observer's gains and on the adaptation law, which allow the reduced order error dynamics to be asymptotically stabilized. A fundamental distinction between this technique and any other approach based on high-gain control is first that the zero-error state of equilibrium is reached exponetially fast, and secondly that although the control contains high-frequency components, its magnitude remains within the input bounds while high-gain controllers may require infinitely large input signals in order to theoretically approach the zero-error state of equilibrium.

# 6. References

*Aizermann, M.A., and Pyatnitskiy, Y.E.S.* : "Theory of Dynamic Systems which incorporate Elements with Incomplete Information and its Relation to the Theory of Discontinuous Systems", Journal of the Franklin Institute, Vol. 306, rf6, dec. 1978.

*Canudas de Wit, C., Aström K.J, Fixot, N.* : "Computed Torque Control via nonlinear Observers", MTNS Conference, Amsterdam, Holland, June, 1989. Also to appear in IJACSP.

*Canudas de Wit, C., Fixot, N., Aström, K.J.* : "Trajectory Tracking in Robot Manipulators via Nonlinear State Estimate Feedback", Submitted on IEEE Trans. Robotics and Automation, 1990.

*Canudas de Wit, C., Slotine, J.J.E.* : "Sliding observers for robot manipulators", IFAC Symposium on nonlinear control systems design, June 1989, Capri , Italy. Also submitted to Automatica.

*Canudas de Wit, C. and Fixot, N.* : "Robot Control via Robust State Estimate Feedback"; Conference on New Trends in Systems Theory", Genova, Italy, July 1990. Also accepted to IEEE-TAC.

*Glad,T.S.* :" Robutsness of Nonlinear State Feedback- A Survey", Automatica, Vol.23, No.4, pp.425 435, 1987.

*Filippov, A.F.,* : "Differential Equations with discontinuous right-hand side, AM. math. Soc. Trans., Vol. 62, p. 199, 1960.

*Nicosia, S. and Tornambè* " High-gain observers in the state and paprameter estimation of robots having elastic joints", Systems and Control Letters, 13, pp. 331-337, 1989., see also Proceedings of the IEEE conference on Robotics and Automation, May,1990, Cincinnati, pp.1423-1430.

*Nicosia, S. and Tomei, P.* : "Robot Control by using only Joint Position Measurements". IEEE Trans.on Automatic Control, Vol. AC-35, n.9 pp. 1058-1061, (1990).

*Leitmann, G.* : "On the Efficacy of nonlinear control in uncertain linear systems", J. Dyn. Sys. Measure and Control, Vol. 103, pp. 95-102, 1981.

*Ortega, R., Spong, M.,* : "Adaptive Motion Control of Rigid Robots : A tutorial, Proceedings of the 27th Conference on Decision and Control, Austin, Texas, Dec. 1988, pp. 1575-1584.

*Samson, C.* : "Robust nonlinear Control of Robotic Manipulators", Proc. 22nd, IEEE CDC, San Antonio, Dec. 1983.

*Spong, M., Vidyasagar* : "Robot Dynamics and Control", John and Wiley and Sons, 1989.

*Slotine, J.J.E., Weiping, L.* : "Adaptive manipulator Control : a Case Study", IEEE Transactions on Automatic Control, vol. 33, n°11, nov. 1988, pp. 995-1003.

*Utkin, V.I.,* : "Variable structure systems with sliding mode : a survey", IEEE Trans. Automatic Control, vol. 22, p. 212, 1977.

# NONLINEAR CONTROL FOR THE NONHOLONOMIC MOTION OF SPACE ROBOT SYSTEMS

Yoshihiko Nakamura and Ranjan Mukherjee

Center for Robotic Systems and Manufacturing
University of California, Santa Barbara, CA 93106

## ABSTRACT

Recent advances in space applications have necessitated the deployment of robotic systems in space. These systems have intrinsic features due to the nonholonomic constraints governing their motion. In this paper we discuss the presence of nonholonomic redundancy, as different from ordinary kinematic redundancy, in space robots. Nonholonomic redundancy can be utilized to increase the workspace of space robots. In this paper we present a path planning scheme using Liapunov functions in hierarchy for the utilization of nonholonomic redundancy.

## 1. INTRODUCTION

In the last few years space robotics has emerged as a new field of study in the robotics community. The advances in space applications and the requirement of robotization to make space missions safe and cost effective, has played an important role in the growth and development of this new field.

A number of researchers like Akin, Minsky, Thiel and Kurtzman (1983), Alexander and Cannon (1987), Vafa and Dubowsky (1987), Vafa (1987), Umetani and Yoshida (1987), Longman, Lindberg and Zedd (1987), Spofford (1988), Miyazaki, Masutani and Arimoto (1988), Koningstein, Ullman and Cannon (1989), Nakamura and Mukherjee (1989 (a), (b), (c)), Papadopoulos and Dubowsky (1989), Yamada (1989), and Mukherjee and Nakamura (1990) have worked on the kinematics, dynamics and control of space manipulators. Nakamura and Mukherjee (1989 (c), 1990) formulated the nonholonomic mechanical structure of space robot systems and provided a mathematical proof of the nonholonomic property of the system. Vafa and Dubowsky (1987) explored the possibility of changing the vehicle orientation without changing the joint configuration by performing cyclic motion of the joints.

Nakamura and Mukherjee (1989 (c), 1990) proposed a path planning scheme known as the bi-directional approach. For a 6-DOF space manipulator the system was described by nine variables; the six joint variables of the manipulator and the three variables of the orientation of the vehicle. It was shown for particular cases that by directly controlling only the six joint variables of the manipulator it was possible to plan a path such that all the nine variables converged to their desired values. When the six variables of the endeffector position and orientation are to be converged to their desired values at the end of motion, the above fact implies that we have an additional three variables that can be selected arbitrarily. This clearly indicated the existence of three degrees of redundancy at the final time in achieving the desired endeffector position and orientation. This redundancy does not exist locally like ordinary kinematic redundancy but exhibits itself only after a global motion. In this paper, we term this redundancy as nonholonomic redundancy and establish a method to utilize it.

Similar to terrestrial robots, space robots will be required to work in the presence of obstacles and with the restrictions imposed by the joint limits. For terrestrial robots, obstacles and or imposed joint limits may be sufficient to limit the workspace. For space robots the workspace is not a fixed region determined by obstacles and joint limits; instead, it is dependent upon the path as it was shown by Papadopoulos and Dubowsky (1989). In order to utilize the complete workspace it is necessary to plan the path carefully.

In this paper we discuss how we can properly plan a path and utilize nonholonomic redundancy to make maximum use of the workspace of a free-flying space robot. While planning the nonholonomic trajectory, we define and use multiple Liapunov functions for the convergence of the endeffector's position and orientation to their desired values, and for the avoidance of joint limits. The input to the system is consistently synthesized by applying the concept of task-priority based control (Hanafusa, Yoshikawa, and Nakamura, 1981; Nakamura, Hanafusa, and Yoshikawa, 1987) established for ordinary kinematically redundant manipulators. The convergence of the endeffector to its desired position and orientation is considered as the first priority task, and avoiding joint limits is considered as the second priority task.

## 2. NONHOLONOMIC REDUNDANCY

### 2.1 Differential Kinematics

In this section we explain the meaning of the term *nonholonomic redundancy*. We assume that the space vehicle-manipulator system is not acted upon by external forces. The discussion is carried out in the context of a 6-DOF manipulator mounted on a passive free-flying space vehicle to show nonholonomic redundancy in the absence of ordinary kinematic redundancy for better understanding. Nonholonomic redundancy exhibits itself in unison with ordinary kinematic redundancy for space robots with more than 6-DOF.

We begin our discussion with the kinematic equations of a free-flying space robot. It is obvious that the kinematic configuration of the system is fully described by twelve variables, namely, six variables of the position and orientation of the vehicle and the six joint variables. If the vehicle were fixed to the inertia frame like a terrestrial robot system, the first six variables would be constant. The absence of this fixture is replaced with the momentum constraints. Since we assume the absence of external forces, the momentum constraints are equivalent to the momentum conservation laws. Assuming zero initial momentum of the system, the linear and angular momentum conservation laws are represented by the single vector equation (Umetani and Yoshida, 1987; Nakamura and Mukherjee, 1989 (a), (b), (c))

$$\dot{\theta}_1 = H\dot{\theta}_2 = \begin{pmatrix} H_L \\ H_A \end{pmatrix} \dot{\theta}_2 \qquad (1)$$

where, $\dot{\theta}_1 \in R^6$ is the dependent variable and represents the vector of the linear and angular velocities of the space vehicle, and $\theta_2 \in R^6$ represents the joint variables of the manipulator. A detailed description of the matrix $H \in R^{6\times6}$ can be found in Nakamura and Mukherjee, 1989 (a), (b), (c)). $H_L \in R^{3\times6}$ and $H_A \in R^{3\times6}$ are the upper and lower halves of $H$ respectively.

Six constraints among twelve variables seem to imply that the behavior of our system should be described by six (twelve minus six) variables. This simple algebra is wrong. This would be true only if all the constraints of Eq.(1) had an analytic integral form that relate $\theta_1$, and $\theta_2$. Unfortunately, Eq.(1) is not completely integrable. Let us now take a little closer look at Eq.(1). The linear momentum conservation is equivalent to the fact that the center of mass of the total system has a constant velocity; with zero initial momentum the center of mass is fixed. Therefore, the linear momentum conservation law has an integrated form. This gives us the scope to eliminate the three position coordinates of the space vehicle. The angular momentum conservation equation is however not integrable analytically. We are therefore required to deal with nine variables and three constraints due to angular momentum conservation. The behavior of the system is represented by a differential equation involving nine variables out of which six are independent. We call the nine variables as the state variable. The derivative of the six independent variables are considered as the input variable. We choose the state variable $x$, and the input variable $u$ as follows:

$$x = \begin{pmatrix} \alpha \\ \beta \\ \gamma \\ \theta_2 \end{pmatrix} \in R^9 \tag{2}$$

$$u = \dot{\theta}_2 \in R^6 \tag{3}$$

where $\alpha$, $\beta$, and $\gamma$ are the z-y-x Euler angles of the vehicle orientation with respect to the inertia frame. The system equation becomes

$$\dot{x} = K u \tag{4}$$

where

$$K \triangleq \begin{pmatrix} N^{-1} H_A \\ E_6 \end{pmatrix} \in R^{9 \times 6} \tag{5}$$

where $E_6 \in R^{6 \times 6}$ is an identity matrix, and $N$ is the matrix that relates the derivative of Euler angles to the angular velocity, and is given as,

$$^I\omega_0 = N \begin{pmatrix} \dot{\alpha} \\ \dot{\beta} \\ \dot{\gamma} \end{pmatrix}, \qquad N \triangleq \begin{pmatrix} 0 & -sin\alpha & cos\alpha\,cos\beta \\ 0 & cos\alpha & sin\alpha\,cos\beta \\ 1 & 0 & -sin\beta \end{pmatrix} \tag{6}$$

where $^I\omega_0 \in R^3$ is the angular velocity of the vehicle with respect to the inertial frame. A mathematical proof of the nonholonomic nature of the system given by Eq.(4), was provided by Nakamura and Mukherjee (1989 (c), 1990). Using the Frobenius theorem from differential geometry, the nonintegrable nature of the system was concluded from the noninvolutivity of the distribution defined by the vector fields of the six column vectors of $K$. It would be possible to conclude that the entire nine dimensional state space is reachable from any initial state if we could show that the dimension of the smallest involutive distribution involving the column vectors of $K$ is nine at every point in the state space. However, due to the complex nature of the $K$ matrix it would be impractical to try to show this by direct computation.

From Eq.(4) we can derive an important nature of the nonholonomic motion. By integrating Eq.(4) with respect to time, we have

$$x(t_2) = x(t_1) + \int_{t_1}^{t_2} K(\theta_1, \theta_2)\dot{\theta}_2 \, dt = x(t_1) + \sum_{i=1}^{6} \int_{\theta_{2i}(t_1)}^{\theta_{2i}(t_2)} K_i(\theta_1, \theta_2) \, d\theta_{2i} \tag{7}$$

where $K_i$ is the $i$-th column of $K$, $\theta_{2i}$ is the $i$-th element of $\theta_2$, and $t_1$ and $t_2$ are the initial and final times respectively. Equation (7) implies that the trajectory of $x$ is dependent on the trajectory of $\theta_2$, but independent of time. Therefore, once the trajectory of $\theta_2$ is determined, changing the velocity along the trajectory does not alter the trajectory of $x$. This nature comes from the fact that the system maintains zero momentum. Otherwise, the trajectory of $x$ becomes time dependent.

The relationship between the endeffector motion $\dot{x}_E$, and $\dot{\theta}_1$ and $\dot{\theta}_2$ is described by the following equation

$$\dot{x}_E = J_1 \dot{\theta}_1 + J_2 \dot{\theta}_2 \tag{8}$$

where, $\dot{x}_E \in R^6$ represents the linear and angular velocities of the endeffector, and $J_1$ and $J_2$ are pure geometrical Jacobian matrices (Nakamura and Mukherjee, 1989 (a), (b), (c)). Umetani and Yoshida (1987) used Eq.(1) to eliminate the dependence of $\dot{x}_E$ on the dependent variable $\dot{\theta}_1$, and arrived at the simpler form

$$\dot{x}_E = \left( J_1 H + J_2 \right) \dot{\theta}_2 = \tilde{J} \dot{\theta}_2 \tag{9}$$

where $\tilde{J} \in R^{6 \times 6}$ is called the *generalized Jacobian matrix*. Equation (9) successfully describes the endeffector motion in terms of the independent variables only. However, it must be noted that Eq.(9) does not have a view of the three dependent non-integrable variables, namely, the vehicle orientation.

To prepare for further discussions, we slightly change Eq.(9) as follows:

$$\dot{h} = \hat{J} \dot{\theta}_2 \tag{10}$$

$$\hat{J} \triangleq \begin{pmatrix} E_3 & 0 \\ 0 & N_E^{-1} \end{pmatrix} \tilde{J}, \qquad N_E \triangleq \begin{pmatrix} 0 & -\sin\alpha_E & \cos\alpha_E \cos\beta_E \\ 0 & \cos\alpha_E & \sin\alpha_E \cos\beta_E \\ 1 & 0 & -\sin\beta_E \end{pmatrix}$$

where, $h \in R^6$ represents the position and the z-y-x Euler angles $(\alpha_E, \beta_E, \gamma_E)$, of the endeffector.

## 2.2 Nonholonomy and Redundancy

If the three constraints of angular momentum were integrable, then, in the nine dimensional space of $x$, the motion of the system would find a hypersurface defined by six independent generalized coordinates. At any point on the hypersurface, the six generalized coordinates would be able to uniquely determine the nine variables of $x$. In such a situation, by controlling six joints of our system, we cannot reach beyond a six dimensional subspace defined by the hypersurface, of the nine dimensional space of $x$.

In reality, the story is quite different because of the non-integrability of the angular momentum constraints. Through various numerical simulations we are convinced that by controlling six joints of our system we can reach beyond a six dimensional space. Now, major questions that arise are: (1) whether a vehicle with a 6-DOF manipulator can reach the whole nine dimensional space, and (2) if so, what are the requirements, if any, for the kinematic and dynamic structure of the system. These are open questions and we do not intend to address these issues in this paper. For similar two dimensional systems, Sreenath (1990) has given an insight into the set of reachable states.

We proposed a scheme that plans a trajectory between given initial and final values of $x$ and named it the *bi-directional approach* (Nakamura and Mukherjee, 1989, 1990). The planned trajectory satisfies the nonholonomic equation, namely, Eq.(4). However, this approach neither guarantees the reachability of the given final state from the initial state, nor does it guarantee the solvability even if the given final state is reachable from the initial state. We can only say that the scheme could successfully plan a trajectory for every example we tried. From our experience through numerical solutions, we have a strong feeling that the system can define the whole nine dimensional space as the reachable space, and the *bi-directional approach* is useful for most cases. However, this conclusion must wait for further studies.

In this paper, we concentrate on how we can utilize the nonholonomic nature merely assuming that the reachable states are significantly large. If we make a hypothesis that the whole nine dimensional space of $x$ is reachable, it is possible to reach any $x$ by carefully planning the trajectory of $\theta_2$. In the case where the position and orientation of the endeffector is specified at the final time, the above statement means that we have three degrees of redundancy at the final time in choosing the nine variables of $x$ that brings the endeffector to the desired configuration, although we only control six joint variables of $\theta_2$. This redundancy is different from the ordinary kinematic redundancy in nature, since it does not exist locally and appears only after a global motion. We call this kind of redundancy as nonholonomic redundancy. It is noteworthy that unlike ordinary kinematic redundancy, nonholonomic redundancy cannot be exhibited by self-motion. This is clear from the fact that $\dot{x}_E = 0$ implies that $\dot{\theta}_2 = 0$, for a nonsingular $\tilde{J}$ from Eq.(9). If the number of joints $n$ of the space robot satisfies $n > 6$, $\tilde{J}$ becomes a $6 \times n$ matrix, and the system exhibits nonholonomic redundancy and kinematic redundancy simultaneously.

The concept of nonholonomic redundancy and its inherent difference with ordinary kinematic redundancy is further explained as follows. Let $u, x$, and $h$, as defined before, denote the input space, the state space, and the endeffector variables respectively. In both ordinary and nonholonomically redundant manipulators the redundancy exists because of the higher dimension of the state space than the number of endeffector variables. The difference between the two types of redundancy is however attributed to the fact that the dimension of the input space is equal to that of the state space for ordinary redundancy whereas for nonholonomic redundancy the input space is smaller than the state space.

If we make a hypothesis that the whole nine dimensional space of $x$ is reachable, the set of reachable points of the endeffector can be computed by eliminating the vehicle linear velocity from Eq.(8) and treating the vehicle angular velocity as virtually independent. The virtual manipulator concept by Vafa and Dubowsky (1987) provides an useful scheme that reduces this computation into a simple geometric problem. However, it must be noted that the trajectory that connects two reachable points of $x_E$ is not computed even with the virtual manipulator. A reachable point $x_E$ can be reached only when an appropriate nonholonomic trajectory is planned. In an extreme case, two neighboring points of $x_E$ may require very long trajectories of $\theta_2$. Papadopoulos and Dubowsky (1989) studied the workspace of free-flying space robots and showed that as a subset of the reachable set there exists a set of the endeffector position and orientation in which any two entries are mutually reachable regardless of the choice of the connecting trajectory. It is also noteworthy that the complementary set of this subset occupies a significant portion of the complete set of reachable points. These facts imply that the workspace of a free-flying space robot is maximized by establishing a planning scheme of nonholonomic trajectory.

## 2.3 The Path Planning Problem

In this section we define the path planning problem. We need to find a path of the state variable that satisfies the nonholonomic equation, namely, Eq.(4), and brings the endeffector to a given position with a given orientation from a given initial state. There are two possible approaches to the problem, namely the *explicit approach*, and the *implicit approach*.

Although we are provided with the initial state variable $x$, we do not know the state variable at the final time. We only have the desired endeffector position and orientation at the final time. Solving for a state variable that satisfies the final endeffector position and orientation is the first subproblem of the explicit approach. This is an inverse kinematics problem for the final configuration. As we discussed in section 2.2, either by eliminating the holonomic linear momentum constraints from Eq.(8) and treating the angular velocity of the vehicle as independent, or with the help of the virtual manipulator, the inverse kinematics problem virtually forms an ordinary kinematic redundancy problem, namely, solving for the nine variables from the given six variables of endeffector position and orientation. The solution can be obtained by using the established schemes for kinematic redundancy (for example,

Liegeois, 1979; Hanafusa, Yoshikawa and Nakamura, 1981; Klein and Huang, 1983; Nakamura and Hanafusa, 1987). While solving the inverse kinematics we have to make sure that the solution does not violate the imposed joint limits or collide with the obstacles. The next subproblem is to plan a path between the initial state and the computed final state by avoiding joint limits and obstacles. This subproblem could be solved by using the bi-directional approach (Nakamura and Mukherjee, 1989 (c)).

The other approach to the problem is the implicit approach. In this approach we do not divide the problem into subproblems. Rather than computing the final state explicitly, we directly compute the path of the state that brings the endeffector to its desired position and orientation at the final time. The final state is obtained as a result when the computation is completed. This is why we call this approach implicit. While planning the path, we need to satisfy imposed joint limits and consider the presence of obstacles.

Although the explicit approach can be done by solving the two subproblems for which schemes have already been established, the kind of computation involved in them are fairly complex. Furthermore, the final state that is chosen from a variety of possible ones, may not be easily reachable. This is because of the fact that the final state is computed without considering the initial state or the connecting trajectory. In the rest of this paper, we explore the implicit approach.

## 3. HIERARCHICAL LIAPUNOV FUNCTIONS

### 3.1 Liapunov functions in hierarchy

To plan a nonholonomic trajectory, we apply the concept of *task-priority based control* that was established for an ordinary kinematically redundant manipulator (Hanafusa, Yoshikawa and Nakamura, 1981; Nakamura, Hanafusa and Yoshikawa, 1987). This method plans the joint trajectory of a kinematically redundant manipulator when an endeffector trajectory is completely specified and the utilization of redundancy is intended. As mentioned in section 2.2, for our space system we have only 6-DOF and no local redundancy. Hence, if an endeffector trajectory is specified, there are no remaining degrees of freedom that can be utilized.

We modify the task-priority based control within its framework as follows. We are only concerned about the convergence of the endeffector position and orientation to their desired values at the final time; we choose this convergence as the first-priority task. At this moment, we do not care about the trajectory the endeffector takes to reach its final configuration. We invoke a Liapunov function (Liapunov, 1892; LaSalle and Lefschetz, 1961), the primary Liapunov function, that takes care of the first-priority task as follows:

$$v_1 = v_1(\theta_1, \theta_2) \tag{11}$$

The time derivative of $v_1$ is given as

$$
\begin{aligned}
\dot{v}_1 &= \frac{\partial v_1}{\partial \theta_1}\dot{\theta}_1 + \frac{\partial v_1}{\partial \theta_2}\dot{\theta}_2 \\
&= \left(\frac{\partial v_1}{\partial \theta_1}H + \frac{\partial v_1}{\partial \theta_2}\right)\dot{\theta}_2 \\
&= -p_1^T u
\end{aligned}
\tag{12}
$$

where

$$p_1 \stackrel{\triangle}{=} -\left(\frac{\partial v_1}{\partial \theta_1}H + \frac{\partial v_1}{\partial \theta_2}\right)^T \tag{13}$$

and where Eq.(1) and $u = \dot{\theta}_2$ were substituted. At this juncture we simply assume that $p_1^T p_1$ is positive definite over $\theta_1$ and $\theta_2$, except at the final destination of the endeffector where it becomes zero. Then it would be a good idea to choose $u$ such that it leads to the following equation

$$\dot{v}_1 = -k_1\, p_1^T p_1 \tag{14}$$

where $k_1$ is a positive scalar. The general form of $u$ that satisfies Eq.(14) is provided by

$$u = k_1\, p_1 + \left(E_6 - p_1 p_1^{\#}\right) w_1 \tag{15}$$

where, $p_1^{\#}$ is the pseudoinverse of $p_1$, $E_6 \in R^{6 \times 6}$ is the identity matrix, and $w_1$ is an arbitrary vector. Indeed, by substituting Eq.(15) into Eq.(12), we get

$$
\begin{aligned}
\dot{v}_1 &= -p_1^T u \\
&= -k_1\, p_1^T p_1 - k_1\, p_1^T \left(E_6 - p_1 p_1^{\#}\right) w_1 \\
&= -k_1\, p_1^T p_1
\end{aligned}
\tag{16}
$$

where we used the relation $p_1^T \left(E_6 - p_1 p_1^{\#}\right) = \left(p_1 - p_1 p_1^{\#} p_1\right)^T = 0$.

Geometrically, Eq.(15) can be explained as follows. The first term gives the steepest direction to reduce the first Liapunov function, by means of $\dot{\theta}_2$. The second term represents the direction in which $\dot{\theta}_2$ can change without affecting $\dot{v}_1$. We could call this second term an equipotential motion. Hence the first term drives the system into the most efficient direction to reduce the Liapunov function. The second term steers the direction of motion by adding the equipotential motion to the first term. The exact nature of the equipotential motion depends upon the choice of $w_1$. The second term does not change the convergence speed locally as seen from Eq.(16).

Note that since $p_1$ is a vector, the rank of the coefficient matrix of $w_1$ in Eq.(16) is five, and we therefore have 5DOF in choosing the equipotential motion. This was made possible by considering the convergence of the endeffector to its desired configuration as the first priority task, not the trajectory of it.

The arbitrary vector $w_1$ in Eq.(16) is now determined by the second priority task. We again use a Liapunov function. We assume that the second priority task can be taken care of by the following secondary Liapunov function:

$$v_2 = v_2(\theta_1, \theta_2) \tag{17}$$

Similar to Eq.(12) we have

$$\dot{v}_2 = -p_2^T u \tag{18}$$

$$p_2 \stackrel{\triangle}{=} -\left(\frac{\partial v_2}{\partial \theta_1} H + \frac{\partial v_2}{\partial \theta_2}\right)^T$$

We also assume that $p_2^T p_2$ is positive definite over $\theta_1$ and $\theta_2$ except at a set of $\theta_1$ and $\theta_2$ where we prefer it not to be. Here we set

$$\dot{v}_2 = -k_2 \, p_2^T p_2 \tag{19}$$

as the goal of choosing $u$. Substituting Eq.(15) into Eq.(18) and comparing with Eq.(19) we obtain

$$p_2^T \left(E_6 - p_1 p_1^\#\right) w_1 = p_2^T \left(k_2 \, p_2 - k_1 \, p_1\right) \tag{20}$$

Unless $\left(E_6 - p_1 p_1^\#\right) p_2 = 0$, the general solution to Eq.(20) is given as

$$w_1 = \left(\overline{p}_2^\#\right)^T p_2^T \left(k_2 \, p_2 - k_1 \, p_1\right) + \left(E_6 - \overline{p}_2 \overline{p}_2^\#\right) w_2 \tag{21}$$

$$\overline{p}_2 \stackrel{\triangle}{=} \left(E_6 - p_1 p_1^\#\right) p_2$$

where $\overline{p}_2^\#$ is the pseudoinverse of $\overline{p}_2$, and $w_2$ is an arbitrary vector. By substituting Eq.(21) into Eq.(15), we obtain

$$u = k_1\, p_1 + \left(E_6 - p_1 p_1^{\#}\right) \left(\overline{p}_2^{\#}\right)^T p_2^T \left(k_2\, p_2 - k_1\, p_1\right)$$
$$+ \left(E_6 - p_1 p_1^{\#}\right)\left(E - \overline{p}_2 \overline{p}_2^{\#}\right) w_2$$
$$= k_1\, p_1 + \left(\overline{p}_2^{\#}\right)^T p_2^T \left(k_2\, p_2 - k_1\, p_1\right)$$
$$+ \left(E_6 - p_1 p_1^{\#}\right)\left(E_6 - \overline{p}_2 \overline{p}_2^{\#}\right) w_2 \tag{22}$$

When $\overline{p}_2 \neq 0$, substituting Eq.(22) into Eq.(18) gets

$$\dot{v}_2 = -k_1\, p_2^T p_1 - \frac{p_2^T \left(E_6 - p_1 p_1^{\#}\right) p_2}{p_2^T \left(E_6 - p_1 p_1^{\#}\right) p_2} p_2^T \left(k_2\, p_2 - k_1\, p_1\right)$$
$$- p_2^T \left(E_6 - p_1 p_1^{\#}\right)\left(E_6 - \overline{p}_2 \overline{p}_2^{\#}\right) w_2$$
$$= -k_1\, p_2^T p_1 - p_2^T \left(k_2\, p_2 - k_1\, p_1\right) - 0$$
$$= -k_2\, p_2^T p_2 \tag{23}$$

where we used $\overline{p}_2^{\#} = \overline{p}_2^T / \left(\overline{p}_2^T \overline{p}_2\right)$ for $\overline{p}_2 \neq 0$. When $\overline{p}_2 = 0$, $\overline{p}_2^{\#}$ becomes zero, and therefore we have

$$\dot{v}_2 = -p_2^T \left(k_1\, p_1 + \left(E_6 - p_1 p_1^{\#}\right) w_2\right)$$
$$= -k_1\, p_2^T p_1 \tag{24}$$

Equation (24) implies that when $\overline{p}_2 = 0$, we may loose the negative definiteness of $\dot{v}_2$. $\overline{p}_2 = 0$ occurs when $p_2$ becomes parallel to $p_1$. Only when $p_1$ and $p_2$ are mutually parallel and the primary and secondary Liapunov functions conflict with each other, $\dot{v}_2$ becomes positive. Equation (23) shows that except in the case of $\overline{p}_2 = 0$, the input $u$ given by Eq.(22) guarantees the negative definiteness of both the primary and secondary Liapunov functions.

Like the task-priority based control of an ordinary kinematically redundant manipulator, we can use the third term of Eq.(22) if there is a third or even lower priority task. If we have only the first and second priority tasks, and if $\overline{p}_2 \neq 0$, by setting $w_2 = 0$ Eq.(22) becomes simplified to

$$u = k_1\, p_1 + k \left(E_6 - p_1 p_1^{\#}\right) p_2 \tag{25}$$

where

$$k \triangleq \frac{p_2^T \left(k_2 p_2 - k_1 p_1\right)}{p_2^T \left(E_6 - p_1 p_1^{\#}\right) p_2} \tag{26}$$

We have developed the hierarchical Liapunov function approach for planning a nonholonomic trajectory. The key idea behind this approach is that we expect that the second term of Eq.(25) implicitly exploits the global nonholonomic redundancy by locally steering the direction based on the secondary Liapunov function. In section 4, the numerical solution will show that this expectation is fulfilled. Note that this approach can be used for the path planning of a fixed base kinematically nonredundant manipulator as well.

In the following subsections of section 3 we show typical examples of the primary and secondary Liapunov functions.

## 3.2 Primary Liapunov function

The following function is chosen as a candidate of the primary Liapunov function

$$v_1 = \frac{1}{2} \Delta h^T A \Delta h \tag{27}$$

$$\Delta h = h_d - h \tag{28}$$

where $A$ is a symmetric positive definite constant matrix and $h$ is defined by Eq.(10). $v_1 = 0$ is attained only when $h_d = h$, where $h_d$ is the constant goal of the endeffector. The time derivative of $v_1$ becomes

$$\dot{v}_1 = -\Delta h^T A \dot{h} = -a^T u, \qquad a \triangleq \left(A \hat{J}\right)^T \Delta h \tag{29}$$

where, Eq.(10) and $u = \dot{\theta}_2$ were substituted. Choosing the input as

$$u = u_1 = k_1 a \tag{30}$$

where $k_1$ is a positive scalar, the rate of change of the Liapunov function becomes

$$\dot{v}_1 = -k_1 a^T a \le 0 \tag{31}$$

If Eq.(31) is negative definite with respect to $\Delta h$, everywhere except at $h = h_d$, then Liapunov's direct method (Liapunov 1892; LaSalle and Lefschetz 1961) can conclude its global stability. In such a case, the Liapunov function $v_1$ shall decrease monotonously and the endeffector shall gradually reach its desired configuration. This is not true only when the matrix $\hat{J}$ is singular. $\hat{J}$ has singularity problems only when $\tilde{J}$ or $N_E$ given by Eq.(10), becomes singular.

A discussion on the singularities of $\tilde{J}$ can be found in Papadopoulos and Dubowsky (1989). The singularity of $N_E$ is however not a physical singularity. It is a singularity of representation of orientation and can be overcome by choosing a different representation.

## 3.3 Secondary Liapunov function

While planning the trajectory, we have to guarantee the physical feasibility of the trajectory, namely, the avoidance of joint limits and of collisions with obstacles. We consider the physical feasibility by using another Liapunov function. The primary Liapunov function defined in section 3.2 is consistently combined with this Liapunov function by defining a hierarchy between them, as it was discussed in the section 3.1. In this section, we explain how we define the secondary Liapunov function for joint limit avoidance. Obstacle avoidance can be done by defining the secondary Liapunov function in the same fashion.

We first assume that each joint of $\theta_2$ has the joint limit $|\theta_{2i}| \leq \theta_{2i\,max}$ ($i = 1, 2, \cdots, 6$), where $\theta_{2i\,max}$ is a positive constant. We choose the following function as the Liapunov candidate:

$$v_{2J} = \sum_{i=1}^{6} \left( \frac{\theta_{2i}}{\theta_{2i\,max}} \right)^2 \tag{32}$$

$v_{2J}$ is always positive except when $\theta_2 = 0$, when $v_{2J}$ becomes zero. The time derivative of Eq.(32) becomes

$$\dot{v}_{2J} = 2\,row.\left( \frac{\theta_{2i}}{\theta_{2i\,max}^2} \right) u \tag{33}$$

where $u = \dot{\theta}_2$, and $row.(*)$ implies a row vector. Therefore if we choose

$$u = u_{2J} = -\theta_2 \tag{34}$$

then $\dot{v}_{2J}$ becomes

$$\dot{v}_{2J} = -2 \sum_{i=1}^{6} \left( \frac{\theta_{2i}}{\theta_{2i\,max}} \right)^2 = -2\,v_{2J} \tag{35}$$

Equation (35) implies that if $|\theta_{2i}| < \theta_{2i\,max}$ for all $i = 1, 2, \cdots, 6$, then $\dot{v}_{2J}$ is negative definite except for $\theta_2 = 0$, when $\dot{v}_{2J}$ becomes zero. Consequently, we can conclude that $v_{2J}$ in Eq.(32) works as a Liapunov function that brings the manipulator back to its home position.

For the sake of simplicity, we assumed the joint limits to be symmetric. Equation (32) can be easily extended to asymmetric joint limits.

Note that the pair of Eqs.(33) and (35) have a different structure from the pair Eqs.(18) and (19). When $\theta_{2i\,max}$ has the same value for all the joints, the two pairs of equations become equivalent. In numerical simulation in section 4, we assume this case.

## 4. SIMULATION

### 4.1 Endeffector trajectory planning

We present here the simulation results of one particular case. The initial state and the final endeffector configuration were set at

$$x = (\,150.0 \quad 0.0 \quad 0.0 \quad -165.0 \quad 10.0 \quad 15.0 \quad 0.0 \quad -15.0 \quad 0.0\,)^T \tag{36}$$

$$x_E = (\,0.217 \quad 0.460 \quad 1.121 \quad -12.45 \quad -39.40 \quad -11.22\,)^T \tag{37}$$

where the units in Eq.(36) are in degrees and those in Eq.(37) are in meters and degrees. We first planned for the endeffector to reach its desired configuration without imposing any joint limits. We used only the primary Liapunov function $v_1$ for path planning. The matrix $A$ in Eq.(27) was chosen as the identity matrix. We chose $p_1$ in Eq.(25) as $a$ given by Eq.(30).

The vector $a$ in Eq.(29) can become very small in magnitude when the matrix $\hat{J}$ is close to becoming singular and the vector $\Delta h$ is in the degenerate direction of $\hat{J}$. In such situations, the input to the system becomes very small. To compensate this we use the scalar positive constant $k_1$ in defining the input by Eq.(30). In this context we would like to mention again that increasing the velocity along the path only reduces the time of travel; it does not change the path. This is because of the time independence of integration of Eq.(7). We used a value of 500.0 for $k_1$. Such a procedure can however result in a very large input to the system when the matrix $\hat{J}$ is well conditioned. To avoid this, we place an upper limit on the norm of the input. If $u^T u > 1.0$, we proportionally reduce the input, by changing $k_1$, such that $u^T u = 1.0$. We concluded convergence and terminated the computation when the Liapunov function $v_1$ was less than 0.0001.

Figure 1 shows the change of the Liapunov function $v_1$ with time, in the absence of joint limits. Figure 2 shows the synthesized trajectory of the joint variables. The behavior of the system for this case is observed in Fig.3 in which the eight figures correspond to the eight marks on the graph in Fig.1. The convergence time was approximately 4.65 seconds.

## 4.2 Endeffector trajectory planning with joint limits

As the next step, we imposed joint limits and using the Liapunov functions $v_1$ and $v_{2J}$ in hierarchy, we carried out a simulation for the same initial and desired configuration given by Eqs.(36) and (37). The joint limit for all the joints were set at 180.0 degrees *i.e.* $\theta_{2i\,max} = 180.0$ for $(i = 1, 2, \cdots, 6)$. From Fig.2 it is clear that the first joint would violate its limit if the trajectory is planned with the primary Liapunov function alone. We chose $p_2$ in Eq.(25) as $u_{2J}$ given by Eq.(34).

The variable $k$ in Eqs.(25) and (26) is set to a constant value for the simplicity of computation. The vectors $p_1$ and $p_2$ in Eq.(25) may be almost parallel in some situations. In such cases the second term of the input of Eq.(25) becomes small. In the case of path planning with joint limits, we alleviate this problem by using a value of 31.4 for the constant $k$. However, when the vectors $p_1$ and $p_2$ are almost perpendicular, a value of 31.4 for $k$ magnifies the second term of the input to a large extent. We therefore place an upper bound on the norm of the second term. When the norm is greater than 3.0, the second term is proportionally reduced to have a norm of exactly 3.0. The gain and the saturation for the primary Liapunov function discussed above was used concurrently.

Figure 4 shows the variation of the Liapunov function $v_1$ with time for the path planning problem with joint limits. Figure 5 shows the synthesized trajectory of the joint variables for the same case. It can be seen from Fig.5, which is very different from Fig.2, that none of the joint limits has been violated. The system behavior is shown in Fig.6 in which the eight figures correspond to the eight marks on the Fig.4. The convergence criterion was kept the same as in the first case, and the time for convergence was noted to be approximately 1.5 seconds.

## 5. CONCLUSION

This paper discussed the presence of nonholonomic redundancy as different from ordinary kinematic redundancy, in space robots. The presence of this special kind of redundancy is exibited only after a global motion unlike ordinary kinematic redundancy that exibits itself locally through self-motion. A path planning scheme was then developed for the utilization of nonholonomic redundancy.

The identification of the reachable set of states for a space robot is still an open problem. A 6DOF space robot is described by nine variables. We are interested to know whether the entire nine-dimensional space of the 6-DOF space robot is reachable. This will be one of the major research problems in this field.

# REFERENCES

Akin, D.L., Minsky, M.L., Thiel, E.D., and Kurtzman, C.R. 1983. Space Applications of Automation, Robotics and Machine Intelligence Systems (ARAMIS) - Phase 2, *NASA Contractor Report 3734, MIT Space Systems Laboratory.*

Alexander, II.L., and Cannon, R.H. 1987 (Seattle). Experiments on the control of a satellite manipulator. *Proc. 1987 American Control Conference.*

IIanafusa, H., Yoshikawa, T., and Nakamura, Y. 1981 (Kyoto, Japan). Analysis and Control of articulated robot arms with redundancy. *In Control Science and Technology for the Progress of Society (Proc. 8th Triennial World Congress of IFAC)*, ed. H.Akashi, Vol.4, pp.1927-1932.

Klein, C.A., and Huang, C.H. 1983. Review of pseudoinverse control for use with kinematically redundant manipulators. *IEEE Trans. Sys., Man, Cyber.* SMC-13:245-250.

Koningstein, R., Ullman, M., and Cannon Jr., R.H. 1989. Computed Torque Control of a Free-Flying Cooperating-Arm Robot. *1989 Proc. NASA Workshop on Space Telerobotics*, JPL, Pasadena, CA.

LaSalle, J., and Lefschetz, S. 1961. *Stability by Liapunov's Direct Method with Applications.* New York: Academic Press.

Liegeois, A. 1977. Automatic supervisory control of the configuration and behavior of multibody mechanisms. *IEEE Trans. Sys., Man, Cyber.* SMC-7:868-871.

Longman, R.W., Lindberg, R.E., and Zedd, M.F. 1987. Satellite-mounted robot manipulators: New kinematics and reaction moment compensation. *International Journal of Robotics Research.* 6 (3): 87-103.

Liapunov, A.M. 1892. On the general problem of stability of motion. *Soviet Union: Kharkov Mathematical Society. (in Russian).*

Miyazaki, F., Masutani, Y., and Arimoto, S. 1988 (Minneapolis). Sensor feedback using approximate Jacobian. *Proc. USA-Japan Symposium on Flexible Automation.* pp. 139-145.

Mukherjee, R., and Nakamura, Y. 1990 (Kobe, Japan). Space Robot Dynamics and its Efficient Computation. *International Symposium on Artificial Intelligence, Robotics and Automation in Space.*

Nakamura, Y., and Hanafusa, H. 1987. Optimal Redundancy Control of Robot Manipulators. *The International Journal of Robotics Research.* Vol.6, No.1, pp.32-42.

Nakamura, Y., Hanafusa, H., and Yoshikawa, T. 1987. Task-Priority based Redundancy Control of Robot Manipulators. *The International Journal of Robotics Research.* Vol.6, No.2, pp.3-15.

Nakamura, Y., and Mukherjee, R. 1989 (a) Redundancy of Space Manipulator on Free-Flying Vehicle and its Nonholonomic Path Planning. *1989 Proc. NASA Workshop on Space Telerobotics*, JPL, Pasadena, CA.

Nakamura, Y., and Mukherjee, R. 1989 (b) (Scottsdale). Nonholonomic path planning of space robots. *1989 Proc. IEEE International Conference on Robotics and Automation.*pp 1050-1055.

Nakamura, Y., and Mukherjee, R. 1989 (c) (Tokyo, Japan). Bi-Directional approach for nonholonomic path planning of space robots. *Proc., Fifth International Symposium of Robotics Research.* pp. 405-416.

Nakamura, Y., and Mukherjee, R. 1990 (Cincinnati). Nonholonomic path planning of space robots via Bi-Directional approach. *Proc. 1990 IEEE International Conference on Robotics and Automation.* pp. 1764-1769.

Papadopoulos, E., and Dubowsky, S. 1989 (San Francisco). On the Dynamic Singularities in the Control of Free-Floating Space Manipulators. *ASME Winter Annual Meeting: Dynamics and Control of Multibody/Robotic Systems with Space Applications.* DSC-Vol.15, pp.45-51.

Spofford, J.R. 1988. Coordinated Control of a Free-Flying Teleoperator. *Ph.D. Dissertation. Mechanical Engineering, MIT.*

Sreenath, N. 1990 (Cincinnati). Nonlinear Control of Multibody Systems in Shape Space. *Proc. 1990 IEEE International Conference on Robotics and Automation.* pp.1776-1781.

Umetani, Y., and Yoshida, K. 1987. Continuous path control of space manipulators mounted on OMV. *Acta Astronautica.* 15 (12): 981-986.

Vafa, Z. 1987. The kinematics, dynamics and control of space manipulators: The virtual manipulator concept. Ph.D. Dissertation. Mechanical Engineering, MIT.

Vafa, Z., and Dubowsky, S. 1987 (Raleigh). On the dynamics of manipulators in space using the virtual manipulator approach. *Proc. 1987 IEEE International Conference on Robotics and Automation.*

Yamada, K. 1989. Formulation of space multi-body systems and its application to control. *Ph.D. Dissertation, Department of Industrial Machinery Engineering, University of Tokyo.* (in Japanese)

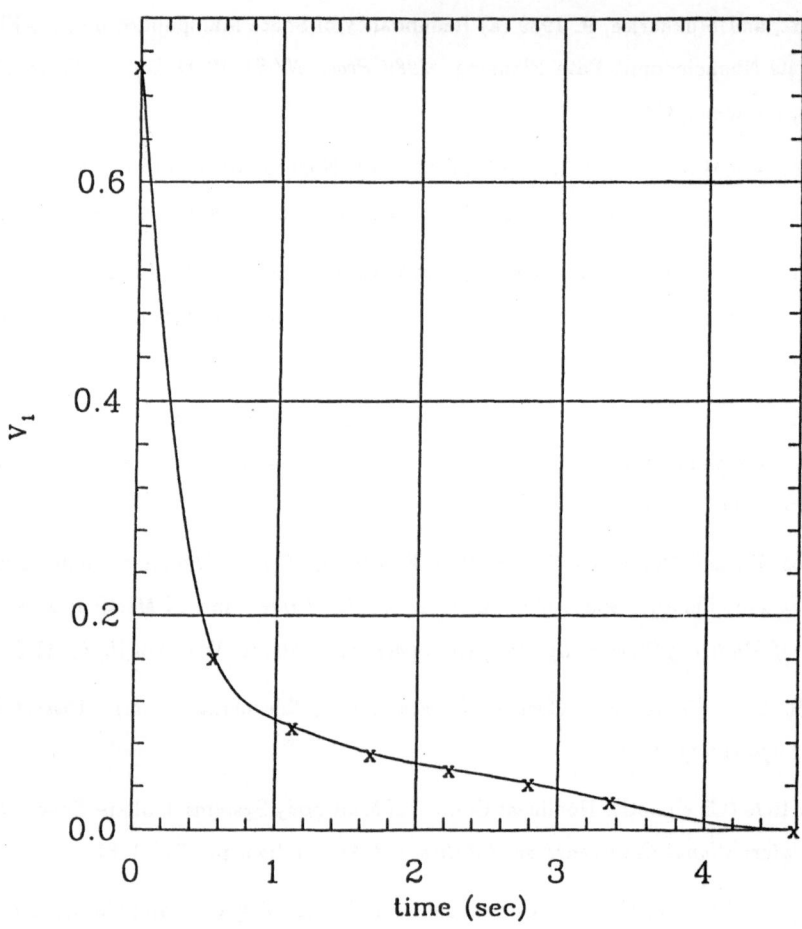

Fig.1. Variation of the Primary Lyapunov function $v_1$ with time for the simulation in the absence of joint limits and obstacles

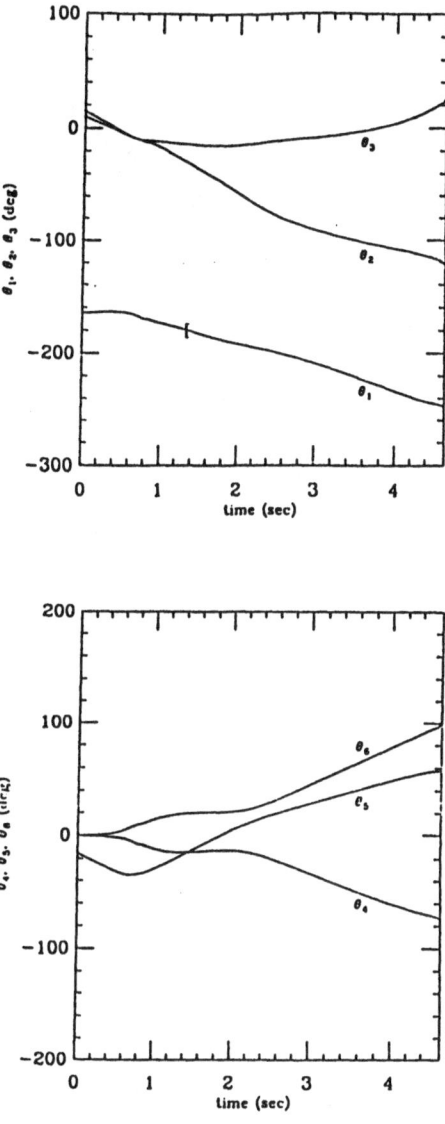

Fig.2. Trajectory of the joint variables for the simulation in the absence of joint limits and obstacles. The region on the curves to the right of the '[' mark and to the left of the ']' mark is where the particular joint would violate the joint limit

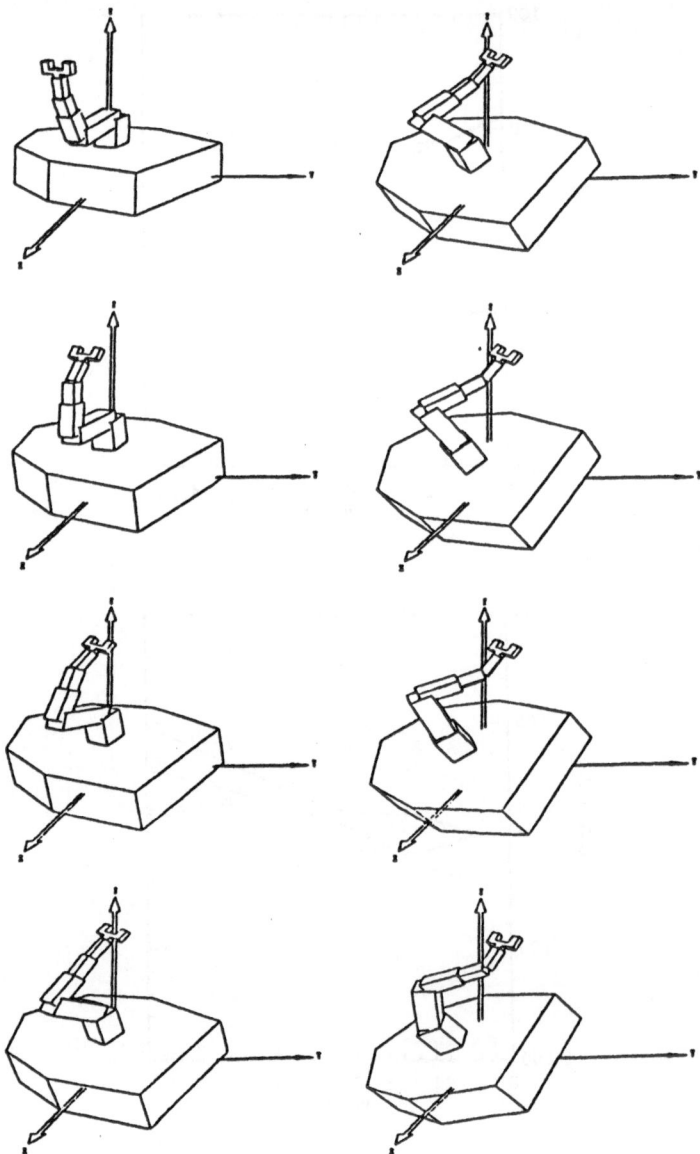

Fig.3. System behaviour for the simulation in the absence of joint limits and obstacles. The eight figures correspond to the eight marks in Fig.1

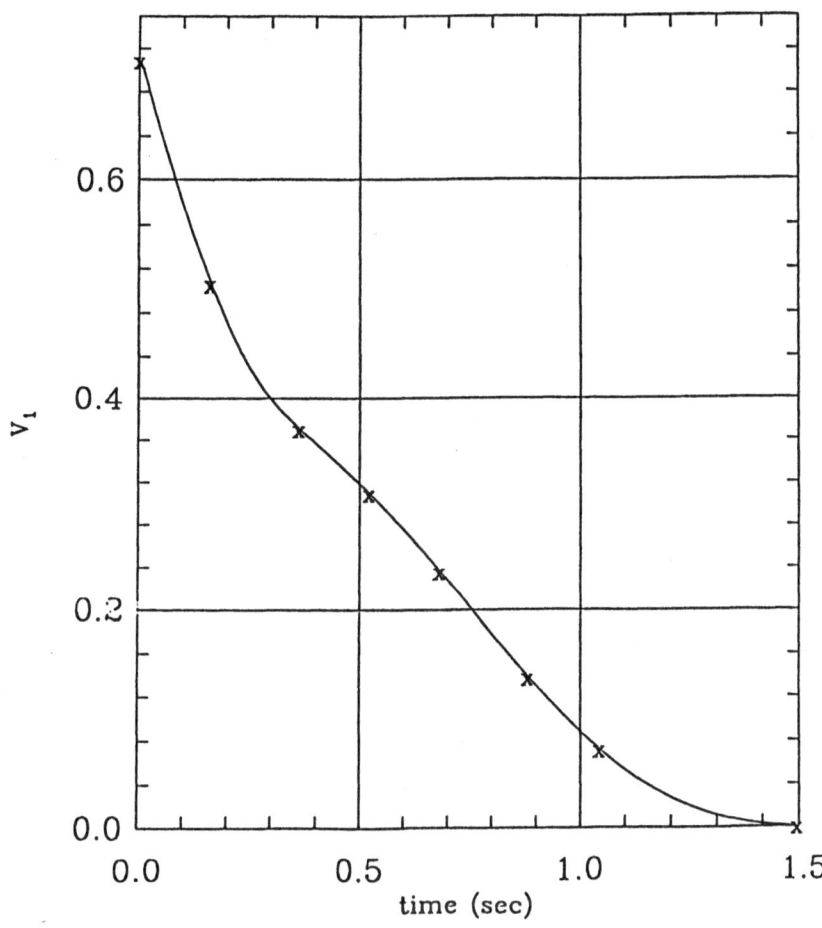

Fig.4. Variation of the Primary Lyapunov function $v_1$ with time for the simulation with joint limits

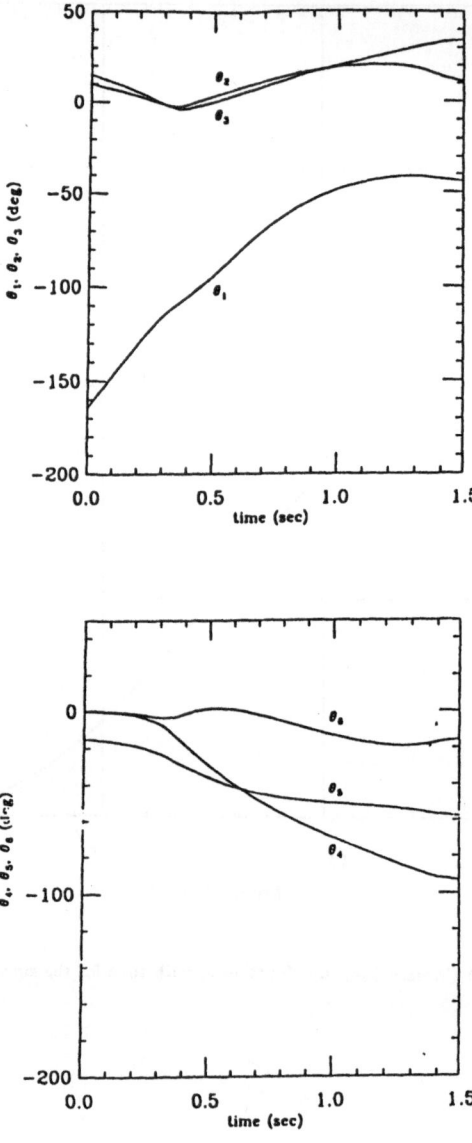

Fig.5. Trajectory of the joint variables for the simulation with joint limits. None of the joints violate their limits

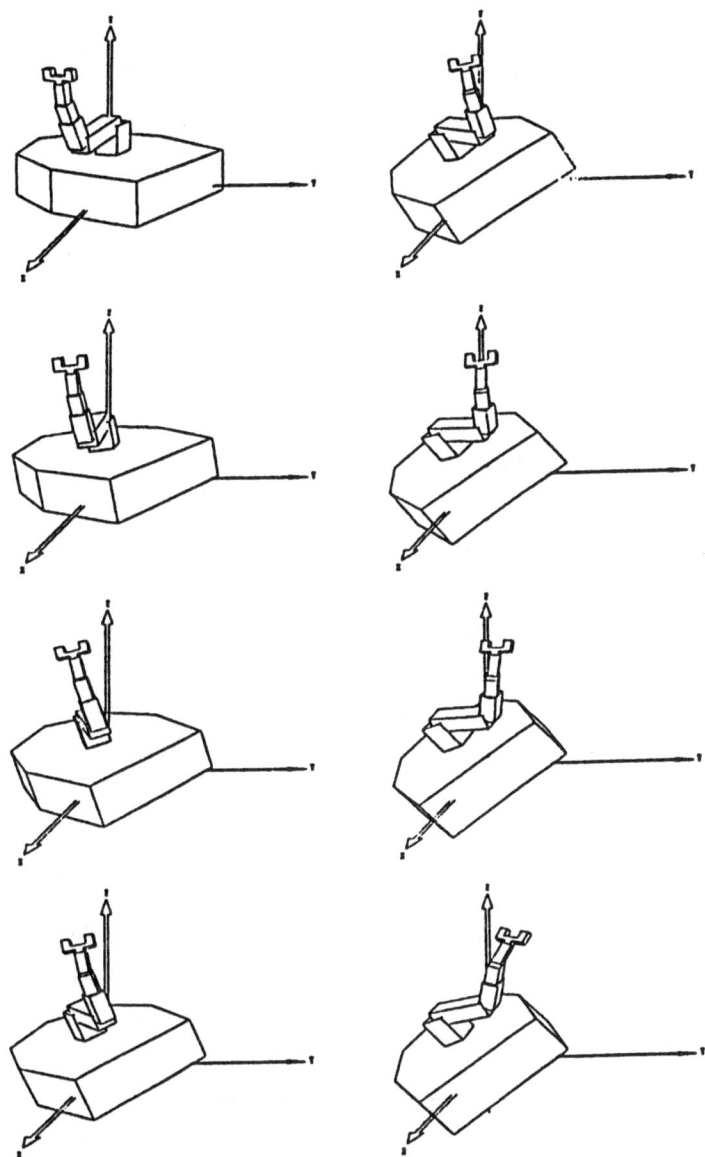

Fig.6. System behaviour for the simulation with joint limits. The eight figures correspond to the eight marks in Fig. 4.

# CONTROLLABILITY AND STATE FEEDBACK STABILIZABILITY OF NON HOLONOMIC MECHANICAL SYSTEMS

## G. Campion[†], B. d'Andrea-Novel[‡], G. Bastin[†]

† Laboratoire d'Automatique, Dynamique et Analyse des Systèmes
Université Catholique de Louvain
Place du Levant, 3-B1348 LOUVAIN-LA-NEUVE (Belgium)

‡ Centre d'Automatique,
Ecole Nationale Supérieure des Mines de Paris
Rue Saint Honoré, 35-F77305 FONTAINEBLEAU (France)

## ABSRACT

The dynamics of non holonomic mechanical system are described by the classical Euler-Lagrange equations subjected to a set of non-integrable constraints. Non holonomic systems are strongly accessible whatever the structure of the constraints. They cannot be asymptotically stabilized by a smooth pure state feedback. However smooth state feedback control laws can be designed which guarantee the global marginal stability of non holonomic systems.

## 1. INTRODUCTION

A mechanical system, whose configuration is completely described by a set of generalized coordinates, can be subjected to kinematic constraints (such as the pure rolling condition of a wheel on a plane), which are expressed by relations between the coordinates and their time derivatives. If these constraints are holonomic (that is integrable) it is possible to characterize the system configuration by a smaller number of coordinates (i.e. to use the constraints in order to eliminate the redundant coordinates) in such a way that the constraints are automatically satisfied in the new coordinates. Unfortunately, in case of non holonomic constraints, this elimination is not possible and the constraints have to be taken into account explicitly in the derivation of the dynamical equations. The theory of mechanical systems with non holonomic constraints has been developped at the end of last century by many authors (e.g. Appell [1], Hamel [2]). The present paper deals with control design of such systems, for which, due to the nonholonomic constraints, the standard control laws developped for holonomic mechanical systems (for instance robotic manipulators) are not applicable .

Control of mechanical systems, with not integrable constraints which are linear in the generalized velocities, has been discussed in the literature through the special case of mobile wheeled robots (see e.g. [3, 4, 5]). In these papers however, the control is designed on the basis of a kinematic state-space model derived from the constraints, but not taking the internal dynamics of the system into account. The purpose of this paper is to derive a full dynamical description of such nonholonomic mechanical systems, including the constraints and the internal dynamics, and to show how a suitable change of coordinates allows to analyse globally the controllability and the state feedback stabilizability of the system. The feedback stabilizability of mechanical systems with constraints (holonomic or not) is also examined by Bloch and McClamroch [9]. However, they use another change of coordinates which is less efficient since it provides only local stability results and is not convenient for a controllability analysis.

The paper is organized as follows. The concept of non holonomic constraints for mechanical systems is introduced in Section 2 within the framework of the theory of nonlinear control systems. The dynamics of non holonomic systems can be partially described by a so-called kinematic state-space model. In Section 3.1, it is shown that this model is completely controllable. The existence of smooth stabilizing state feedback controls is then addressed in Section 3.2. It is shown that the origin of the generalized coordinates cannot be asymptotically stabilized by a smooth pure state feedback but can nevertheless be globally maginally stabilized. A general dynamical state-space model of non holonomic systems is then derived in Section 4, using the classical Euler-Lagrange formalism. By a suitable change of coordinates,this model can be partially linearized in such a way that the remaining nonlinearities only depend on the structure of the constraints. On this basis it is then shown, in Section 4.1, that non holonomic systems are strongly accessible whatever the structure of the constraints. Furthermore, as shown in Section 4.2, the stabilizability results of Section 3.2 can be extended to the general case: non holonomic systems cannot be asymptotically stabilized by a smooth pure feedback control but can be globally marginally stabilized. The design of the stabilizing control law is explicited.

## 2. NONHOLONOMIC CONSTRAINTS

We are concerned, in this paper, with mechanical systems whose configuration space is an n-dimensional simply connected manifold $\mathcal{M}$ and whose dynamics are described, in local coordinates, by the so-called Euler-Lagrange equations of motion. Usually, the local coordinates used for the description of these systems are termed "generalized coordinates" and denoted $q_1, q_2, ..., q_n$. Each configuration of the system is represented by the vector of these generalized coordinates and is denoted:

$$q \equiv [q_1, q_2, ... , q_n]^T$$

The configuration manifold $\mathcal{M}$, which is the set of all possible configurations, is represented in local coordinates by an open set $\Omega \subseteq \mathbb{R}^n$. The position of each material point of the system is a function of the generalized coordinates. A motion of the system is represented in the q coordinates by a smooth time function q(t). The corresponding trajectory is a one-dimensional immersed submanifold of $\mathcal{M}$. The tangent vector at a point of the trajectory is then represented by the vector $\dot{q} \equiv [\dot{q}_1, \dot{q}_2, ... , \dot{q}_n]^T$ whose components $\dot{q}_1, \dot{q}_2, ... , \dot{q}_n$ are termed generalized velocities.

In many instances, the motion of mechanical systems is subjected to various constraints which are permanently satisfied during the motion and which take the form of algebraic relationships between the positions and the velocities of particular material points of the system. Two kinds of constraints can be distinguished: geometric constraints and kinematic constraints.

## Geometric constraints.

These constraints are represented by analytical relations between the generalized coordinates. When the system is subjected to m such constraints, there exists an m-dimensional vector function $\rho(q) : \Omega \rightarrow \mathbb{R}^m$ such that $\rho(q) = 0$ for all q in $\Omega$. The m (<n) constraints are said *independent* when the jacobian matrix of $\rho(q)$ has full rank for all q. In that case m generalized coordinates can be eliminated and n - m generalized coordinates are sufficient to provide a full description of the configurations of the system.

## Kinematic constraints.

These constraints are represented by analytical relations between the generalized coordinates and velocities. In most applications, these relations are linear with respect to the generalized velocities and written as:

$$a_j^T(q)\dot{q} = 0 \tag{1}$$

where $a_1^T, a_2^T, ... , a_m^T$ are smooth n-dimensional covector fields on $\mathcal{M}$. In matrix form, the constraints (1) are written :

$$A^T(q)\dot{q} = 0$$

where A(q) is the (n x m) matrix made up of the vector functions $a_j(q)$ as follows:

$$A(q) \equiv [a_1(q), a_2(q), \ldots, a_m(q)]$$

The m (<n) constraints are said *independent* when this matrix has full rank for all q. Unlike geometric constraints, the kinematic constraints do not necessarily lead to the elimination of generalized coordinates from the system description. The elimination is possible only when the constraints are *holonomic* (that is: integrable). Our concern in this paper will precisely be to discuss the controllability and the feedback stabilization of mechanical systems with *nonholonomic* constraints.

Hence, without loss of generality, we can consider that all the redundant generalized coordinates associated to the geometric constraints have been eliminated and restrict our attention to mechanical systems subjected to m *independent* kinematic constraints only. These constraints are assumed to have the form (1).

We assume that the annihilator of the codistribution spanned by the covector fields $a_1^T, a_2^T, \ldots, a_m^T$, is an (n-m)-dimensional *smooth* nonsingular distribution $\Delta$ on $\mathcal{M}$. This distribution $\Delta$ is spanned by a set of (n-m) smooth vector fields $s_1, s_2, \ldots, s_{n-m}$ :

$$\Delta \equiv \text{span}\{s_1, s_2, \ldots, s_{n-m}\}$$

which satisfy, in local coordinates, the following relations:

$$a_j^T(q)s_i(q) = 0 \quad \forall q \in \Omega \quad j = 1,\ldots,m \quad i = 1,\ldots,n-m$$

Since $\Delta$ is nonsingular, any vector field $\tau$ of $\Delta$ can be expressed in the form:

$$\tau(q) = \sum_{i=1}^{n-m} c_i(q)s_i(q)$$

where $c_1(q), c_2(q), \ldots, c_{n-m}(q)$ are smooth functions on $\Omega$ (see e.g. Isidori [6], Chapter 1, Section 1.3).

We introduce also the full rank matrix S(q) made up of the vector functions $s_i(q)$:

$$S(q) \equiv [s_1(q),s_2(q), \ldots,s_{n-m}(q)]$$

It is then clear that the constraints (1) may be expressed as:

$$\dot{q} \in \Delta(q) \text{ or equivalently } \dot{q} \in \text{Im}[S(q)]$$

Consider now the involutive closure of $\Delta$, denoted $\Delta^*$, and defined as the smallest involutive distribution containing $\Delta$. Assume that this distribution is regular (that is has constant dimension on $\mathcal{M}$). Clearly:

$$n - m \leq \dim(\Delta^*) \leq n$$

Let $(n - m^*)$ denote the dimension of $\Delta^*$, with $m^* \leq m$. From Fröbenius Theorem, at each $q$ in $\Omega$, there exists a set of $m^*$ independent smooth functions denoted $\mu_1(q)$, $\mu_2(q)$, ..., $\mu_{m^*}(q)$, such that, for each vector field $\tau \in \Delta^*$ the following relations hold :

$$L_\tau \mu_i(q) = 0 \qquad i = 1, ..., m^* \tag{2}$$

where $L_\tau \mu_i$ denotes the Lie derivative of $\mu_i$ along $\tau$.

Let us now define a change of coordinates $\xi = \Phi(q)$, with $\Phi(0) = 0$, with $m^*$ coordinates being the functions $\mu_1(q)$, $\mu_2(q)$, ..., $\mu_{m^*}(q)$, and the remaining $n-m^*$ coordinates being chosen to complete the diffeomorphism:

$$\xi = \Phi(q) = \begin{bmatrix} \varphi_1(q) \\ \varphi_2(q) \\ \cdot \\ \cdot \\ \varphi_{n-m^*}(q) \\ \mu_1(q) \\ \mu_2(q) \\ \cdot \\ \cdot \\ \mu_{m^*}(q) \end{bmatrix}$$

Hence the tangent vector to the trajectory at the point $\xi = \Phi(q)$ is represented in the $\xi$ coordinates by $\dot{\xi}$, with :

$$\dot{\xi} \in \mathrm{Im}\left[\left(\frac{\partial \Phi}{\partial q}\right)_{\Phi^{-1}(\xi)} S(\Phi^{-1}(\xi))\right]$$

It then follows from (2) that, since $\dot{q} \in \Delta(q)$, the last $m^*$ components of $\dot{\xi}$ are identically zero:

$$\dot{\xi}_{n-m^*+1} = \dot{\xi}_{n-m^*+2} = ... = \dot{\xi}_n = 0$$

This means that the $m^*$ coordinates $\xi_{n-m^*+1}, \xi_{n-m^*+2}, ..., \xi_n$ which are identical to the $m^*$ functions $\mu_i$ are constant along the motions of the system.

Then, depending on the dimension of $\Delta^*$, several situations may arise:

a) If $m^* = m$ (that is if $\Delta$ is involutive) the system is said to be *holonomic* . The configuration space can be characterized with $(n-m)$ coordinates only, namely $\xi_1, \xi_2, ..., \xi_{n-m}$. The configuration space is thus an $(n-m)$-dimensional manifold.

b) If $m^* = 0$ (that is if $\dim(\Delta^*) = n$) the constraints are completely nonintegrable and the system is said to be *nonholonomic*. The characterization of the configuration space requires n coordinates.

c) If $0 < m^* < m$ it is possible to eliminate $m^*$ coordinates. The configuration space is a manifold of dimension $n - m^*$.

Without loss of generality, we can thus assume that all the geometric and all the integrable kinematic constraints have been eliminated from the system description and restrict our attention to the situation b) that is to nonholonomic mechanical systems evolving in an n-dimensional configuration manifold and subjected to m independent nonintegrable constraints.

## 3. THE KINEMATIC STATE-SPACE MODEL : CONTROLLABILITY AND FEEDBACK STABILIZATION

The dynamics of nonholonomic mechanical systems are partially described by a state-space model which is associated to the kinematic constraints and referred to as the *kinematic state-space model*. Our purpose, in this section is to examine the controllability properties of this model and to discuss its state feedback stabilization.

Along the motions of the system, the constraints (1) imply the existence of a vector time function $w(t) \in \mathbb{R}^{n-m}$ for all t, such that:

$$\dot{q} = S(q)w(t) \tag{3}$$

where $S(q)$ is the matrix defined above. Conversely, for any initial condition $q(0)$ and any time function $w(t)$, the solution $q(t)$ of (3) will satisfy the constraints (1) and be a possible motion of the system.

Hence the model (3) can be interpreted as an n-dimensional state space representation of the motion of a nonholonomic mechanical system with state q and control input w. Obviously, for a given choice of the generalized coordinates, this representation is not unique since it depends on the particular selection of the basis (i.e. the vector fields $s_i$) of the distribution $\Delta$.

## 3.1. Controllability.

It follows immediately from the property of nonholonomy of the constraints that the *strong accessibility rank condition* (see [10]) is satisfied for all $q \in \Omega$ and, therefore, that the system (3) is strongly accessible from any configuration. Furthermore, since equation

(3) does not contain a drift vector field, strong accessibility implies controllability (see e.g. Nijmeijer and van der Schaft [7], Chapter 3, Section 3.1). We thus have the following result.

**Lemma 1.** The kinematic state space model of a nonholonomic system is controllable.

In practice, this means that for any two configurations $q^{(1)}$ and $q^{(2)}$ in $\Omega$, there exists a finite time T and an input function w(t) such that if $q(0) = q^{(1)}$ then $q(T) = q^{(2)}$. It is however worth noting that this does not mean that any velocity can be achieved since the generalized velocities are constrained to belong to the (n-m)-dimensional space spanned by the columns of S(q).

## 3.2. State feedback control.

In this section, we are concerned by the question of the existence of smooth pure state feedback stabilizing control laws for the kinematic state space model (3). More precisely, we would like to stabilize the system at a particular configuration which may be taken, without loss of generality, as the origin of the generalized coordinates (i.e. q = 0).

A smooth pure state feedback control law for the system (3) is defined as a smooth mapping:

$$w: \Omega \to \mathbb{R}^{n-m} : q \to w(q)$$

with the property that w(0) = 0. The application of this control law to the kinematic model (3) yields closed loop dynamics of the form:

$$\dot{q} = S(q)w(q) \tag{4}$$

which have the origin q = 0 as equilibrium point. Our concern is to find feedback controls w(q) that make this equilibrium point stable. Several definitions of the stability of equilibrium points are however in order here.

**Definitions.**

The equilibrium point q = 0 is Lagrange stable if, for any initial condition $q(0) = q_0$, there exist a bound $b(q_0)$ such that $\|q(t)\| \leq b(q_0)$ for all t.

The equilibrium point q = 0 is asymptotically stable (in the sense of Lyapunov) if there exists a positive constant $\varepsilon$ such that if $\| q(0) \| \leq \varepsilon$, then q(t) is bounded and converges to zero as time tends to infinity.

The equilibrium state q = 0 is (globally) marginally stable if it is Lagrange stable but not asymptotically stable.

It follows from the controllability of the system (Lemma 1) that there exist control laws which ensure the convergence of q(t) to zero. However the controllability does not imply the existence of a *smooth feedback* control law which can make the origin asymptotically stable and which can be synthetized as a smooth function of the state q only. In fact, it is easily shown that such smooth feedback stabilizing controls do *not* exist for nonholonomic systems.

**Lemma 2.** The equilibrium point q = 0 of the closed loop system (4) cannot be made asymptotically stable by a smooth state feedback w(q).
*Proof.* From the smoothness of A(q) and the independence of the constraints, it results that there exists a neighbourhood of the origin in $\mathbb{R}^n$, say $\mathcal{U}_0$, such that a given set of m rows of A(q) are independent on $\mathcal{U}_0$. Without loss of generality, we assume that the first m rows of A(q) are independent on $\mathcal{U}_0$, and we partition A(q) as follows:

$$A(q) = \begin{pmatrix} A_1(q) \\ A_2(q) \end{pmatrix}$$

where $A_1(q)$ is a square matrix, non singular on $\mathcal{U}_0$.
Define a neighbourhood $\mathcal{U}_1$, in $\mathbb{R}^{n-m}$, containing the origin, and $\mathcal{U}$ as the cartesian product of $\mathcal{U}_0$ by $\mathcal{U}_1$. Consider the following mapping, inspired by Eq.(4):

$$(q,w) \rightarrow g(q,w) = S(q)w$$

and denote $\mathcal{V}$, the image of $\mathcal{U}$ by this mapping g.
Then, for any $\sigma$ belonging to $\mathcal{V}$, there exists q such that $\sigma$ belongs to Im(S(q)) and therefore that

$$A_1^T(q)\sigma_1 + A_2^T(q)\sigma_2 = 0$$

where $\sigma$ is partitioned in a m-subvector $\sigma_1$ and a (n-m)-subvector $\sigma_2$. This implies that any $\sigma$, with $\sigma_1$ not equal to zero and $\sigma_2$ equal to zero, does not belong to $\mathcal{V}$, and therefore that $\mathcal{V}$, the image of the open set $\mathcal{U}$, is not an open neighbourhood of the origin. The result then follows from a necessary condition for the existence of smooth stabilizing feedback(see Brockett [8]).

*Remarks*:
-It must be noted that this proof is not based on the nonholonomy of the constraints and Lemma 2 holds therefore also for holonomic systems.

-Stabilization of non holonomic systems can however be achieved by open loop control, by non smooth state feedback control (see an example in Bloch and McClamroch[12]),or by using smooth state-feedback control depending explicitly on time, i.e. of the form w(q,t). Samson ([13]) proposes such a control ensuring the stability of the closed-loop, in the case of a mobile wheeled robot.

-However, as shown in the next theorem, there exists a smooth pure state-feedback control which can globally marginally stabilize the closed loop at the origin.

**Theorem 1.** With the smooth feedback control law:

$$w(q) = -S^T(q)q$$

the equilibrium point q = 0 of the closed loop system (4) is globally marginally stable. Precisely :

a) the state q(t) is bounded as follows for all t : $\| q(t) \| \leq \| q(0) \|$

b) the state q(t) converges to the invariant set U:

$$U \equiv \{ q \mid S^T(q)q = 0 \}$$

*Proof.* Straightforward by considering the Lyapunov function candidate $V(q) = q^T q$ whose time derivative along the closed loop trajectories is:

$$\dot{V} = -2\, q^T S(q) S^T(q) q$$

**Comment.** We notice that:

$$\text{rank} \left[ \frac{\partial}{\partial q} \{ S^T(q)q \} \right]_{q=0} = n - m$$

This implies that, at least locally around the origin, the invariant set defined in the statement of Theorem 1 is an m - dimensional manifold.

## 4. THE DYNAMICAL STATE-SPACE MODEL / CONTROLLABILITY AND FEEDBACK STABILIZATION.

In Section 3, the kinematic state-space model has been advocated to analyse the controllability of nonholonomic systems. It is however worth noting that this model does not provide a full description of the dynamics of mechanical systems. The variables w considered as inputs in this model are actually internal states which are dynamically related to the physical inputs that is to the generalized forces and torques applied to the system by the actuators. Our purpose in this section is to examine the controllability properties of the dynamical state-space model of nonholonomic systems and to discuss its state feedback stabilization.

Using the Lagrange formalism, the dynamics of a mechanical system are described by the following differential equations:

$$\frac{d}{dt}\left(\frac{\partial L}{\partial \dot{q}}\right) - \frac{\partial L}{\partial q} = A(q)\lambda + B(q)u \qquad (5)$$

with the following notations and definitions:

a) $L(q, \dot{q}) = T(q, \dot{q}) - W(q)$ is the Lagrangian of the system with $T(q, \dot{q})$ the kinetic energy and $W(q)$ the potential energy.

b) $B(q)u$ is the set of generalized forces applied to the system with $B(q)$ a (n x p) kinematic matrix and u the p-vector of external forces and torques applied to the system by the actuators.

c) $A(q)$ is the matrix associated to the constraints (see Section 2); $\lambda$ is the m-vector of Lagrange multipliers.

The n-dimensional vector function $A(q)\lambda$ is the vector of the generalized forces acting on the system in order to satisfy the constraints. These forces are said "ideal" which means that their potential power is zero for any potential velocity field compatible with the constraints.

The kinetic energy $T(q, \dot{q})$ is defined as:

$$T(q, \dot{q}) = \frac{1}{2}\dot{q}^T M(q)\dot{q}$$

where $M(q)$ is the (n x n) definite positive symmetric inertia matrix. We define also the matrix $C(q,\dot{q})$ and the vector $g(q)$ as follows:

$$C(q, \dot{q}) \equiv \frac{dM(q)}{dt} - \frac{1}{2}\frac{\partial}{\partial q}[\dot{q}^T M(q)]$$

$$g(q) \equiv \frac{\partial W(q)}{\partial q}$$

With these definitions, the model (5) is rewritten as follows:

$$M(q)\ddot{q} + C(q, \dot{q})\dot{q} + g(q) = A^T(q)\lambda + B(q)u \qquad (6)$$

This equation, together with the constraints (1) written in matrix form as:

$$A^T(q)\dot{q} = 0 \qquad (7)$$

provide a full description of the dynamics of the nonholonomic system.

We note that the following equality is a consequence of the definitions of section 1:

$$S^T(q)A(q) = 0 \qquad\qquad \forall q \in \Omega$$

Using this expression, we eliminate the Lagrange multipliers by premultiplying equation (6) by $S^T(q)$ to obtain:

$$S^T(q)[M(q)\ddot{q} + C(q, \dot{q})\dot{q} + g(q)] = S^T(q)B(q)u \qquad (8)$$

Moreover, the constraints (7) imply the existence of a vector time function $\eta(q, \dot{q})$ smooth in q and linear in $\dot{q}$ which satisfies the following equality along the trajectories of the system:

$$\dot{q} = S(q)\eta(q, \dot{q}) \qquad (9)$$

This is precisely the kinematic state-space model introduced in the previous section which appears now as a part of the system dynamics.

By differentiating (9), one obtains:

$$\ddot{q} = S(q)\dot{\eta} + R(q, \dot{q})\eta \qquad (10)$$

with:

$$R(q, \dot{q}) \equiv \frac{dS(q)}{dt} = \sum_{i=1}^{n} \frac{\partial}{\partial q_i}[S(q)]\dot{q}_i$$

Substituting (9) and (10) into (6) then leads to the following alternative state space description of the system:

$$\Sigma(q)\dot{\eta} = S^T(q)\{-[M(q)R(q,S(q)\eta)\eta + C(q,S(q)\eta)S(q)\eta + g(q)] + B(q)u\} \qquad (11.a)$$

$$\dot{q} = S(q)\eta \qquad (11.b)$$

where $\Sigma(q) = S^T(q)M(q)S(q)$ is a definite positive symmetric matrix. The state vector $\{\eta, q\}$ of this model, referred to as "the dynamical state-space model" of the system, has dimension $(2n - m)$. It shows clearly that $\eta$ is an internal state instead of being regarded as a fictitious input function $w(t)$ in the kinematical model.

As a first step towards the analysis of the controllability of this system, we have the following property.

**Lemma 3.** If $p \geq n - m$ (recall that p is the number of inputs) and if $S^T(q)B(q)$ has full rank for all q in $\Omega$, the dynamical state-space model (11) is partially feedback linearizable with a control law $u(\eta,q)$ chosen such that:

$$S^T(q)B(q)u = \Sigma(q)v + S^T(q)[M(q)R(q,S(q)\eta)\eta + C(q,S(q)\eta)S(q)\eta + g(q)] \qquad (12)$$

where v denotes an (n-m) - dimensional external input. Indeed, with such a control law, the closed loop is written:

$$\dot{\eta} = v \qquad (13.a)$$

$$\dot{q} = S(q)\eta \qquad (13.b)$$

Thus it appears that the static state feedback (12) allows to reduce the system (11) to the simple form (13) whose structure only depends on the nonholonomic constraints. Our concern is now to discuss the controllability properties of this model and the design of a second state feedback loop $v(\eta,q)$ to stabilize the system around the origin.

## 4.1. Controllability.

Due to the presence of a drift vector field in the model, the controllability of the system (13) cannot be analyzed without an explicit knowledge of the matrix S(q). However, we know that a necessary controllability condition is that the strong accessibility rank of the system be equal to the state dimension (2n - m). As a matter of fact, this condition holds for nonholonomic systems whatever the structure of S(q) as is shown in the following theorem.

**Theorem 2.** The strong accessibility rank of a nonholonomic system evolving in an n-dimensional configuration manifold and subjected to m constraints is (2n - m).

*Proof.*

Results directly from the fact that, if a system is strongly accessible from an input, then it is also strongly accessible from the derivative of this input.

## 4.2. State feedback control.

In this section, we are concerned with the design of smooth state feedback stabilizing controls for the dynamical state space model (13). When a smooth state feedback control law $v(\eta, q)$, such that $v(0, 0) = 0$, is applied to the system (13), the closed loop dynamics :

$$\dot{\eta} = v(\eta, q) \qquad (16.a)$$

$$\dot{q} = S(q)\eta \qquad (16.b)$$

have the origin $(\eta, q) = (0, 0)$ as an equilibrium point.

We have properties quite similar to those that have been emphasized for the kinematic state-space model, namely that the equilibrium $(\eta, q) = (0, 0)$ of the closed loop cannot be made asymptotically stable by pure state feedback, but can be marginally stabilized.

**Lemma 4.** The equilibrium point $(\eta, q) = (0, 0)$ of the closed loop (16) cannot be made asymptotically stable by a smooth state feedback $v(\eta, q)$.

*Proof.* Similar to that of Lemma 2.

**Theorem 3.** With the smooth state feedback control law:

$$v(\eta, q) = - S^T(q)S(q)\eta - D(\eta, q)q - \Lambda[S^T(q)q + \eta] - S^T(q)q \qquad (17)$$

where:

$$D(\eta, q) \equiv \frac{d}{dt} S^T(q)$$

the equilibrium point $(\eta, q) = (0, 0)$ of the closed loop system (16)-(17) is Lagrange stable. Precisely:

    a) the state $\eta(t), q(t)$ is bounded for all t

    b) the state $\eta(t), q(t)$ converges to the invariant set U:

$$U = \{ (\eta, q) \mid \eta = 0 \text{ and } S^T(q)q = 0 \}$$

*Proof.* We define

$$\tilde{\eta} \equiv - S^T(q)q - \eta$$

The closed loop (16)-(17) is then easily shown to be equivalent to:

$$\dot{\tilde{\eta}} = - D(\eta, q)q - S^T(q)S(q)\eta - v(\eta, q) = - \Lambda\tilde{\eta} + S^T(q)q \qquad (18.a)$$

$$\dot{q} = - S(q)S^T(q)q - S(q)\tilde{\eta} \qquad (18.b)$$

The theorem follows by considering the following Lyapunov function candidate:

$$V(\tilde{\eta}, q) = \frac{1}{2} [\tilde{\eta}^T \tilde{\eta} + q^T q]$$

whose time derivative along the solutions of (18) is given by:

$$\dot{V} = -\frac{1}{2} \tilde{\eta}^T (\Lambda + \Lambda^T)\tilde{\eta} - q^T S(q) S^T(q) q \leq 0$$

## 5. CONCLUSIONS.

Our main contribution in this paper has been to show that non holonomic systems: (i) are strongly accessible whatever the structure of the constraints; (ii) cannot be asymptotically stabilized by a smooth pure state feedback; (iii) can nevertheless be globally marginally stabilized by a smooth state feedback. Furthermore the design of these stabilizing controls has been explicited.

An application of the foregoing theory to mobile wheeled robots can be found in reference [11]. A brief sketch of this application is given in Appendix as a matter of illustration.

## 6. ACKNOWLEDGEMENTS.

We thank M. Fliess and C. Samson for their interesting suggestions and comments concerning the dynamics of non holonomic systems.

The results presented in this paper have been obtained within the framework of the Belgian Program on Concerted Research Actions and on Interuniversity Attraction Poles initiated by the Belgian State, Prime Minister's Office, Science Policy Programming. The scientific responsibility rests with its authors.

## 7. REFERENCES.

[1]    P. Appell , "Traité de mécanique rationnelle",Gauthier-Villars, Paris, 1903.

[2]    G. Hamel, "Die Lagrange-Euler'schen Gleichungen der Mechanik", Teubner, Leipzig, 1903.

[3]    J.P. Laumond, "Feasible trajectories for mobile robots with kinematic and environment constraints", Int. Conf. on Intelligent Autonomous Systems, Amsterdam, 1986, pp .346,354.

[4]     J. Barraquand, J.C. Latombe, "On non holonomic mobile robots and optimal manoeuvring", Revue d'intelligence Artificielle, Vol 3-2, 1989, Ed. Hermes, pp. 77,103.

[5]     C. Samson, K. Ait-Abderrahim, "Mobile robot control. Part 1: Feedback control of non holonomic wheeled cart in Cartesian space", INRIA Report, N°1288, 1990

[6]     A. Isidori, "Nonlinear control systems", Springer Verlag, 1989.

[7]     H. Nijmeijer and A.J. van der Schaft, "Nonlinear Dynamical Control Systems", Springer Verlag, 1990.

[8]     R.W. Brockett, "Asymptotic stability and feedback stabilization", in Differential Geometric Control Theory (eds. Brockett, Millmann, Sussmann), Birkhauser, Boston, pp.181-191, 1983.

[9]     Bloch and Mc Clamroch, "Control of mechanical systems with classical non holonomic constraints", Proc. 28th IEEE Conf.on Decision and Control, Tampa, 1989, pp.201-205.

[10]    H.J. Sussmann and V. Jurdjevic, "Controllability of nonlinear systems", J. Diff. Eqns., 12, pp.95-116, 1972.

[11]    B. d'Andrea, G.Bastin and G.Campion, "Modelling and Feedback Control of Nonholonomic Systems: The case of mobile robots", to be presented at the IEEE Conf. on Robotics and Automation, Sacramento, 1991.

[12]    A. M. Bloch and N. H. McClamroch, "Controllability and stabilizability properties of a nonholonomic control system", Proc. 29th Conference on Decision and Control, Hawaii, pp.1312-1314, 1990.

[13]    C. Samson,"Velocity and torque feedback control of a nonholonomic cart" Int. Workshop in Adaptive and Nonlinear Control: Issues in Robotics, Grenoble, 1990.

## APPENDIX: Simplified model of a mobile wheeled robot.

We consider a mobile robot moving on an horizontal plane, constituted by a rigid trolley equipped with non deformable wheels. During the motion, the plane of each wheel remains vertical and the wheel rotates around its (horizontal) axis. The orientation of 2 wheels with respect to the trolley is fixed, while the orientation of the third wheel is varying (see Figure 1). The contact between the wheels and the ground satisfied the *pure rolling* and *non slipping* conditions. The motion of the robot is achieved by 2 motors which provide torques acting on the rotation of the 2 wheels whose orientation is fixed.

In order to characterize the position of the trolley , we define an inertial reference frame in the plane of motion $\{0, I_1, I_2\}$, a reference point Q on the trolley and a basis $\{x_1, x_2\}$ attached to the trolley. The position of the trolley in the plane is therefore characterized by 3 variables:

- x, y : the coordinates of the reference point Q in the inertial frame,

-$\theta$ : the orientation of the basis $\{x_1, x_2\}$ with respect to the inertial frame.

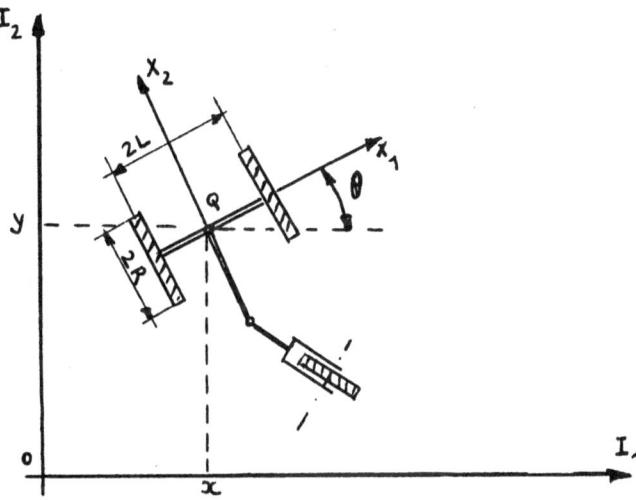

Figure 1: Mobile robot configuration

The configuration of the robot is described by 7 variables : ( x, y, $\theta$ ) for the position of the trolley, 3 angles characterizing the rotations of the 3 wheels, and 1 angle describing the orientation of the mobile wheel. A complete description can be found in [11]. In the present simplified illustrative analysis we restrict ourself in describing the motion of the robot in the plane and we define therefore the generalized coordinates vector q as:

$$q = ( x \; y \; \theta )^T$$

There is one constraint involving the time dervative of q only , namely the non slipping condition of the axis of the 2 front wheels. This constraint is written as:

$$\dot{x} \cos\theta + \dot{y} \sin\theta = 0$$

The matix A(q) is therefore defined as follows:

$$A(q) = \begin{pmatrix} \cos\theta \\ \sin\theta \\ 0 \end{pmatrix}$$

A particular choice for S(q) is the following:

$$S(q) = (s_1(q) \; s_2(q)) = \begin{pmatrix} -\sin\theta & 0 \\ \cos\theta & 0 \\ 0 & 1 \end{pmatrix}$$

The Lie brackett of the vector fields associated with the colums of S(q) is computed as:

$$[s_1(q) , s_2(q)] = \begin{pmatrix} \cos\theta \\ \sin\theta \\ 0 \end{pmatrix}$$

Since this new vector field does not belong to the distribution represented by S(q), we conclude that the system is nonholonomic.

## A1. Kinematic model.

According to Eq.(4) the kinematic model is written as follows:

$$\dot{x} = -w_1 \sin\theta$$
$$\dot{y} = w_1 \cos\theta$$
$$\dot{\theta} = w_2$$

The 2 inputs $w_1$ and $w_2$ have a physical interpretation: they are respectively the velocity of the robot in the $x_2$ direction and its angular velocity.

The state feedback control of Theorem 1 is given by:

$$w_1 = x \sin\theta - y \cos\theta$$
$$w_2 = -\theta$$

and the invariant set U is described by:

$$-x \sin\theta + y \cos\theta = 0$$
$$\theta = 0$$

or, equivalently by:

$$y = \theta = 0$$

## A2. Dynamical model

Neglecting the masses and inertias of the wheels, the kinetic energy reduces to:

$$T(q,\dot{q}) = \frac{1}{2} (\dot{x} \ \dot{y} \ \dot{\theta}) \begin{pmatrix} m & 0 & 0 \\ 0 & m & 0 \\ 0 & 0 & I_0 \end{pmatrix} \begin{pmatrix} \dot{x} \\ \dot{y} \\ \dot{\theta} \end{pmatrix}$$

where m is the mass of the robot, and $I_0$ is its inertia moment around the vertical axis at point Q.

We consider now as inputs $(u_1, u_2)$ the torques provided by the 2 motors. The corresponding generalized forces are given by :

$$B(q) \, u = \frac{1}{R} \begin{pmatrix} -\sin\theta & -\sin\theta \\ \cos\theta & \cos\theta \\ L & -L \end{pmatrix} \begin{pmatrix} u_1 \\ u_2 \end{pmatrix}$$

where R is the radius of the wheels and 2L the length of the axis of the front wheels.

Equation (6) takes the following form:

$$m\ddot{x} = \lambda \cos\theta - \frac{1}{R}(u_1 + u_2) \sin\theta$$

$$m\ddot{y} = \lambda \sin\theta + \frac{1}{R}(u_1 + u_2) \cos\theta$$

$$I_0\ddot{\theta} = \frac{L}{R}(u_1 - u_2)$$

According to Eq.(9) we define $\eta$ by:

$$\eta_1 = -\dot{x} \sin\theta + \dot{y} \cos\theta$$

$$\eta_2 = \dot{\theta}$$

After elimination of the Lagrange multiplier $\lambda$, we obtain the following dynamical model of the robot:

$$m\dot{\eta}_1 = \frac{1}{R}(u_1 + u_2)$$

$$I_0\dot{\eta}_2 = \frac{L}{R}(u_1 - u_2)$$

$$\dot{x} = -\eta_1 \sin\theta$$

$$\dot{y} = \eta_1 \cos\theta$$

$$\dot{\theta} = \eta_2$$

The following static state feedback allows to reduce these equations to the form of Eq.(13):

$$u_1 = \frac{R}{2}\left(m v_1 + \frac{I_0}{L} v_2\right)$$

$$u_2 = \frac{R}{2}\left(m v_1 - \frac{I_0}{L} v_2\right)$$

The stabilizing state feedback control (17) takes the following form:

$$v_1 = -(1 + k_1)(\eta_1 + x \sin\theta + y \cos\theta) + \eta_2 (x \cos\theta + y \sin\theta)$$

$$v_2 = -(1 + k_2)(\theta + \eta_2)$$

where the gains $k_1$ and $k_2$ are non negative design parameters.

This choice of v, combined with the first state feedback, ensures the convergence of the closed-loop to the invariant set characterized by $\eta = 0$, $y = 0$ and $\theta = 0$.

# VELOCITY AND TORQUE FEEDBACK CONTROL OF A NONHOLONOMIC CART

## Claude SAMSON

INRIA, Centre de Sophia-Antipolis
Route des Lucioles, 06565 VALBONNE, FRANCE.

### Abstract

A framework for designing and analysing velocity and torque feedback controls for a nonholonomic wheeled-cart is presented. A stability analysis of a set of nonlinear systems, the equations of which encompass all stable linear invariant systems, is first proposed. This analysis is then applied to the design and analysis of feedback controls for the wheeled-cart. The control inputs are either the cart's motorized wheels angular velocities (velocity control) or the torques applied to those wheels (torque control). Various control objectives are discussed and sufficient conditions for asymptotic convergence of the proposed controls are given. Among other results, the existence of smooth feedback controls that stabilize the cart at a desired position and orientation is established. This result does not contradict earlier non-existence results because the stabilizing controls depend not only on the robot's configuration variables but also on the exogeneous time variable.

**Key words:** mobile robots, nonholonomy, nonlinear systems, controllability, feedback stabilization.

## 1 Introduction

Feedback control of nonholonomic mobile robots is currently arising the interest of several research groups within the Robotics and Automatic Control communities [1]-[6]. While the topic may be seen as a logical extension of the work done in the case of holonomic robot manipulators, it turns out that this extension is not as straightforward as it may appear at first glance. In particular, a nonholonomic mobile robot may be strongly controllable in its configuration space, as established by various authors focussing on path planning issues [7]-[12], and yet not be stabilisable by using smooth pure-state feedback [4]. In some cases, global stability results can be proven by using either linear or nonlinear control laws, although convergence may, or may not, be exponential depending on the choice of the state vector which is to be regulated (in relation to the control objectives) [4]. Moreover, due to the structural diversity of nonholonomic mobile robots, it still is not clear whether it is possible to prove general results applicable to a wide set of systems (see for example [5] and [6] in this volume). Thus, even at a very basic control level, much remains to be done...

In this article, an attempt to derive a methodology for the design and analysis of stabilizing feedback control laws for a wheeled mobile robot is presented. The article is organized as follows. In Section 2, the stability properties of a set of nonlinear systems, the equations of which encompass stable linear invariant systems, are studied via classical Lyapunov techniques, and results are gathered in a lemma which may be seen as an adaptation of a Lasalle's theorem [15]. Application to the considered mobile robot is done in Section 3 where it is shown how the lemma can be utilized for the design and analysis of stabilizing feedback controls. Firstly, the control inputs are assumed to be the motorized wheels' angular velocities (velocity control) and several control objectives are discussed: stabilisation of a cart's point at a desired position which may either be fixed or moving, stabilisation of the cart's position and orientation, and tracking of preplannified reference trajectories. This part may in fact be seen as a generalisation of the study performed in [4]. In addition, it is shown that there exist smooth feedback controls able to make the cart asymptotically converge to any desired configuration. In view of another result derived in [4], it appears that such controls laws have to be functions of not only the cart's configuration variables but also of the independent time variable. Then, in order to take into account second order effects due to the cart's dynamics, the torques applied to the motorized wheels are chosen as control inputs (torque control). By using again the analysis of Section 2, stabilizing feedback controls are derived for the first two objectives considered previously in the case of velocity control, and simple relations between velocity and torque control are explicited. Robustness with respect to errors in the modelling of the cart's inertia matrix is pointed out, but is not fully explored here. The extension to the other control objectives is not difficult and left to the interested reader.

## 2 Stability analysis of a class of nonlinear systems

Let $S$ denote the set of matrix valued functions $f(X, t)$ defined on $R^k \times R^+$ ($k \in N$), of class $C^\infty$, uniformly bounded with respect to the independent time variable $t$, and with successive partial derivatives also uniformly bounded with respect to $t$.

By choosing functions in this set, many technicalities which are not essential to the forthcoming analysis will be avoided.

We will consider in this section a class of systems defined .by an equation in the following form:

$$\dot{X} = -P^{-1}(Q(X,t) + \tilde{C}(X,t))X \qquad (1)$$

where:

- $X \in R^n$

- $\dot{X}(t) = \frac{d}{dt}X(t)$

- $P$ is a positive-definite symmetric ($p.d.s.$) matrix (meaning that $P = P^T$ and $X^T P X > 0$ for all $X \in R^n$ and $X \neq 0$)

- $Q(X,t) \in S$ is a positive symmetric ($p.s.$) matrix valued function (meaning that $Q(X,t) = Q(X,t)^T$ and $X^T Q(X,t) X \geq 0$ for all $(X,t) \in R^n \times R^+$)

- $\tilde{C}(X,t) \in S$ is a skew-symmetric ($s.s.$) matrix valued function (meaning that $\tilde{C}(X,t) + \tilde{C}(X,t)^T = 0$).

We may already notice that the set of systems defined by 1 encompasses all *stable* linear systems in the form:

$$\dot{X} = AX \tag{2}$$

where $A$ is a *stability* matrix (i.e. a matrix the eigenvalues $\lambda_i(A)$ of which have a strictly negative real part).

Indeed, it is known (Lyapunov) [13] that $A$ is a stability matrix if and only if there exists two $p.d.s.$ matrices $P$ and $Q$ such that:

$$A^T P + PA = -2Q \tag{3}$$

It is then easy to verify that the matrix $A$ may also be written:

$$A = -P^{-1}(Q + \tilde{C}) \tag{4}$$

with:

$$\tilde{C} = \frac{1}{2}(A^T P - PA) \tag{5}$$

Conversely, if a matrix $A$ can be written as 4, with $P$ and $Q$ being $p.d.s.$ and $\tilde{C}$ being $s.s.$, then $A$ is a stability matrix because it satisfies the equation 3.

In other words, we have the following result:

**Lemma 2.1** *A matrix $A$ is a stability matrix if and only if $(-A)$ is the product of a* p.d.s. *matrix (denoted as $P^{-1}$) with a positive-definite matrix (denoted as $Q + \tilde{C}$).*

Obviously, this product decomposition is not unique since relation 3 can be satisfied for any $p.d.s.$ matrix $Q$ when $A$ is a stability matrix.

Moreover, the positive matrix $Q$ does not have to be nonsingular, provided $X(t)^T Q X(t)$ is not identically zero along any nonzero solution of $\dot{X}(t) = AX(t)$ (Kalman) [13]. This is an *observability* type condition which yields the following result:

**Lemma 2.2** *If:*

- $P$ is a $(n \times n)$ p.d.s. *matrix*

- $Q$ is a $(n \times n)$ *(semi)* p.s. *matrix*

- $\tilde{C}$ is a $(n \times n)$ s.s. *matrix*

then the matrix:

$$A = -P^{-1}(Q + \check{C})$$

is a stability matrix if and only if:

$$\text{rank} \begin{bmatrix} Q \\ QP^{-1}\check{C} \\ \vdots \\ Q(P^{-1}\check{C})^{n-1} \end{bmatrix} = n$$

## Proof of 2.2:

i) $A$ is a stability matrix if the rank condition is satisfied.

Consider any solution $X(t)$ to the equation $\dot{X} = AX$, and the (Lyapunov) function $V = 1/2X^T PX$. From now on, we will use the notation $\|X\|_P^2 = X^T PX$. We have, from the definition of $A$:

$$\dot{V} = -X^T QX \quad (\leq 0) \tag{6}$$

and therefore:

$$\lim_{t \to \infty} QX(t) = 0 \tag{7}$$

Moreover, since $V(t)$ is decreasing, $\|X(t)\|$ is bounded, and so are the derivatives $X^{(k)}(t)$ for $k \in [1, n]$. Since $QX(t)$ tends to zero and its second derivative is bounded, its first derivative tends to zero:

$$\lim_{t \to \infty} Q\dot{X}(t) = \lim_{t \to \infty} -QP^{-1}(Q + \check{C})X(t) = 0 \tag{8}$$

Thus, again from 7:

$$\lim_{t \to \infty} QP^{-1}\check{C}X(t) = 0 \tag{9}$$

Repeating the same procedure a few more times, we finally obtain that $QX(t)$, $QP^{-1}\check{C}X(t)$, $Q(P^{-1}\check{C})^2X(t)$, ..., $Q(P^{-1}\check{C})^{n-1}X(t)$ tend to zero when $t$ tends to infinity. The rank condition of 2.2 then implies that $X(t)$ tends to zero, and thus that $A$ is a stability matrix.

ii) the rank condition is satisfied if $A$ is a stability matrix.

Assume that the rank condition is not satisfied. Then there exists a vector $X_0 \neq 0$ such that (by the Cayley-Hamilton theorem):

$$Q(P^{-1}\check{C})^k X_0 = 0 \quad for \ k \in N \tag{10}$$

Let us then consider the solution $X(t)$ to the following equation:

$$\dot{X} = -P^{-1}\check{C}X \quad ; \quad X(0) = X_0 \tag{11}$$

Since $\frac{d}{dt}X(t)^T PX(t) = 0$, we know that $X(t)$ does not tend to zero.

We also have:

$$X(t) = exp(-P^{-1}\tilde{C}t)X_0 \tag{12}$$

and:

$$
\begin{aligned}
QX(t) &= Qexp(-P^{-1}\tilde{C}t)X_0 \\
&= QX_0 - tQP^{-1}\tilde{C}X_0 + \tfrac{t^2}{2}Q(P^{-1}\tilde{C})^2X_0 + ..... \\
&= 0
\end{aligned} \tag{13}
$$

Therefore $X(t)$ is also a solution to the equation:

$$
\begin{aligned}
\dot{X} &= -P^{-1}\tilde{C}X - P^{-1}QX \\
&= AX
\end{aligned} \tag{14}
$$

and $A$ cannot be a stability matrix, since $X(t)$ would otherwise tend to zero. (end of proof)

Lemma 2.2 provides us with a matrix rank test which is usually simpler to perform than the calculation of the eigenvalues of the matrix $A = -P^{-1}(Q + \tilde{C})$. It is thus of interest when the system under study is directly given in the form $\dot{X} = -P^{-1}(Q + \tilde{C})X$ and the matrices $P$ and $(Q + \tilde{C})$ are known to be positive.

An extension of this lemma, that may be used to study the stability properties of nonlinear systems in the form 1, is given in the next lemma which may be seen as an adaptation of Lasalle's theorem [15].

**Lemma 2.3** *Consider the subset:*
$\mathcal{D} = \{Y \in S \mid \forall t \in R^+ : \|Y(t)\|_P = constant\}$
*of vector valued functions defined from $R^+$ to $R^n$, and a sequence of matrices $\{Q_i\}_{i \in N}$ such that:*

- *for $i = 0$:*

$$
\left.\begin{aligned}
&Y \in \mathcal{D} \\
&\lim_{t \to \infty} Q(Y(t), t)Y(t) = 0
\end{aligned}\right\} \Rightarrow \lim_{t \to \infty} Q_0 Y(t) = 0
$$

- *for $i \geq 1$:*

$$
\left.\begin{aligned}
&Y \in \mathcal{D} \\
&\lim_{t \to \infty} Q_{i-1} Y(t) = 0 \\
&\lim_{t \to \infty} Q_{i-1} P^{-1}\tilde{C}(Y(t), t)Y(t) = 0
\end{aligned}\right\} \Rightarrow \lim_{t \to \infty} Q_i Y(t) = 0
$$

*then the solutions $X(t)$ to the system 1 are such that:*
$\lim_{t \to \infty} Q_i X(t) = 0 \quad ; \forall i \in N$

**Proof of Lemma 2.3**
Consider the positive function:

$$V(X) = 1/2\|X\|_P^2 \tag{15}$$

and a particular solution $X(t)$ of equation 1. We have:

$$\dot{V}(X(t)) = -X(t)^T Q(X(t), t) X(t) \quad (\leq 0) \tag{16}$$

Thus:

$$\|X(t)\|_P \leq \|X(0)\|_P \tag{17}$$

The solutions $X(t)$ are uniformly bounded with respect to the initial conditions and thus exist on $R^+$. Moreover, from 15 and 16:

$$\exists l \geq 0 \; ; \; \lim_{t \to \infty} \|X(t)\|_P = l \quad (with \;\; \frac{l}{\|X(t)\|_P} \leq 1) \tag{18}$$

and:

$$\lim_{t \to \infty} Q(X(t), t) X(t) = 0 \tag{19}$$

Moreover, the function $t \mapsto X(t)$ belongs to the set $S$ because the function:
$F(X, t) = -P^{-1}(Q(X, t) + \check{C}(X, t)) X$
belongs to this set.
Consider now the function:
$t \longmapsto Y(t) = \frac{l}{\|X(t)\|_P} X(t)$
It belongs to $S$ because $X(t) \in S$. Also:

$$\|Y(t)\|_P = l \quad ; \quad \forall t \in R^+ \tag{20}$$

It is thus an element of the set $\mathcal{D}$. Moreover, because of 18:

$$\lim_{t \to \infty} \|X(t) - Y(t)\| = 0 \tag{21}$$

By using an argument of (uniform) continuity, 19 and 21 yield:

$$\lim_{t \to \infty} Q(Y(t), t) Y(t) = 0 \tag{22}$$

and therefore:

$$\lim_{t \to \infty} Q_0 Y(t) = 0 \tag{23}$$

which in turn implies:

$$\lim_{t \to \infty} Q_0 X(t) = 0 \tag{24}$$

Let us now finish the proof by induction and show that $\lim_{t \to \infty} Q_{i+1} X(t) = 0$ if $\lim_{t \to \infty} Q_i X(t) = 0$.

Since $\|\ddot{X}(t)\|$ is bounded ($\|X(t)\|$ being an element of $S$), the convergence of $Q_i X(t)$ to zero implies the convergence to zero of the derivative $Q_i \dot{X}(t)$. Premultiplication by $Q_i$ of both members of the equality 1 gives:

$$Q_i\dot{X}(t) = -Q_iP^{-1}Q(X(t),t)X(t) - Q_iP^{-1}\tilde{C}(X(t),t)X(t) \tag{25}$$

and, because of the convergence of $Q_i\dot{X}(t)$ and $Q(X(t),t)X(t)$ to zero:

$$\lim_{t\to\infty} Q_iP^{-1}\tilde{C}(X(t),t)X(t) = 0 \tag{26}$$

By (uniform) continuity:

$$\lim_{t\to\infty} Q_iP^{-1}\tilde{C}(Y(t),t)Y(t) = 0 \tag{27}$$

and therefore:

$$\lim_{t\to\infty} Q_{i+1}Y(t) = 0 \tag{28}$$

which in turn implies:

$$\lim_{t\to\infty} Q_{i+1}X(t) = 0 \tag{29}$$

(end of proof).

Note that, although the conditions of Lemma 2.3 are always satisfied by the null matrix, the practical purpose of the determination of the sequence $\{Q_i\}_{i\in N}$ is clearly to reach a matrix $Q_i$ of maximum rank.

**Corollary 1 (of Lemma 2.3)** *If one of the matrices $Q_i$ is nonsingular then:*
$\lim_{t\to\infty} X(t) = 0$
*If, in addition, one of the following conditions is satisfied:*

- *i) $Q(0,t) > \alpha I_n$ for some strictly positive number $\alpha$ (meaning that $X^TQ(0,t)X > \alpha\|X(t)\|^2$, $\forall X \in R^n$, $X \neq 0$)*

- *ii) $Q(0,t)$ and $\tilde{C}(0,t)$ are constant matrices and the matrix $-P^{-1}(Q(0,t)+\tilde{C}(0,t))$ is a stability matrix*

*then the convergence is exponential in the neighbourhood of zero.*

**Proof of the Corollary 1**
The first part of the corollary is obvious.
Concerning the second part, the convergence of $X(t)$ to zero implies ( because of the uniform continuity of $Q(X,t)$ on $C \times R^+$, where $C$ is any compact set in $R^n$) the existence of an instant $t_0$ such that:
$\|Q(X(t),t) - Q(0,t)\| < \alpha/2$    for $t > t_0$
Moreover, since $\|X(t)\|_P$ is decreasing, we have $t_0 = 0$ for all solutions starting close enough to zero.
In view of 16, and the assumption according to which $Q(0,t)$ is bounded from below by $\alpha I_n$, we thus have:

$$\frac{d}{dt}\frac{1}{2}\|X(t)\|_P^2 < -X(t)^T Q(0,t)X(t) + \frac{\alpha}{2}\|X(t)\|^2 \qquad for \ \ t > t_0$$
$$< -\frac{\alpha}{2}\|X(t)\|^2$$

This yields:

$$\|X(t)\|_P \le \|X(t_0)\|_P \ exp(-\frac{\alpha}{2}\lambda_{min}(P)\ (t-t_0)) \qquad for \ \ t > t_0$$

which establishes the exponential convergence of $X(t)$ to zero in the case where $Q(0,t) > \alpha I_n$.

Let us now consider the case where $Q(0,t)$ and $\tilde{C}(0,t)$ are constant matrices and the matrix $-P^{-1}(Q(0,t) + \tilde{C}(0,t))$ is a stability matrix.

We introduce the following notations:

$$A = -P^{-1}(Q(0,t) + \tilde{C}(0,t))$$
$$f(X,t) = -P^{-1}(Q(X,t) + \tilde{C}(X,t))X$$
$$\eta(X,t) = f(X,t) - f(0,t) - \frac{\partial f}{\partial X}(0,t)X$$

Application of Taylor's second order expansion formula to the $n$ components $f_i(X,t)$ of the function $f(X,t)$, and uniform boundedness of the second partial derivatives $\frac{\partial^2 f_i}{\partial X^2}$ with respect to the variable $t$, yields the existence of a real number $M_\alpha$ such that:

$$\|X\| < \alpha \Rightarrow \|\eta(X,t)\| < M_\alpha \|X\|^2 \quad \forall t \in R^+$$

Since $f(0,t) = 0$ and $\frac{\partial f}{\partial X}(0,t)X = AX$, we thus have:

$$\|X\| < \alpha \Rightarrow \|f(X,t) - AX\| < M_\alpha \|X\|^2 \quad \forall t \in R^+ \tag{30}$$

Now, since $A$ is a stability matrix, there are (according to Lemma 2.1) a *p.d.s.* matrix $P_1$ and a *s.s* matrix $\tilde{C}_1$ such that:

$$A = -P_1^{-1}(\beta I_n + \tilde{C}_1) \quad with \ \beta > 0$$

Consider the positive function:

$$V(X) = 1/2 \ \|X\|_{P_1}^2$$

We have:

$$
\begin{aligned}
\dot{V}(X) &= X^T P_1 f(X,t) \\
&= -\beta\|X\|^2 + X^T P_1(f(X,t) - AX) \\
&\le -\beta\|X\|^2 + \lambda_{max}(P_1) \ \|f(X,t) - AX\| \ \|X\|
\end{aligned}
\tag{31}
$$

Since $\|X(t)\|$ converges to zero, there is also a time $t_0$ such that:

$$\|X(t)\| < inf(\alpha; \beta/2M_\alpha\lambda_{max}(P_1)) \quad for \ t > t_0 \tag{32}$$

By using 30 and 32 in 30, we obtain:

$$\dot{V}(X(t)) \le -\frac{\beta}{2}\|X(t)\|^2 \ (\le 0) \quad for \ t > t_0$$

which clearly yields the exponential convergence of $\|X(t)\|$ to zero.
(end of proof).

# 3 Application to the feedback control of a cart

The cart that is considered is schematized in Fig.1 and Fig.2. It is equipped with two motorized wheels that share the same rotation axis. To simplify, it is assumed that it moves on a horizontal ground and that gravity forces do not influence its motion. The model equations of the cart's motion will also be derived under the usual rolling-without-slippage assumption.

## 3.1 Kinematic equations

These equations are obtained by expressing the fact that the point of each motorized wheel in contact with the ground has zero velocity.

The following notations will be used:

- $M_0$: the cart's point located at mid-distance of the motorized wheels

- $M_d$: the cart's point located on the $(M_0, \vec{i_1})$ axis at a distance $|d|$ of the point $M_0$ $(\vec{M_0M_d} = d\vec{i_1})$

- $\dot{q_1}, \dot{q_2}$: the motorized wheels' angular velocities

- $\mathcal{F}_0 = (O; \vec{i_0}, \vec{j_0})$: a fixed reference frame such that the point $M_0$ belongs to the plane $(O; \vec{i_0}, \vec{j_0})$

- $\mathcal{F}_1 = (M_0; \vec{i_1}, \vec{j_1})$: a frame rigidly linked to the cart

- $X_l(P) = [x_l(P), y_l(P)]^T$: the matrix of coordinates of a point $P$ in the frame $\mathcal{F}_l$ $(l = 0$ or $l = 1)$

- $\vec{V}_{P/\mathcal{F}_l}$: the velocity vector of the point $P$ in the frame $\mathcal{F}_l$

and also:

$$\dot{q} = \begin{bmatrix} \dot{q_1} \\ \dot{q_2} \end{bmatrix} \tag{33}$$

$$D = \begin{bmatrix} r/2 & r/2 \\ r/2R & -r/2R \end{bmatrix} \tag{34}$$

with $r$: the radius of the motorized wheels, and $2R$: the distance between the two wheels

$$B_1(d) = \begin{bmatrix} 1 & 0 \\ 0 & d \end{bmatrix} \tag{35}$$

$$R_1(\theta) = \begin{bmatrix} c\theta & s\theta \\ -s\theta & c\theta \end{bmatrix} \quad (c\theta = \cos\theta, \ s\theta = \sin\theta) \tag{36}$$

$$\tilde{C}_1 = \begin{bmatrix} 0 & 1 \\ -1 & 0 \end{bmatrix} \tag{37}$$

$$U = \begin{bmatrix} v \\ \dot{\theta} \end{bmatrix} \tag{38}$$

with $v$: the cart's advancement velocity $(\vec{V}_{M_0/\mathcal{F}_0} = v\vec{i_1})$.

We then obtain the following relations (see [4], for example):

$$U = D\dot{q} \tag{39}$$

and:

$$\dot{X}_0(M_d) = R_1^T(\theta)B_1(d)U \tag{40}$$

If, instead of the coordinates $(x_0(M_d), y_0(M_d))$ of vector $\vec{OM_d}$ in the frame $\mathcal{F}_0$, we choose to work with its coordinates $(x_1(M_d), y_1(M_d))$ in the frame $\mathcal{F}_1$, we have:

$$X_1(M_d) = R_1(\theta)X_0(M_d) \tag{41}$$

$$\dot{X}_1(M_d) = R_1(\theta)\dot{X}_0(M_d) + \dot{R}_1(\theta)X_0(M_d) \tag{42}$$

$$\dot{R}_1(\theta) = \dot{\theta}\tilde{C}_1 R_1(\theta) \tag{43}$$

and, from 40, 42 and 43:

$$\dot{X}_1(M_d) = \dot{\theta}\tilde{C}_1 X_1(M_d) + B_1(d)U \tag{44}$$

40 and 44 are basic kinematic equations which relate the velocity of the point $M_d$ in the frame $\mathcal{F}_0$ to the wheels' angular velocities (contained in the vector $U$).

## 3.2 Velocity control

In this section, $U$ is taken as the control vector. According to 39, and since the matrix $D$ is known and nonsingular, this is equivalent to choosing the wheels' angular velocities as the control inputs. The aim is to determine $U$ in order to have the cart achieve some objective. Several possible objectives will be considered and studied in view of the analysis performed in Section 2.

### 3.2.1 Objective 1: bring the cart's point $M_d$ (with $d \neq 0$) to the point $O$ by using state feedback control

This objective is equivalent to having the vector:

$$X = X_1(M_d) \tag{45}$$

converge to zero.
According to 44, we have:

$$\dot{X} = -\tilde{C}_2(U)X + B_1(d)U \tag{46}$$

with:

$$\tilde{C}_2(U) = -([01]U)\tilde{C}_1 \tag{47}$$

We may already notice that the matrix $\tilde{C}_2$ is $s.s..$.
The analysis of Section 2 suggests trying a feedback control in the form:

$$U = -B_1(d)^{-1}[Q(X,t) + \tilde{C}_3(X,t)]X \tag{48}$$

where $Q(X,t)$ is p.s. and $\tilde{C}_3(X,t)$ is s.s..

Indeed, by using 48 in 46, the equation of the controlled system is then in the form 1 with:

$$\begin{aligned} P &= I_2 \\ \tilde{C}(X,t) &= \tilde{C}_2(U(X,t)) + \tilde{C}_3(X,t) \end{aligned} \tag{49}$$

Let us explore a few possibilities for the choice of $Q(X,t)$ and $\tilde{C}_3(X,t)$.

**possibility 1:**

$Q(X,t)$ is taken equal to a p.d.s. constant matrix $Q$. Then, whatever the choice of $\tilde{C}_3(X,t)$, the application of Lemma 2.3 with $Q_0 = Q$ indicates that $X(t)$ converges to zero. Moreover, in view of Corollary 1, the convergence is exponential in the neighbourhood of zero. In fact, since $\frac{d}{dt}1/2||X||^2 = -||X||_Q^2$, it appears that the rate of convergence to zero is exponential from the beginning.

**possibility 2:**

$$Q(X,t) = \frac{1}{(||X||^2 + \xi)^{1/2}}Q \quad with \ Q: \ p.d.s. \ and \ \xi > 0 \tag{50}$$

The norm of the control vector is then unconditionnally bounded, and Lemma 2.3 again applies with $Q_0 = Q$. Thus $X(t)$ converges to zero. Moreover, since $Q(0,t) = \xi^{-1/2}Q$, the convergence is exponential in the neighbourhood of zero by application of Corollary 1.

**possibility 3:**

$$\begin{aligned} Q(X,t) &= R_1(\theta(t))Q_0 R_1^T(\theta(t)) \quad with \ Q_0: \ p.d.s. \ matrix \\ \tilde{C}_3(X,t) &= R_1(\theta(t))\tilde{C}_0 R_1^T(\theta(t)) \quad with \ \tilde{C}_0: \ s.s. \ matrix \end{aligned} \tag{51}$$

In these functions $\theta(t)$ denotes the value taken by the orientation angle $\theta$ at time $t$ when the control is applied to the cart. Since the boundedness of $||X(t)||$ yields the boundedness of $|\dot{\theta}(t)|$, the matrix valued function $R_1(\theta(t))$ belongs to the set $S$, and so do the functions $Q(X,t)$ and $\tilde{C}_3(X,t)$ defined above.

Lemma 2.3 may again be applied to show the convergence of $X(t)$ to zero. Convergence is in fact exponential because we have $\frac{d}{dt}1/2||X||^2 = -||X||_{Q_0}^2$.

Notice that the control thus obtained may also be written:

$$U = -B_1(d)^{-1}R_1(\theta)(Q_0 + \tilde{C}_0)X_0(M_d) \tag{52}$$

and that, in view of 40, this control yields the following linear closed-loop equation:

$$\dot{X}_0(M_d) = -(Q_0 + \tilde{C}_0)X_0(M_d) \tag{53}$$

This control thus linearizes the system in the coordinates $X_0(M_d)$, and decoupling is additionnally obtained by choosing $\tilde{C}_0 = 0$ and $Q_0$ diagonal. As explained in [4], linearization can also be performed in the coordinates $X_1(M_d)$, but in this case singularities arise at points where $x_1(M_d) = d$.

**Remark:**

From 40, it is easy to see that a linearizing stable control expression more general than 52 is:

$$U = -B_1(d)^{-1}R_1(\theta)AX_0(M_d) \qquad A: \text{ stability matrix} \tag{54}$$

Simulation results are given in [4].

### 3.2.2 Objective 2: bring the cart's point $M_0$ to the point $O$ by using state feedback control

The objective is to have:

$$X = X_1(M_0) \tag{55}$$

converge to zero.

An important difference with the previous case is that the matrix $B_1(0)$ is now singular. As a consequence, the p.l. system associated with the nonlinear system 46 (with $d = 0$):

$$\dot{X} = B_1(0)U \qquad with \ B_1(0) = \begin{bmatrix} 1 & 0 \\ 0 & 0 \end{bmatrix} \tag{56}$$

is not controllable, or even stabilisable. Therefore, the study of the p.l. system is, in this case, of no help for finding a locally stabilizing feedback control. Despite of this, globally stabilizing feedback controls exist as shown below.

We will use the fact that $B_1(0)$ satisfies the following equality:

$$(I_2 - K(X,t))B_1(0) = 0 \tag{57}$$

for any matrix valued function $K(X,t)$ ($\in \mathcal{S}$) in the form:

$$K(X,t) = \begin{bmatrix} 1 & k_1(X,t) \\ 0 & k_2(X,t) \end{bmatrix} \tag{58}$$

The analysis of Section 2 then suggests trying a control in the form:

$$U = -[g(X,t)K(X,t)^T + \tilde{C}_3(X,t)B_1(0)]X$$
$$g(X,t) \in \mathcal{S}: \text{ scalar positive function} \tag{59}$$
$$\tilde{C}_3(X,t) \in \mathcal{S}: \text{ s.s. matrix valued function}$$

so as to obtain, by using 59 in 46, a controlled system in the form 1 with:

$$\begin{array}{rcl} P & = & I_2 \\ Q(X,t) & = & g(X,t)B_1(0) \\ \tilde{C}(X,t) & = & \tilde{C}_2(U(X,t)) \end{array} \tag{60}$$

To simplify the stability analysis, let us choose the gain $g(X,t)$ equal to a constant positive number $g$. Then:

$$Q(X,t) = \begin{bmatrix} g & 0 \\ 0 & 0 \end{bmatrix} \tag{61}$$

and a matrix $Q_0$ which satisfies the condition of Lemma 2.3 clearly is:

$$Q_0 = B_1(0) \tag{62}$$

Thus:

$$Q_0 \tilde{C}(X,t) = \begin{bmatrix} 0 & \dot{\theta}(X,t) \\ 0 & 0 \end{bmatrix} \quad with \quad \dot{\theta}(X,t) = -g[k_1(X,t)|k_2(X,t)]X \tag{63}$$

Consider now an element $Y(t)$ of the set $\mathcal{D}$ such that $||Y(t)|| = l$. We have:

$$\left. \begin{array}{l} \lim_{t\to\infty} Q_0 Y(t) = 0 \\ \lim_{t\to\infty} Q_0 \tilde{C}(Y(t),t)Y(t) = 0 \end{array} \right\} \Rightarrow \left\{ \begin{array}{l} \lim_{t\to\infty} Y(t) = Y_{lim} = \begin{bmatrix} 0 \\ \pm l \end{bmatrix} \\ \lim_{t\to\infty} k_2(Y_{lim}, t)l = 0 \end{array} \right. \tag{64}$$

We thus obtain the following result: if $k_2([0, x]^T, t)$ does not converge to zero ($\forall x \in R$), then: i) $l = 0$, ii) a matrix $Q_1$ which satisfies the condition of Lemma 2.3 is the identity matrix $I_2$, and iii) the solutions $X(t)$ of the nonlinear system converge to zero.

One may for example choose $k_2(X,t)$ equal to any constant nonzero number. The choice of $k_1(X,t)$ is free.

However, since $Q + \tilde{C}(0,t) = Q$ and $Q$ is singular, Corollary 1 cannot be used to show that the convergence is exponential. In fact, simulation results given in [4] indicate that the convergence is *not* exponential and that the angle $\theta(t)$ *does not converge* to a limit value in general. This clearly reduces the practical usefulness of this type of control.

### 3.2.3 Objective 3: have the cart's point $M_d$ track a moving reference point $M_{d,r}(t)$

This is a slight generalisation of Objective 1. Let $X_{l,r}(t)$ denote the coordinates, expressed in the basis of the frame $\mathcal{F}_l$, of the vector $\overrightarrow{OM_{d,r}}$, and define:

$$W(0,t) = R_1(0)\dot{X}_{0,r}(t) \tag{65}$$

$$X = X_1(M_d) - X_{1,r}(t) \qquad : coordinates\ of\ \overrightarrow{M_{d,r}M_d}\ in\ \mathcal{F}_1 \tag{66}$$

It is simple to show that we now have (see [4] also):

$$\dot{X} = -\tilde{C}_2(U)X + B_1(d)U - W(0,t) \tag{67}$$

with $\tilde{C}_2(U)$ defined in 47.

Then, by introducing the auxiliary control vector:

$$U' = U + B_1(d)^{-1}W(0,t) \tag{68}$$

equation 67 may also be written:

$$\dot{X} = -\tilde{C}_2(U)X + B_1(d)U' \tag{69}$$

Since this equation is similar to equation 46, we are brought back to the study of Objective 1 for which globally exponentially stabilizing feedback control laws have been derived.

### 3.2.4  Objective 4: have the cart track a reference cart in both position and orientation

So far, the objectives that we have considered have not involved the orientation $\theta$ of the cart. This angle was not actively controlled.

We now assume that an *ideal reference trajectory* $(X_{0,r}(t), \theta_r(t)) \in S$ is known, and we would like the cart to track this trajectory. In other words, we would like our cart to track a similar *reference cart* the position and orientation of which is given by $(X_{0,r}(t), \theta_r(t))$.

Because of the cart's nonholonomy, we already know that the variations of $X_{0,r}(t)$ are not independent of the variations of $\theta_r(t)$. In fact, according to 40 and assuming that $X_{0,r}(t)$ is the vector of coordinates (in the fixed frame $\mathcal{F}_0$) of the reference cart's point $M_{0,r}$ located at mid-distance of the motorized wheels, we have:

$$\dot{X}_{0,r}(t) = R_1{}^T(\theta_r(t))B_1(0)U_r(t) \tag{70}$$

with:

$$U_r(t) = \begin{bmatrix} v_r(t) \\ \dot{\theta}_r(t) \end{bmatrix} \qquad v_r(t) \; : \; \textit{reference cart's advancement velocity} \tag{71}$$

Relation 70 indicates that the reference cart's trajectory is entirely determined once the initial condition $(X_{0,r}(0), \theta_r(0))$ is given and the velocities $v_r(t)$ and $\dot{\theta}_r(t)$ are chosen. We define:

$$\begin{aligned} \tilde{v} &= v - v_r \\ \tilde{\theta} &= \theta - \theta_r \\ \tilde{U} &= U - U_r \end{aligned} \tag{72}$$

and:

$$X = \begin{bmatrix} X_1(M_0) - X_{1,r} \\ \tilde{\theta}/k \end{bmatrix} \qquad ; \; k \neq 0 \tag{73}$$

The problem consists of finding (when it exists) a feedback control $\tilde{U}$ that ensures the asymptotic convergence of $X$ to zero.

We leave to the reader the task of verifying (see also [4]) that $X$ satisfies the equation:

$$\dot{X} = \tilde{C}_2(X, \tilde{U}, t)X \; + \; B\,(\tilde{U} + K_1(X, t)X) \tag{74}$$

with:

$$\tilde{C}_2(X,\tilde{U},t) = \begin{bmatrix} \tilde{C}_{2,1}(\tilde{U},t) & kC_{2,2}(X,t) \\ -kC_{2,2}{}^T(X,t) & 0 \end{bmatrix} \tag{75}$$

$$\tilde{C}_{2,1}(\tilde{U},t) = (\dot{\theta}_r(t) + [01]\tilde{U})\tilde{C}_1 = \begin{bmatrix} 0 & \dot{\theta} \\ -\dot{\theta} & 0 \end{bmatrix} \tag{76}$$

$$C_{2,2}(X,t) = (1/\tilde{\theta})(I_2 - R_1(\tilde{\theta}))B_1(0)U_r(t) = \begin{bmatrix} \frac{1-\cos\tilde{\theta}}{\tilde{\theta}}v_r(t) \\ \frac{\sin\tilde{\theta}}{\tilde{\theta}}v_r(t) \end{bmatrix} \tag{77}$$

$$B = \begin{bmatrix} B_1(0) \\ 0 & 1/k \end{bmatrix} = \begin{bmatrix} 1 & 0 \\ 0 & 0 \\ 0 & 1/k \end{bmatrix} \tag{78}$$

$$K_1(X,t) = \begin{bmatrix} 0 & 0 \\ k^2C_{2,2}{}^T(X,t) & 0 \end{bmatrix} = \begin{bmatrix} 0 & 0 & 0 \\ k^2\frac{1-\cos\tilde{\theta}}{\tilde{\theta}}v_r(t) & k^2\frac{\sin\tilde{\theta}}{\tilde{\theta}}v_r(t) & 0 \end{bmatrix} \tag{79}$$

We notice that the matrix $\tilde{C}_2(X,\tilde{U},t)$ is *s.s.*.
The analysis of Section 2 then suggests a control in the form:

$$\tilde{U}(X,t) = -[K_1(X,t) + (Q_1(X,t) + \tilde{C}_3(X,t))B^T]X \tag{80}$$

with:

- $Q_1(X,t) \in \mathcal{S}$: *p.d.s.* matrix valued function
- $\tilde{C}_3(X,t) \in \mathcal{S}$: *s.s.* matrix valued function

so as to obtain a controlled system in the form 1 with:

$$\begin{array}{rcl} P & = & I_3 \\ Q(X,t) & = & BQ_1(X,t)B^T \\ \tilde{C}(X,t) & = & -\tilde{C}_2(X,\tilde{U}(X,t),t) + B\tilde{C}_3(X,t)B^T \end{array} \tag{81}$$

From there, we may use Lemma 2.3 in order to analyse the stability properties of the controlled system. Let us illustrate this possibility with an example.

**example:**
Let us take:

$$\begin{array}{rcl} Q_1 & = & \begin{bmatrix} q_1 & 0 \\ 0 & q_2 \end{bmatrix} \qquad ; \; q_1 > 0, \; q_2 > 0 \\[12pt] \tilde{C}_3(X,t) & = & \begin{bmatrix} 0 & k^2\frac{1-\cos\tilde{\theta}}{\tilde{\theta}}v_r(t) \\ k^2\frac{\cos\tilde{\theta}-1}{\tilde{\theta}}v_r(t) & 0 \end{bmatrix} \end{array} \tag{82}$$

The corresponding control is:

$$\tilde{U} = -\begin{bmatrix} q_1 & 0 & -k\frac{\cos\tilde{\theta}-1}{\tilde{\theta}}v_r(t) \\ 0 & k^2\frac{\sin\tilde{\theta}}{\tilde{\theta}}v_r(t) & q_2/k \end{bmatrix} X \tag{83}$$

By setting: $g_1 = q_1$, $g_2 = k$ and $g_3 = q_2/k^2$, this control may also be written:

$$\begin{cases} v &= v_r(t)\cos\tilde{\theta} - g_1(x_1(M_0) - x_{1,r}(t)) \quad\quad ; g_1 > 0 \\ \dot{\theta} &= \dot{\theta}_r(t) - g_2 v_r(t)\frac{\sin\tilde{\theta}}{\tilde{\theta}}(y_1(M_0) - y_{1,r}(t)) - g_3\tilde{\theta} \quad\quad ; g_2 > 0,\ g_3 > 0 \end{cases} \tag{84}$$

We also have in this case:

$$\tilde{C}(X,t) = \begin{bmatrix} 0 & -\dot{\theta}(X,t) & 0 \\ \dot{\theta}(X,t) & 0 & -k\frac{\sin\tilde{\theta}}{\tilde{\theta}}v_r(t) \\ 0 & k\frac{\sin\tilde{\theta}}{\tilde{\theta}}v_r(t) & 0 \end{bmatrix} \tag{85}$$

with $\dot{\theta}(X,t)$ given by the second equation of 84, and:

$$Q = \begin{bmatrix} q_1 & 0 & 0 \\ 0 & 0 & 0 \\ 0 & 0 & q_2/k^2 \end{bmatrix} \tag{86}$$

Obviously, a matrix $Q_0$ which satisfies the condition of Lemma 2.3 is:

$$Q_0 = \begin{bmatrix} 1 & 0 & 0 \\ 0 & 0 & 0 \\ 0 & 0 & 1 \end{bmatrix} \tag{87}$$

so that we already know, from Lemma3, that the first and third components of $X$ tend to zero.

We also have:

$$Q_0\tilde{C}(X,t) = \begin{bmatrix} 0 & -\dot{\theta}(X,t) & 0 \\ 0 & 0 & 0 \\ 0 & kv_r(t)\frac{\sin(\tilde{\theta}(X,t))}{\tilde{\theta}(X,t)} & 0 \end{bmatrix}. \tag{88}$$

Consider an element $Y(t)$ of the set $\mathcal{D}$ with norm $l$. It is simple to verify:

$$\left.\begin{array}{l} \lim_{t\to\infty} Q_0 Y(t) = 0 \\ \lim_{t\to\infty} Q_0\tilde{C}(Y(t),t)Y(t) = 0 \end{array}\right\} \Rightarrow \begin{cases} \lim_{t\to\infty} \dot{\theta}_r(t)l = 0 \\ \lim_{t\to\infty} v_r(t)l = 0 \end{cases} \tag{89}$$

Therefore, if either $v_r(t)$ or $\dot{\theta}_r(t)$ does not tend to zero, then i) $l = 0$, ii) a matrix $Q_1$ which satisfies the condition of Lemma 2.3 is the identity matrix $I_3$, and iii) all solutions $X(t)$ tend to zero.

Moreover, if $v_r(t)$ and $\dot{\theta}_r(t)$ are constant and not both equal to zero, then the matrix:

$$\begin{bmatrix} Q \\ Q\tilde{C}(0,t) \end{bmatrix} = \begin{bmatrix} q_1 & 0 & 0 \\ 0 & 0 & 0 \\ 0 & 0 & q_2/k^2 \\ 0 & -q_1\dot{\theta}_r & 0 \\ 0 & 0 & 0 \\ 0 & q_2 v_r/k & 0 \end{bmatrix} \tag{90}$$

is a full-rank matrix, and, according to Lemma 2.2 and Corollary 1, the convergence of $X$ to zero is exponential in the neighbourhood of zero.

### 3.2.5 Objective 5: have the cart converge to a desired position/orientation

In Sections 3.2.1 and 3.2.2, we saw that there are smooth feedback controls able to make any cart's point converge to a desired location.

In the previous section, we saw that there are smooth feedback controls able to make the cart converge to the position/orientation of a reference cart...as long as the reference cart keeps moving.

The question addressed in this section is the following: is there a smooth feedback control, which does not rely on any trajectory plannification, able to make the cart converge to a desired configuration?

We may assume, without restricting the problem, that the desired configuration to be reached is $(X_0(M_d) = 0, \theta = 0)$.

It is well known that the considered cart is *controllable* in both position and orientation ( [12], [4], [5],...). This means that, for any initial configuration of the cart, there are *open-loop* controls $U(t)$ able to make the cart converge to the desired configuration (assuming that the model of the cart is correct). However, we showed in [4], by using a theorem from Brockett ([14]), that no smooth feedback control $U(X_0(M_d), \theta)$, function of the cart's position/orientation variables only, could ensure such a convergence whatever the initial configuration of the cart.

We show now that the problem has solutions when the feedback control is also allowed to be a function of the independent time index.

Defining the state vector:

$$X = \left[ \begin{array}{c} X_1(M_0) \\ \theta \end{array} \right] \tag{91}$$

where $X_1(M_0) = [x_1(M_0), y_1(M_0)]^T$ is the matrix of coordinates of the vector $\vec{OM}$ in the basis of the frame $\mathcal{F}_1$, the control objective is to have $X$ converge to zero.

According to 44, $X$ satisfies the following equation:

$$\dot{X} = \left[ \begin{array}{ccc} 0 & \dot{\theta} & 0 \\ -\dot{\theta} & 0 & 0 \\ 0 & 0 & 0 \end{array} \right] X + \left[ \begin{array}{cc} 1 & 0 \\ 0 & 0 \\ 0 & 1 \end{array} \right] U \tag{92}$$

Obviously, the *p.l.* system associated with 92 at the stationnary point $(X = 0, U = 0)$ is not stabilizable.

Let us consider the following control:

$$U = - \left[ \begin{array}{ccc} k_1 & 0 & 0 \\ k_2(X_1(M_0), t) & k_3(X_1(M_0), t) & k_4 \end{array} \right] X \tag{93}$$

with:

- $k_1$ and $k_4$: positive scalars

- $k_2(X_1(M_0), t)$: a scalar function in $\mathcal{S}$

- $k_3(X_1(M_0), t)$: a scalar function in $\mathcal{S}$ such that $\forall y \in R$, $\frac{\partial k_3}{\partial t}([0, y]^T, t)$ *does not converge to zero when $t$ tends to infinity.*

From 92 and 93, we obtain the following closed-loop system:

$$\dot{X} = \begin{bmatrix} -k_1 & \dot{\theta}(X,t) & 0 \\ -\dot{\theta}(X,t) & 0 & 0 \\ -k_2(X_1,t) & -k_3(X_1,t) & -k_4 \end{bmatrix} X \tag{94}$$

with:

$$\dot{\theta}(X,t) = -\begin{bmatrix} k_2(X_1,t) & k_3(X_1,t) & k_4 \end{bmatrix} X \tag{95}$$

From 94, we already see that the partial state vector $X_1(M_0)$ satisfies the equation:

$$\dot{X}_1(M_0) = -(\tilde{C}(X,t) + Q)X_1(M_0) \tag{96}$$

with:

$$\tilde{C}(X,t) = \begin{bmatrix} 0 & -\dot{\theta}(X,t) \\ \dot{\theta}(X,t) & 0 \end{bmatrix} \quad : \ s.s.\ matrix$$
$$Q = \begin{bmatrix} k_1 & 0 \\ 0 & 0 \end{bmatrix} \quad : \ p.s.\ matrix \tag{97}$$

This equation would be in the form 1 if $\dot{\theta}(X,t)$ were not a function of the variable $\theta$. This is not the case here because of the non-zero control gain $k_4$. However, we still have:

$$\frac{d}{dt} 1/2\|X_1(M_0)\|^2 = -X_1(M_0)^T Q X_1(M_0) \quad (\leq 0) \tag{98}$$

Thus, $\|X_1(M_0)\|$ decreases to some limit value and $QX_1(M_0)$ tends to zero. This in turn yields, assuming that the solutions of 96 exist on $R^+$:

$$\lim_{t\to\infty} X_1(M_0) = \begin{bmatrix} 0 \\ y_{lim} \end{bmatrix} \tag{99}$$

The solutions of 94 and 96 indeed exist on $R^+$ because the uniform boundedness of $\|X_1(M_0)\|$ (with respect to initial conditions) yields , in view of 95, the uniform boundedness of $\theta(t)$. This is due to the choice of a positive gain $k_4$ and the properties of functions in $S$. From there, it follows that $\|\dot{X}(t)\|$, as well as $\|\ddot{X}(t)\|$ (by using again the properties of functions in $S$), are also bounded.

Now, since $X_1(M_0)$ converges and $\|\ddot{X}_1(M_0)\|$ is bounded, $\dot{X}_1(M_0)$ tends to zero. If we now remark that $x_1(M_0)$ satisfies the equation:

$$\dot{x}_1 = -k_1 x_1 + \dot{\theta} y_1 \tag{100}$$

it comes from what precedes:

$$\lim_{t\to\infty} \dot{\theta} y_{lim} = 0 \tag{101}$$

Assume that $y_{lim} \neq 0$, then $\dot{\theta}(t)$ tends to zero and, in view of 95:

$$\lim_{t\to\infty} (k_3(X_1(t),t)y_1(t) + k_4\theta(t)) = 0 \tag{102}$$

Thus, from 99 and by using an argument of (uniform) continuity:

$$\lim_{t \to \infty} (k_3([0, y_{lim}]^T, t)y_{lim} + k_4\theta(t)) = 0 \qquad (103)$$

By using again the argument according to which the first derivative of a converging function tends to zero when its second derivative is bounded, we obtain:

$$\lim_{t \to \infty} (\frac{\partial k_3}{\partial t}([0, y_{lim}]^T, t)y_{lim} + k_4\dot\theta(t)) = 0 \qquad (104)$$

and, since $\dot\theta(t)$ tends to zero and $y_{lim} \neq 0$:

$$\lim_{t \to \infty} \frac{\partial k_3}{\partial t}([0, y_{lim}]^T, t) = 0 \qquad (105)$$

This contradicts the initial assumption made on the choice of $k_3(X_1(M_0), t)$. Therefore, $y_{lim} = 0$ and, from 95:

$$\lim_{t \to \infty} (\dot\theta - k_4\theta) = 0 \qquad (106)$$

which implies that $\theta(t)$ also tends to zero.

We thus have shown that the control 93 makes all solutions $X(t)$ of the system 92 converge to zero and brings the cart to the desired configuration.

We may notice that this solution does not require any trajectory plannification since the dependence of the control gains $k_2$ and $k_3$ upon the time index is not *a priori* connected to initial conditions. For example, one may choose $k_2 = 0$ and $k_3(t) = sint$. However, simulations of the corresponding control indicate that the convergence of $X(t)$ to zero may be very slow. The possibility of finding more efficient gains, or more efficient feedback control strategies, remains to be explored.

## 3.3   Dynamic equations

In the next section, the control inputs will no longer be the motorized wheels' velocities $\dot q_i$, but the torques $\gamma_i$ ($i = 1, 2$) applied those wheels. In order to be able to perform a stability analysis similar to the one performed previously in the case of velocity control, a model of the effects of those torques on the cart's motion is needed. Such a model, called *dynamic model* of the cart, is now derived under the rolling-without-slippage assumption and other simplifying assumptions.

If we assume that the cart is an assembly of rigid bodies rigidly linked either to the motorized wheels (assumed to be perfectly circular and balanced) or to the cart's platform (itself mechanically linked to the wheels' axis), and if the ground is ideally flat, then it is simple to show that, *under the rolling-without-slippage assumption*, all velocity screws associated with these bodies depend linearly on the wheels' angular velocities $\dot q_i$. This was already established in relation 39 for the velocity screw associated with the cart's platform. As a consequence, the cart's kinetic energy, denoted as $\mathcal{T}$, is a quadratic form in those velocities, and there exists a *p.d.s inertia* matrix $\mathcal{M}$ so that:

$$\mathcal{T} = 1/2\dot q^T \mathcal{M} \dot q \qquad (107)$$

Notice that, in writing this relation, the nonholonomic constraints are explicitly used to eliminate redundant velocity variables in the expression of the kinetic energy. In the general case, this should not be done before applying Lagrange equations ([16]) because this would usually lead to incorrect equations. However, this is of no consequence here because all velocity screws depend linearly on the independent wheels' velocities $\dot{q}_i$ while they do not depend on the cart's configuration. If we also assume that the ground is horizontal (so that gravity has no effect on the cart's motion) and that friction terms can be neglected, then Lagrange equations may, in this particular case, be formally written as follows:

$$\frac{d}{dt}(\frac{\partial T}{\partial \dot{q}_i}) = \gamma_i \quad ; \; i = 1, 2 \tag{108}$$

From 107 and 108, we then obtain the following dynamic equation:

$$\mathcal{M}\ddot{q} = \Gamma \tag{109}$$

with $\Gamma = [\gamma_1, \gamma_2]^T$.

A complete dynamic model of the cart is thus formed of the dynamic relation 109 together with the following kinematic relations (derived previously):

$$U = D\dot{q} \tag{110}$$

$$\dot{X}_1(M_d) = \dot{\theta}\tilde{C}_1 X_1(M_d) + B_1(d)U \tag{111}$$

## 3.4 Torque control

In this section, the torque vector $\Gamma$ is the control vector to be determined in order to achieve some objective. We will show how the analysis of Section 2 may again be applied to the design and analysis of stabilizing feedback controls. This will be done for the first two objectives considered in Section 3.2.

### 3.4.1 Objective 1: bring the cart's point $M_d$ (with $d \neq 0$) to the point $O$ by using state feedback

Let us set as state vector:

$$X = \begin{bmatrix} X_1(M_d) \\ U \end{bmatrix} \tag{112}$$

From 109-111, we have:

$$\dot{X} = f(X, \Gamma)$$
$$= \begin{bmatrix} \dot{\theta}\tilde{C}_1 & B_1(d) \\ 0 & 0 \end{bmatrix} X + \begin{bmatrix} 0 \\ D\mathcal{M}^{-1} \end{bmatrix} \Gamma \tag{113}$$

The analysis of Section 2 suggests (for example) a control in the form:

$$\Gamma = -\mathcal{M}D^{-1}G[B_1(d)X_1(M_d) + RU] \tag{114}$$

where $G$ and $R$ are two *p.d.s.* matrices. Indeed, this type of control leads to a closed-loop system equation in the form 1 with:

$$P = \begin{bmatrix} I_2 & 0 \\ 0 & G^{-1} \end{bmatrix} \tag{115}$$

$$Q = \begin{bmatrix} 0 & 0 \\ 0 & R \end{bmatrix} \tag{116}$$

$$\tilde{C}(X) = \begin{bmatrix} -\dot{\theta}\tilde{C}_1 & -B_1(d) \\ B_1(d) & 0 \end{bmatrix} \tag{117}$$

A matrix $Q_0$ which satisfies the condition of Lemma 2.3 is:

$$Q_0 = \begin{bmatrix} 0 & 0 \\ 0 & I_2 \end{bmatrix} \tag{118}$$

Thus:

$$Q_0 P^{-1} \tilde{C}(X) = \begin{bmatrix} 0 & 0 \\ GB_1(d) & 0 \end{bmatrix} \tag{119}$$

Consider an element $Y(t)$ of the set $\mathcal{D}$. Obviously, if $Q_0 Y(t)$ and $Q_0 P^{-1}\tilde{C}(Y(t))Y(t)$ tend to zero, then $Y(t)$ must also tend to zero. Therefore, a matrix $Q_1$ which satisfies the condition of Lemma 2.3 is the identity matrix $I_4$, and, according to Corollary 1, all solutions $X(t)$ converge to zero.

Morerover, since the matrix:

$$\begin{bmatrix} Q \\ QP^{-1}\tilde{C}(0) \end{bmatrix} = \begin{bmatrix} 0 & 0 \\ 0 & R \\ 0 & 0 \\ RGB_1(d) & 0 \end{bmatrix} \tag{120}$$

is a full-rank matrix, the convergence is, according to Lemma 2.2 and Corollary 1, exponential in the neighbourhood of zero.

We may also notice that the particular choice:

$$G = kD\mathcal{M}^{-1}D^T \qquad ; k > 0 \tag{121}$$

gives the following control:

$$\Gamma = -kD^T[B_1(d)X_1(M_d) + RU] \tag{122}$$

and that the calculation of this stabilizing control *does not require the knowledge of the inertia matrix* $\mathcal{M}$. This may be interpreted as a robustness property with respect to modelling errors.

Another way of writing 122 is:

$$\Gamma = -kD^T R[U - U_{des.}] \tag{123}$$

with:

$$U_{des.} = -B_1(d)^{-1}Q'X_1(M_d)$$
$$Q' = B_1(d)R^{-1}B_1(d) \quad : p.d.s \; matrix \tag{124}$$

This writing points out clearly how a stabilizing torque control $\Gamma$ can simply be obtained from a *desired* stabilizing (as shown in Section 3.2.1) velocity control $U_{des.}$.

### 3.4.2 Objective 2: bring the cart's point $M_0$ to the point $O$ by using state feedback control

Since $d = 0$ in this case, the matrix $B_1(d)$ is now singular and the pseudo-linearized system associated with the nonlinear system 113 is not controllable, as already pointed out in the case of velocity control.

In order to find a stabilizing control, we will again make use of the following equality:

$$(I_2 - K_1)B_1(0) = 0 \tag{125}$$

with:

$$K_1 = \begin{bmatrix} 1 & k_1 \\ 0 & k_2 \end{bmatrix} \quad ; k_1 \in R, \; k_2 \in R \tag{126}$$

Let us then consider the following linear state-feedback control:

$$\Gamma = -\mathcal{M}D^{-1}G[K_1{}^T X_1(M_0) + Q_\Gamma U] \tag{127}$$

with:

$$G = \begin{bmatrix} g_1 & 0 \\ 0 & g_2 \end{bmatrix} \quad (g_1 > 0, \; g_2 > 0)$$
$$Q_\Gamma = \begin{bmatrix} q_1 & 0 \\ 0 & q_2 \end{bmatrix} \quad (q_1 > 0, \; q_2 > 0) \tag{128}$$
$$k_2 \neq 0$$

The closed-loop system satisfies the equation:

$$\dot{X} = \begin{bmatrix} I_2 & 0 \\ 0 & G \end{bmatrix} \begin{bmatrix} \dot{\tilde{C}}_1 & B_1(0) \\ -K_1{}^T & -Q_\Gamma \end{bmatrix} X \tag{129}$$

Let us then introduce the following vector:

$$X' = \begin{bmatrix} X_1(M_0) \\ B_1(0)U \end{bmatrix} \tag{130}$$

From 129, and by using 125 and the fact that the products of $B_1(0)$ with $G$ and $Q_\Gamma$ commute, we obtain:

$$\dot{X}' = -P^{-1} \begin{bmatrix} -\dot{\theta}\tilde{C}_1 & -B_1(0) \\ B_1(0) & Q_\Gamma \end{bmatrix} X' \tag{131}$$

with:

$$P = \begin{bmatrix} I_2 & 0 \\ 0 & G^{-1} \end{bmatrix} \tag{132}$$

The system 131 would be in the form 1 if each solution $\dot{\theta}(t)$ were shown to have the properties of a function in $S$. Since this has not yet been established, Lemma 2.3 does not apply directly to this system. However, convergence of $X(t)$ to zero may still be proven as shown below.

From 131, we have:

$$\begin{aligned} 1/2\tfrac{d}{dt}X'^T P X' &= -X'^T \begin{bmatrix} 0 & 0 \\ 0 & Q_\Gamma \end{bmatrix} X' \\ &= -q_1 U^T B_1(0) U \quad (\le 0) \end{aligned} \tag{133}$$

Thus $||X'||$ decreases to some value and $B_1(0)U = [v,0]^T$ converges to zero. Now, since $U = D\dot{q}$, we have from 110 and 127:

$$\dot{U} = -G[K_1^T X_1(M_0) + Q_\Gamma U] \tag{134}$$

According to this equation, and since $-GQ_\Gamma$ is clearly a stability matrix, the boundedness of $||X_1(M_0)||$ yields the boundedness of $||U||$. Thus $||X||$ is bounded, and so is $||\dot{X}||$. By differentiating 129 once, we also check that $||\ddot{X}||$ is bounded. The convergence of $B_1(0)U$ to zero then yields the convergence of $B_1(0)\dot{U}$ to zero. Moreover, premultiplication of 134 by $B_1(0)$ gives:

$$B_1(0)\dot{U} = -GB_1(0)[X_1(M_0) + q_1 U] \tag{135}$$

Therefore:

$$\lim_{t\to\infty} B_1(0)X_1(M_0) = 0 \tag{136}$$

The first component $x_1(M_0)$ of $X_1(M_0)$ thus converges to zero.

Now, since $||X'||$ converges, and since the first and third components of $X'$ were shown to converge to zero, while the fourth one is zero, the second component $y_1(M_0)$ also converges:

$$\lim_{t\to\infty} y_1(M_0) = y_{lim} \tag{137}$$

Also, since $||\ddot{X}'||$ is bounded, the derivative $B_1(0)\dot{X}_1(M_0) = \dot{\theta}S\tilde{C}_1 X_1(M_0) + B_1(0)U$ of $B_1(0)X_1(M_0)$ tends to zero. We thus obtain:

$$\lim_{t\to\infty} (\dot{\theta}y_1(M_0)) = 0 \tag{138}$$

Let us assume that $y_{lim}$ is different from zero. Then, according to 138, $\dot{\theta}$ must tend to zero. Since $||\dot{X}||$ is bounded, $\ddot{\theta}$ must also tend to zero. Now, the second equation of relation 134 is:

$$\ddot{\theta} = -g_2(k_1 x_1(M_0) + k_2 y_1(M_0) + q_2\dot{\theta}) \tag{139}$$

Since $k_2 \neq 0$ by assumption, the convergence of $x_1(M_0)$, $\dot{\theta}$ and $\ddot{\theta}$ to zero thus yields the convergence of $y_1(M_0)$ to zero. But this contradicts our previous assumption about $y_{lim}$. Therefore, this assumption is not correct and we must have $y_{lim} = 0$. The convergence of $\dot{\theta}$ and $\ddot{\theta}$ to zero then follows from 139.

We thus have shown that the control 127-128 makes the solutions $X(t)$ of the closed-loop system converge to zero.

**Remarks:**

- The preceding convergence proof may clearly be adapted to the case where the matricial gains $K_1$ and $Q_\Gamma$ are not chosen constant.

- As in the case of velocity control studied in Section 3.2.2, the convergence of the cart's point $M_0$ to the point $O$ may not be exponential and the cart's orientation $\theta(t)$ may not converge to any limit value.

- With the choice: $Q_\Gamma = 1/g\, I_2$, the stabilizing torque control 127-128 may be written:

$$\Gamma = -\frac{1}{g}\mathcal{M}D^{-1}G[U - U_{des}] \tag{140}$$

where:

$$U_{des} = -gK_1^T X_1(M_0) \tag{141}$$

is a stabilizing velocity control, as shown in Section 3.2.2.. This remark is to emphasize the correspondance existing between velocity and torque control.

## 3.5 Concluding comments

From there, we will leave to the interested reader the task of verifying that the other objectives considered in Section 3.2, in the case of velocity control, can be treated in the same manner and that convergence properties, similar to those obtained in the case of velocity control, can be derived. This correspondance between velocity control and torque control may be seen as an extension of the one already known and thoroughly used by roboticians in the case of holonomic robot manipulators. From the Automatic Control point of view, it may be interpreted as a consequence of the fact that if a system can be controlled by a vector $U$ (a velocity vector in the present case), then it can be controlled in the same way by the derivative $\dot{U}$ of this vector (accelerations or torques). It seems also that the *robustness* of torque control with respect to model uncertainties, which was only briefly pointed out in this paper, is again much related to positivity properties of mechanical systems (positivity of the inertia matrix, for example) and, by extension, to passivity properties of these systems. Of course, these general ideas remain to be closely examined and more precisely stated. They could lead, as in the case of robot manipulators, to interesting developments.

Acknowledgment: The author would like to thank G. Campion for his insightful reviewing comments.

# References

[1] P.F. Muir, C.P. Neuman: **Kinematic Modeling for Feedback Control of an Omnidirectional Mobile Robot**, Proc. of the 1987 IEEE Int. Conf. on Robotics and Automation, Raleigh, N. Carolina, pp. 1772-1778, 1987.

[2] W.L. Nelson, I.J. Cox: **Local Path Control for an Autonomous Vehicle**, Proc. of the 1988 IEEE Int. Conf. on Robotics and Automation, Philadelphia, Penn., pp. 1504-1510, 1988.

[3] Y. Kanayama, Y. Kimura, F. Miyazaki, T. Noguchi: **A Stable Tracking Control Method for an Autonomous Mobile Robot**, 1990 IEEE Int. Conf. on Robotics and Automation, Cincinnati, Ohio, pp. 384-389, 1990.

[4] C. Samson, K. Ait-Abderrahim: **Mobile Robot Control. Part 1: Feedback Control of a Nonholonomic Wheeled Cart in Cartesian Space**, INRIA report, No. 1288, 1990.

[5] G. Campion, B. d'Andrea-Novel, G. Bastin: **Controllabilty and State Feedback Stabilisation of Non Holonomic Mechanical Systems**, Int. Workshop in Adaptive and Nonlinear Control: Issues in Robotics, Grenoble, France, 1990.

[6] Y. Nakamura, R. Mukherjee: **Nonlinear Control for the Nonholonomic Motion of Space Robot Systems**, Int. Workshop in Adaptive and Nonlinear Control: Issues in Robotics, Grenoble, France, 1990.

[7] J.P. Laumond: **Feasible Trajectories for Mobile Robots with Kinematic and Environment Constraints**, Int. Conf. on Intelligent Autonomous Syst ems, Amsterdam, pp. 346-354, 1986.

[8] J.P. Laumond: **Finding Collision-Free Smooth Trajectories for a Non-Holonomic Mobile Robot**, 10th Int. Joint Conf. on Artificial Intelligence, Milano, Italy, pp. 1120-1123, 1987.

[9] P. Tournassoud, O. Jehl: **Motion Planning for a Mobile Robot with a Kinematic Constraint**, IEEE Int. Conf. on Robotics and Automation, Philadelphia, Penn., pp. 1785-1790, 1988.

[10] J.F. Canny: **The Complexity of Robot Motion Planning**, MIT Press, 1988.

[11] Z. Li, J.F. Canny, G. Heinzinger: **Robot Motion Planning with Nonholonomic Constraints**, Memo UCB/ERL M89/13, Electronics Research University, University of California, Berkeley CA.

[12] J. Barraquand, J.C. Latombe: **On Nonholonomic Mobile Robots and Optimal Maneuvering**, Revue d'Intelligence Artificielle, vol. 3 - n.2/1989, Ed. Hermes, pp. 77-103, 1989.

[13] T. Kailath: **Linear Systems**, Prentice-Hall Information and Systems Sciences Series, 1980.

[14] R.W. Brockett: **Asymptotic Stability and Feedback Stabilization**, Differential Geometric Control Theory, Proc. of the conf. held at Michigan Technological University in June-July 1982, Progress in Math., vol.27, Birkhauser, pp. 181-208, 1983.

[15] A. Lasalle, S. Lefschetz: **Stability by Lyapunov's Direct Method**, Academic Press, 1961.

[16] Y. Bamberger: **Mécaniques de l'ingénieur I. Systèmes de corps rigides**, Hermann, 1981.

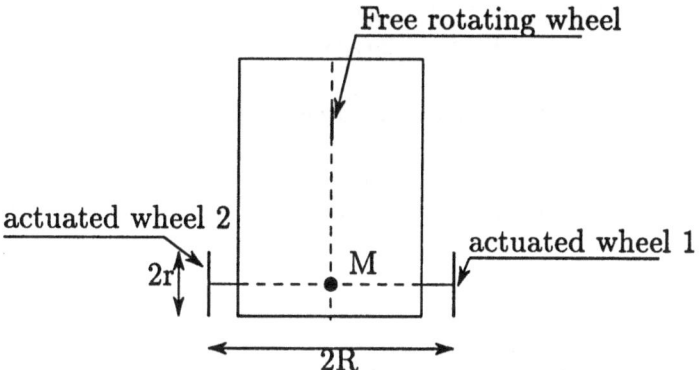

**Fig.1** : Above view of the cart

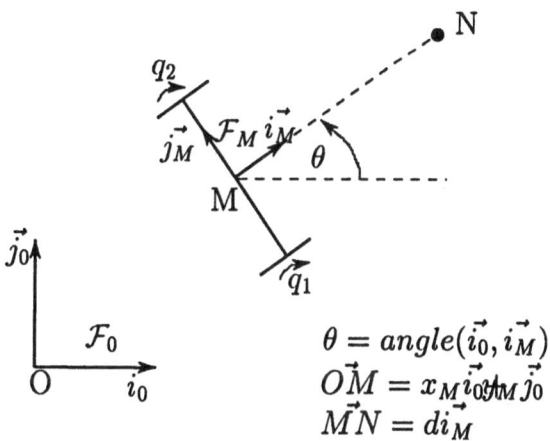

$$\theta = angle(\vec{i_0}, \vec{i_M})$$
$$\vec{OM} = x_M \vec{i_0} + y_M \vec{j_0}$$
$$\vec{MN} = d\vec{i_M}$$

**Fig.2** : Cart's parameters

# SOME ISSUES IN THE CONTROL OF RIGID ROBOTS IN A SENSORY SPACE

Bernard Espiau*, François Chaumette**

(*) ISIA-ENSMP, Rue Claude Daunesse, 06565 Valbonne Cedex, France
(**) IRISA-INRIA Rennes, Campus de Beaulieu, 35042 Rennes, France

**Abstract:** The aim of this paper is to examine some problems related to robot control in an output sensory space, with focus on the particular case of visual sensors. A single sensor is here considered as a mapping from $SE_3$ to $\mathbb{R}$. This assumption allows us to set that the related jacobian is a particular set of screws, which authorizes some unification in the analysis. Then, we define the concept of virtual linkage as a way of expressing the constraints induced by the sensors. We then use the redundancy approach of C. Samson in order to design a correct task function, specifying in that way the right output working space. Some facts in the control are then recalled.

All the analysis, originally suited for proximity, range or force sensors, is applied to the case of visual sensors. The various involved mappings and the needed assumptions are precised; the design of adequate task functions which use special image features is presented, with some indications on the practical derivation of the jacobian models. Some results and comments are finally given.

# 1 Introduction

## 1.1 Notation

The following notation will be employed:

Let $E$ be the three-dimensional affine euclidean space, the related vector space being $\mathbb{R}^3$. The configuration space of a rigid body, which is also the frame configuration space, is the Lie group of displacements, $SE_3$ (Special Euclidean Group), isomorphic to $\mathbb{R}^3 \times SO_3$ where $SO_3$ is the group of rotations. It is a six-dimensional differential manifold. An element of $SE_3$, called a 'position' (i.e. location and attitude owing to the previous isomorphism) is denoted as $\bar{r}$. The tangent space to $SE_3$ at identity is denoted as $se_3$, and its dual, or cotangent space, $se_3^*$. $se_3$ is a Lie algebra isomorphic to the Lie algebra of equiprojective fields of $E$ in $\mathbb{R}^3$, which means that any element (field) of $se_3$ is no more than the classical velocity screw. A screw $H$ is also defined by its vector $u$ and the value of its field in a point $P$ of $E$. We may therefore write: $H = (H(P), u)$.

Frames are denoted as $F$, with origin $O$. A given screw expressed in $F$ is then also a vector in $\mathbf{R}^6$. Finally, the velocity of a frame $F_i$ with respect to a frame $F_j$ is denoted as $V_{ij}$.

The screw product is the bilinear mapping associated with $\begin{bmatrix} 0 & I_3 \\ I_3 & 0 \end{bmatrix}$. It may be written, for any considered point $P$:

$$H_1 \bullet H_2 = <u_1, H_2(P)> + <u_2, H_1(P)> \tag{1}$$

where $<, >$ is the usual scalar product between two vectors of $\mathbf{R}^3$. Let $S$ be a screw space. The screw product induces an isomorphism between $S$ and its dual $S^*$, which is itself a screw space.

We will denote the skew-symmetric matrix associated with a vector of $\mathbf{R}^3$ as $As(.)$; a matrix $A$ $(n \times n)$ will be said positive if $x^T A x > 0$, $\forall x \neq 0 \in \mathbf{R}^n$.

We will also consider a rigid robot, the state equation of which is given by

$$\Gamma = M(q)\,\ddot{q} + N\,(q, \dot{q}, t)\ , \ \dim(q) = \dim(\Gamma) = n \tag{2}$$

where $\Gamma$ is the vector of applied external forces (actuator torques), $M$ is the kinetics energy matrix, $N$ gathers gravity, centrifugal, Coriolis and friction forces, and $(q, \dot{q})$, the joint position and velocity, is the state vector of the system.

It is assumed that an actuator is associated to every degree of freedom of the robot. We will also assume here for simplicity that $n = 6$. Let $F_6$ be a frame linked to the 'last' body, and $F_0$ a reference frame. The robot jacobian, $J$, is the jacobian associated to the mapping from $q \in \mathbf{R}^6$ to the position of $F_6$ with respect to $F_0$. We do not consider here the case where $J(q)$ falls singular. Some techniques to cope with this problem are presented in [11].

## 1.2 A Flavour of the Problem

Decoupling and linearizing (2) in the joint space is trivial as soon as the dynamics is known and computed. However, control in joint space is generally of little interest for the user: it is at least wished to control the position of $F_6$. The ideal decoupling and linearizing control takes then the form:

$$\Gamma = M(q)\,J^{-1}(q)\,u + N(q,\dot{q}) - M(q)\,J^{-1}(q) \begin{bmatrix} \vdots \\ \dot{q}^T W_i(q,t)\dot{q} \\ \vdots \end{bmatrix} \tag{3}$$

where $W_i(q,t)(i = 1, \cdots, n)$ is the partial derivative of the i-th row of $J^T(q)$ with respect to $q$, and $u$ the new control vector. The need for nonsingularity of $J(q)$ appears here. Nevertheless, this kind of control, in $SE_3$, is not suitable in more complex (and interesting) applications, especially when exteroceptive sensors are used. Another working space is then required. This situation is a particular case of the more general 'control in task space', developed in [8] and [11], which will be briefly stated in section 4.

We will therefore try in the present paper to show how a control in sensory space, extending in some way the scheme (3), may be designed, and we will apply this approach to the case of visual sensors (other cases are examined in [11] and [4]). It should be emphasized that, in robotics, this area, known as 'visual servoing', is not as largely investigated as classical robot vision. Some relevant references are [2], [5], [6], [7], [13], [14], [12]. The related works will not be discussed here, since done in [3], to which we refer the reader.

In fact, modelling aspects and design of the adequate task function (i.e. of the output space associated to (2)) are the most delicate points, and we shall focus the development on these aspects. Section 2 will be therefore devoted to general considerations on sensor-environment interactions; the concept of hybrid task (i.e combining tasks expressed in $SE_3$ and in a sensory space) is presented in section 3, while section 4 examines the specific case of visual features. Finally experimental results are presented in section 5, followed by few remarks on the need for on-line estimation schemes.

# 2 Modelling of Sensor-environment Interaction

## 2.1 The Interaction Screw

We restrict our study to the case where, formally:

- *A sensor is completely defined by a $C^2$ mapping from $SE_3$ to $\mathbb{R}^k$.*

This assumption implies in particular that, for a given sensor, relative environmental modifications of the geometrical kind are the only ones allowed to make the sensor output varying. This is true for many kinds of proximity, range force and visual sensors. Let us now link a frame $F_T$ to the part of environment observed by the sensor, and another, $F_S$, to the sensor itself. The reference frame may be $F_S$ or $F_T$, or, even, when the environment is time-invariant, any associated frame $F_o$. The sensor output $s$ may then be written: $s(F_S, F_T)$. Furthermore, let us assume that the sensor mobility is got through a generalized coordinate system, $q$, which constitutes a local chart of $SE_3$. Then, when the observed objects are autonomously mobile themselves, $s$ may be also written $s = s(q,t)$, the independent time variable $t$ representing this contribution of the objects motion. The six variables $q_i$ are for example the joint angular positions of a rigid manipulator which handles a camera.

Let us now examine a one-dimensional component $s_j$ of $s$. Owing to the above preliminaries and to the definition of $s$, we know that its differential at $\bar{r}$, $ds_{j|_r}$, is a linear mapping from $se_{3|_r}$ to $\mathbf{R}$. It is also known that the differential of any analytic function from a manifold $M$ to $\mathbf{R}$ may be identified with an element of the cotangent space. In our case, this implies that the differential of $s_j$ at $\bar{r}$ is simply an element of $se_3^*$, that is to say a screw. Recalling that an element of $se_3$ is the velocity screw $V$, we may finally write at $\bar{r}$ in $SE_3$ the basic screw product:

$$\dot{s}_j = H_j \bullet V_{ST} \tag{4}$$

where $V_{ST}$ is the velocity of the frame $F_T$ with respect to the frame $F_S$, $\bullet$ is the screw product defined above, and $H_j$ is a screw, the expression of which depends both on the environment characteristics and on the sensor itself. It therefore fully characterizes the interaction between a sensor and its environment, and we thus call it *Interaction Screw*.

## 2.2 The Concept of Virtual Linkage

A set of *compatible* and *independent* constraints, $s(\bar{r}) - s^* = 0$, where $s^*$ is stationary, constitutes a **virtual linkage** between the sensor $(S)$ and the objects of the environment $(T)$. Let thus $V^*$ be a virtual motion at $\bar{r}$ keeping constant the sensor output component $s_j$, i.e. preserving the satisfaction of the $j$th constraint. $V^*$ is solution of the equation:

$$H_j \bullet V^* = 0 \tag{5}$$

and is therefore a screw *reciprocal* to $H_j$. Let us now return to the full sensor output vector, $s$, with dimension $k$. The set of the motions $V^*$ leaving $s$ invariant is $S^*$, the subspace reciprocal to the screw subspace $S$ spanned by the set $\{H_1 \cdots H_j \cdots H_k\}$ in $se_3$.

In a position where these constraints are satisfied, the dimension, $N$, of $S^*$ may be called the **class** of the virtual linkage in $\bar{r}$.

- **Remark:** With an obvious breach of notation, equation (4) may also be written:

$$\dot{s}_j = L_j^T V_{ST} \quad \text{where} \quad L_j^T = H_j \begin{bmatrix} 0 & I_3 \\ I_3 & 0 \end{bmatrix} \tag{6}$$

$L_j^T$ is the matrix-form of the interaction screw $H_j$, in a given frame $F$ and in a chosen point $O$. In the same way, the matrix form of the set $\{H_1 \cdots H_k\}$ is called *Interaction Matrix*, and is denoted as $L^T$. With a similar breach of notation, we may write $S^* = \text{Ker } L^T$.

The interest of this approach lies both in its generality and in the unification it realizes. Let us simply emphasize two aspects of these advantages:

- **Computation:** Knowing that the robot direct kinematics is a mapping from $\mathbb{R}^n$ to $SE_3$, it is easy to see that every column of the robot jacobian matrix, $J$, is the matrix representation of a screw. In most of the computations needed in practice, the three useful transformations are therefore simply:

  - the change of basis ($i$ to $j$), with matrix $R_{ij}$ (rotation matrix);

  - the change of frame ($F_i$ to $F_j$), with homogeneous matrix $\bar{R}_{ij} = \begin{bmatrix} R_{ij} & [O_iO_j]_{F_i} \\ 0 & 1 \end{bmatrix}$

  - the change of reference frame for screws, with matrix $Ad_{ij} = \begin{bmatrix} R_{ij} & R_{ij}As(O_iO_j)_{F_j} \\ 0 & R_{ij} \end{bmatrix}$
  which is required for transforming the robot jacobian expression.

  Since the assumption made in sensor modelling leads also to consider sets of screws, it appears that the three transformations above are also the only ones to be used. This finally allows to obtain some unity in the computational issues.

- **Virtual linkage:** This concept may be related to the basic kinematics of contacts, as classically used in the theory of mechanisms. The idea of virtual linkage, which may include the physical linkage when contact sensors are used, will allow us to design the wished sensor-referenced robotics tasks in a simple way. This will also establish a connection with the approach known in the litterature as 'hybrid control', which is traditionnaly used in control schemes involving contact force sensors. This finally shows that many types of sensors may be used within a single framework: the one of *hybrid tasks* which realize *virtual linkages*.

# 3 Tasks and Control Design

## 3.1 The concept of task function

The dynamic behaviour of a rigid manipulator is described by equation (2). The task to be performed may then be specified as an *output function* associated to (2). More precisely, it may be shown ([11]) that the user's objective may in general be expressed as the regulation to zero of some $n$-dimensional $C^2$ function, $e(q, t)$, called *task function*, during a time interval $[0, T]$. An immediate example of task function is

$$e(q, t) = x(q) - x_d(t) \tag{7}$$

where $x_d(t)$ is for example a parametrization of the desired position of a robot wrist in $SE_3$. Many other cases are presented in [11]. When sensors are used, it appears that the sensor vector $s(q, t)$ has to contribute to the design of the task function, in a way explained later.

As detailed in [8] and [11], the problem of regulating $e$ is well-posed if $e$ has some specific properties. One of them is the existence and the unicity of a $C^2$ *ideal trajectory*, $q_r(t)$, such that $e(q_r(t), t) = 0$, $t \in [0, T]$ and $q_r(0) = q_0$, where $q_0$ is a given initial condition. Another one, very important, is the non-singularity of the task-jacobian matrix $\frac{\partial e}{\partial q}(q, t)$, around $q_r(t)$. When all the required conditions are satisfied, which will be implicitly assumed in the following, the task function is said to be 'admissible'. Efficient control laws may then be designed.

## 3.2 Control and stability

We only give here an intuitive idea of the used approach and of the obtained results. All the related developments may be found in [8] and [11]. Let us consider the exact decoupling and feedback linearization in the task space: in a way similar to (3), it is easy to see that an adequate control is:

$$\Gamma = M \left(\frac{\partial e}{\partial q}\right)^{-1} u' + N - M \left(\frac{\partial e}{\partial q}\right)^{-1} f \tag{8}$$

with:

$$f(q, \dot{q}, t) = \begin{bmatrix} \vdots \\ \dot{q}^T W_i(q, t) \dot{q} \\ \vdots \end{bmatrix} + 2 \frac{\partial^2 e}{\partial q \partial t}(q, t) \, \dot{q} + \frac{\partial^2 e}{\partial^2 t}(q, t) \tag{9}$$

where $W_i(q, t)(i = 1, \cdots, n)$ is the partial derivative of the i-th row of $\left(\frac{\partial e}{\partial q}\right)^T (q, t)$ with respect to $q$. We may choose a PD feedback of the form:

$$u' = -\lambda G \left(\mu D e + \dot{e}\right) \tag{10}$$

$G$ and $D$ being positive matrices, $\lambda$ and $\mu$ being positive scalars, all to be tuned by the user.

The ideal control scheme (8) (10) requires a perfect knowledge of all its components, which is neither possible, nor even wished. A more realistic approach consists in generalizing the previous control as:

$$\Gamma = -\lambda \hat{M} \left(\frac{\widehat{\partial e}}{\partial q}\right)^{-1} G \left(\mu D e + \frac{\widehat{\partial e}}{\partial q} \dot{q} + \frac{\widehat{\partial e}}{\partial t}\right) + \hat{N} - \hat{M} \left(\frac{\widehat{\partial e}}{\partial q}\right)^{-1} \hat{f} \tag{11}$$

where the carets point out that models (approximations, estimates) are used instead of the true terms. In this general expression, all the terms but $\mu, D$ and $G$ are allowed to be functions of $q$ and $t$, even of $\dot{q}$ for $\lambda, \hat{f}$ and $\hat{N}$.

A stability analysis of the system (2) with control (11) was done by Samson ([11]) in a nonlinear framework. Two main classes of sufficient stability conditions (in the sense of the boundedness of $\|e(t)\|$) were then exhibited: **gain** conditions (these tuning parameters leave more or less possibilities to the user) and **modelling** conditions. Among them, the most critical concerns the task itself, and has the form:

$$\frac{\partial e}{\partial q}\left(\widehat{\frac{\partial e}{\partial q}}\right)^{-1} > 0 \tag{12}$$

This essential condition allows to characterize the robustness of the task itself with regard to uncertainties and approximations.

It may already be noticed that, when we are interested in the motion of the end effector, we may write $\frac{\partial e}{\partial q} = \frac{\partial e}{\partial \bar{r}} \frac{\partial \bar{r}}{\partial q}$, where $\frac{\partial \bar{r}}{\partial q}$ is the robot jacobian matrix, $J$. When it is known and nonsingular, as we shall assume afterwards, the choice $\widehat{\frac{\partial e}{\partial q}} = \widehat{\frac{\partial e}{\partial \bar{r}}}J$ allows condition (12) to be reduced to:

$$\frac{\partial e}{\partial \bar{r}}\left(\widehat{\frac{\partial e}{\partial \bar{r}}}\right)^{-1} > 0 \tag{13}$$

## 3.3 Hybrid Tasks

Regulating sensor signals is generally not the unique user's objective; very often, this task has to be combined with another such that a trajectory tracking.

Generally, the problem specification leads in a first step to defining a sensor-based task vector, $e_1(q, t)$, with $m \leq n$ independent components, the regulation of which constitutes the part of the global task which requires the use of exteroceptive sensors. How to derive such a vector when using visual sensors will be described later. A second objective, for example a desired sensor motion, might me represented in a first glance by a second vector $e_2(q, t)$. However, $e_1$ and $e_2$ would be gathered in a single task vector $e(q, t)$ **admissible**, such that the two tasks are compatible and independent.

It may indeed be shown that a more efficient way of setting the problem consists in embedding it in the framework of *task redundancy*. In this approach, $e_1$ is considered as priority, and $e_2$ is defined as the representation of the constrained minimization of a secondary cost function.

### 3.3.1 The Redundancy Framework

Let us assume that $J = \frac{\partial r}{\partial q}$ is known and nonsingular everywhere needed. Let $e_1$ be a $m$-dimensional main task, with jacobian matrix $J_1^r \left(= \frac{\partial e_1}{\partial r}\right)$ in $SE_3$, and let $h_s$, with gradient $g_s^r = \frac{\partial h_s}{\partial r}$, be a secondary cost function to be minimized (the choice of $h_s$ is discussed in [11]). Minimizing $h_s$ under the constraint $e_1 = 0$ requires the subspace of motions left free by this constraint to be determined. This comes back to knowing the null space of $J_1^r$, Ker $J_1^r$ (or the range of $J_1^{rT}$, $R(J_1^{rT})$) along the ideal trajectory. In other words, it has to be found any $m \times n$ full rank matrix $W$, such that:

$$R(W^T) = R\left(J_1^{rT}\right) \tag{14}$$

along the robot's ideal trajectory, $q_r(t)$.

Once this matrix is determined, it may rather easily be shown ([9], [11]) that a task function minimizing $h_s$ under the constraint $e_1 = 0$ is:

$$e = W^+ e_1 + \alpha \left(I_n - W^+ W\right) g_s^{rT} \tag{15}$$

where $\alpha$ is a positive scalar, $W^+$ is the pseudo-inverse of $W$, and $(I_n - W^+ W)$ is an orthogonal projection operator on the null space of $W$, i.e. on that of $J_1^r$.

- **Remark:** When $e_1$ is made from sensor signals and $h_s$ expresses a trajectory tracking task in $SE_3$, the task represented by (15) is then called 'hybrid task'.

It clearly appears that the computation of the jacobian matrix related to (15), possibly required in the control scheme, may be complex. The positivity condition (13) may then be of some interest. It may indeed be shown that if, in addition to (14), $W$ satisfies the property:

$$J_1^r W^T > 0 \tag{16}$$

along $q_r(t)$, then, under 'normal circumstances' (see [11]) the jacobian matrix of $e$ in $SE_3$ is such that

$$\frac{\partial e}{\partial r} \left(I_n + \gamma \left(I_n - W^+ W\right)\right) > 0 \tag{17}$$

along $q_r(t)$, and $\forall \gamma \geq \gamma_m(\alpha) \geq 0$. The condition (13) is therefore satisfied by taking:

$$\left(\frac{\widehat{\partial e}}{\partial r}\right)^{-1} = \left(I_n + \gamma \left(I_n - W^+ W\right)\right) \tag{18}$$

More, when $\alpha$ is 'small enough', then $\gamma_m = 0$, $\frac{\partial e}{\partial r}$ is positive, and we may choose:

$$\frac{\widehat{\partial e}}{\partial r} = I_n \tag{19}$$

### 3.3.2 The Specific Case of Sensor Signals

Let us know apply this approach to the use of sensor signals as defined in section 2.1. Let us recall that the vector $s$ is of dimension $k$. Recall that the jacobian of $s$ in $SE_3$ corresponds to the interaction matrix $L^T$. The dimension of $L$ is $6 \times k$ and its rank is $m$, $N = 6 - m$ being the class of the associated virtual linkage.

We are interested in regulating $s$ around a desired value or trajectory $s^*(t)$. Let $C(t)$ be a 'combination matrix', with dimension $m \times k$, such that the matrix $CL^T$ is of full rank $m$ along $q_r(t)$. The main task may then be written ([10]):

$$e_1 = C(t)\left(s(\bar{r}, t) - s^*(t)\right) \tag{20}$$

One of the advantages of the existence of $C$ is the possibility of taking into account more sensors $(k)$ than the actual dimension of the constraints they specify $(m)$.

The jacobian matrix of $e_1$ in $SE_3$ is then $J_1^{\bar{r}} = \frac{\partial e_1}{\partial \bar{r}} = CL^T$ and we may easily show that $R(J_1^{\bar{r}^T}) = R(L)$. Owing to (15), the task to be regulated may finally be written:

$$e = W^+ C\left(s(\bar{r}, t) - s^*(t)\right) + \alpha\left(\mathbf{I_6} - W^+ W\right)g_s^{\bar{r}^T} \tag{21}$$

$W$ must ideally satisfy property (14), which then becomes $R\left(W^T\right) = R(L)$ ; this also means that the rows of $W$ are made from basis vectors of $S$. Finally, property (16) becomes

$$CL^T W^T > 0 \tag{22}$$

which prevents $C$ and $W$ from being chosen independently. For example, (22) may be satisfied by selecting: $C = WL$ or $C = WL^{T^g}$ where $L^{T^g}$ is the generalized inverse of $L^T$. In many cases, $C$ and $W$ may be chosen constant, as assumed in the following.

### 3.3.3 Model of $\frac{\partial e}{\partial t}$

In the control equation (11), an expression of the term $\frac{\widehat{\partial e}}{\partial t}$ is needed. Considering again the task function given by (21), we then have:

$$\frac{\partial e}{\partial t} = W^+ \frac{\partial e_1}{\partial t} + \alpha\left(\mathbf{I_n} - W^+ W\right)\dot{g}_s^{\bar{r}^T} \tag{23}$$

Vector $\frac{\partial e_1}{\partial t}$ represents, when $C$ is constant, the contribution of a possible autonomous target motion and is in general unknown. The choice made in many cases is $\frac{\widehat{\partial e_1}}{\partial t} = 0$. If the target moves, this choice may lead to a tracking error, the size of which decreases with $\lambda$. On the other hand, if, as in trajectory tracking, the used secondary cost function allows to know $\dot{g}_s^{\bar{r}}$, we may choose:

$$\frac{\widehat{\partial e}}{\partial t} = \alpha\left(\mathbf{I_n} - W^+ W\right)\dot{g}_s^{\bar{r}^T} \tag{24}$$

# 4 Case of a Visual Sensor

## 4.1 Framework

Let us reduce a camera to a perspective projection model (Figure 1). All the used variables, for example the camera velocity screw $(v(O), \omega)$, denoted in the following as $V_c$, the point coordinates, the interaction screws, will be assumed to be expressed in the frame $(O, \vec{x}, \vec{y}, \vec{z})$ linked to the camera.

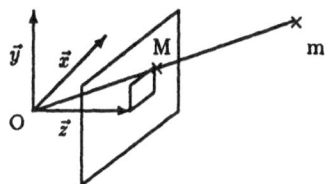

Figure 1: A simple camera model

Without loss of generality, the focal length is assumed to be equal to 1, such that any point $m$ of $E$ with coordinates $\underline{x} = (x\,y\,z)^T$ is projected on the image plane as a point $M$ with coordinates $\underline{X} = (X\,Y\,1)^T$ with:

$$\underline{X} = \frac{1}{z}\,\underline{x} \tag{25}$$

- **Remark:** It would seem to be necessary to complete this geometrical model with a photometric model. Some arguments given in [3] justify the absence of such a model in the present development.

Let us consider a single rigid solid in $E$, to which tridimensional primitives (points, lines, vertices...) may be associated. A set of such primitives is called a 'scene feature'. A configuration of the scene feature is an element $p$ of the set $\mathcal{P}_s$ of all possible configurations. When a scene feature fully characterizes the position (location and attitude) of the associated rigid body, the dimension $n'$ of $\mathcal{P}_s$ is 6. Otherwise, it is smaller.

Let us denote as $f$ the perspective projection mapping, with $f(p) = P \in \mathcal{P}_i$ ; $\dim \mathcal{P}_i = m < n'$. $P$ is called an 'image feature'. It is assumed that $p$ belongs to an open subset $U \subseteq \mathcal{P}_s$ such that $P = f(p)$ is not a degenerated element of $\mathcal{P}_i$ (case where, for example, a line projects onto the image plane as a point, a circle as a segment,...). This restriction implies that it exists a complete parametrization $\underline{P}$ of $P$ on the open set $V = f(U) \subseteq \mathcal{P}_i$. Moreover, we assume that this parametrization is differentiable and minimal (the dimension of $\underline{P}$ is $m$).

Let $\underline{p}$ a complete and unique parametrization of $p$ in $U$ (in the sense that a single parametrization is necessary and sufficient to represent any configuration of the scene feature $p$ in $U$. The dimension of $p$ is $n \geq n'$. If we write $\underline{p} = \phi(p)$ and $\underline{P} = \psi(P)$, we have (see Figure 2):

$$\underline{P} = \psi \circ f \circ \phi^{-1}(\underline{p}) \qquad (26)$$

$$
\begin{array}{ccccc}
W \subseteq SE3 & \xrightarrow{\phantom{xx}} & U \subseteq P_s & \xrightarrow{\phantom{xx}} & V \subseteq P_i \\
(\bar{r}) & \delta & (p) & f & (P) \\
 & & \Big\downarrow \phi & & \Big\downarrow \psi \\
 & & \mathbb{R}^n & \xrightarrow{\phantom{xx}} & \mathbb{R}^m & \xrightarrow{\phantom{xx}} & \mathbb{R} \\
 & & (\underline{p}) & \psi \circ f \circ \phi^{-1} & (\underline{P}) & \sigma & (s)
\end{array}
$$

Figure 2: From $\bar{r}$ to $s$

Finally the group of displacements acts on $P_s$ through the mapping $\delta$ and we assume that the displacements are restricted to an open set $W$ of $SE_3$ such that $\delta(W) \subseteq U$ in order to prohibit any degenerated case. We then have:

$$\underline{p} = \phi \circ \delta(\bar{r}) \qquad (27)$$

whence:

$$\underline{P} = \psi \circ f \circ \delta(\bar{r}) \qquad (28)$$

The components $s_j$ of the 'signal sensor' will then be chosen as a function $s_j = \sigma(\underline{P})$ such that $\sigma$ is differentiable, the most frequent situation being to choose for $s_j$ a component of $\underline{P}$.

The derivation of the interaction matrix reduces therefore to the computation of the expression $\frac{\partial s}{\partial \underline{P}} \frac{\partial \underline{P}}{\partial P} \frac{\partial P}{\partial p} \frac{\partial p}{\partial \bar{r}}$ , and, often practically to the computation of $\frac{\partial s}{\partial \underline{P}} \frac{\partial \underline{P}}{\partial p} \frac{\partial p}{\partial \bar{r}}$. This form is not always the most adequate for an analytic computation of $L^T$, and another method may sometimes be preferred which allows to directly obtain $\frac{\partial \underline{P}}{\partial \bar{r}}$.

Indeed, the $i^{th}$ primitive of a scene feature may generally be described by an equation of type:

$$h_i(\underline{x}, \underline{p}_i) = 0 \qquad (29)$$

where $h_i$ defines the kind of the primitive and the value of $\underline{p}_i$ corresponds to one of its configurations.

In practice, the interaction matrix associated to the parametrization $\underline{P}_i$ of each primitive is computed, and the global interaction matrix associated to $\underline{P}$ is obtain by concatenation of all the elementary interaction matrices. In the sequel, $h_i$, $p_i$, $\underline{p}_i$, $P_i$ and $\underline{P}_i$ related to the $i^{th}$ primitive will be respectively denoted $h$, $p$, $\underline{p}$, $P$ and $\underline{P}$. By using (25), equation (29) becomes:

$$h(z\underline{X}, \underline{p}) = 0 \tag{30}$$

i.e.

$$h'(\underline{X}, z, \underline{p}) = 0 \tag{31}$$

Under the condition $\frac{\partial h'}{\partial z} \neq 0$ which is ensured in all the non-degenerated cases, the implicit function theorem ensures the existence of a unique function $\mu$ around a solution $\underline{x}_0$ of (31), such that:

$$z = \mu(\underline{X}, \underline{p}) \tag{32}$$

Let us give two simple examples of basic primitives. For each of them, we give the function $\mu$ and the projection in the image of $h(\underline{x}, \underline{p}) = 0$, which will be written under the form:

$$g(\underline{X}, \underline{P}) = 0 \tag{33}$$

- *Case of a point:* Let $m_i$ be a point of $E$ with coordinates $\underline{x}_i = (x_i \, y_i \, z_i)^T$. We have:

$$h(\underline{x}, \underline{p}) = \begin{cases} h_1 = x - x_i = 0 \\ h_2 = y - y_i = 0 \\ h_3 = z - z_i = 0 \end{cases} \tag{34}$$

In this simple case, $\mu$ is obtained from $h_3$: $z = z_i$ and $g$ by $g(\underline{X}, \underline{P}) = \begin{cases} X - X_i = 0 \\ Y - Y_i = 0 \end{cases}$ where $X_i$ and $Y_i$ are the coordinates of the projection of $m_i$ in the image.

- *Cases of a straight line and of plane primitives:* $h$ is then two-dimensional ($h = (h_1 \, h_2)^T$). We may for example choose the equation of the plane in which the primitive lies as a function $h_2$ (in the line case, $h_2$ is not unique). $\mu$ is then obtained from $h'_2(\underline{X}, z, \underline{p}) = 0$. With (32), $h'_1(\underline{X}, z, \underline{p}) = 0$ gives $\tilde{h}(\underline{X}, \underline{p}) = 0$, with dim $\tilde{h} = 1$, which, after change of parametrization, may be written $g(\underline{X}, \underline{P}) = 0$.

The case of tri-dimensional primitives is treated in [3], [1].

Knowing that the rigidity assumption implies $\dot{g} = 0$, $\forall \underline{X} \in P$, we may now compute the interaction matrix $L^T$ associated to $\underline{P}$. Differentiation of (33) gives:

$$\frac{\partial g}{\partial \underline{P}}(\underline{X}, \underline{P}) \, \dot{\underline{P}} = -\frac{\partial g}{\partial \underline{X}}(\underline{X}, \underline{P}) \, \dot{\underline{X}}, \, \forall \underline{X} \in P \tag{35}$$

Differentiating (25) leads to the wellknown optic flow equations, which may be written:

$$\dot{\underline{X}} = L_{of}^T(\underline{X}, z) \, V_c \tag{36}$$

where $V_c$ is the velocity of the camera with respect to the scene, and with:

$$L_{of}^T = \begin{pmatrix} -1/z & 0 & X/z & XY & -(1+X^2) & Y \\ 0 & -1/z & Y/z & 1+Y^2 & -XY & -X \end{pmatrix} \quad (37)$$

Using (32) in (37), gives:

$$L_{of}^T(\underline{X}, z) = L_{of}^{\prime T}(\underline{X}, \underline{p}) \quad (38)$$

Finally, (35), (36) and (38) lead to:

$$\frac{\partial g}{\partial \underline{P}}(X, \underline{P}) \, \dot{\underline{P}} = -\frac{\partial g}{\partial \underline{X}}(X, \underline{P}) \, L_{of}^{\prime T}(\underline{X}, \underline{p}) \, T \;, \; \forall \underline{X} \in P \quad (39)$$

This equation may be solved, either by explicitly using (33) in (39) and identifying, or by choosing $m$ points $\bar{X}_i$ of $Q$ and solving (see [1]).

## 4.2 Two examples of usual features

### 4.2.1 Points

Consider a point $m_i$ of $E$ with coordinates $\underline{x}_i$ ; then, $\underline{p} = (x_i\,y_i\,z_i)$ and $\underline{P} = (X_i\,Y_i)$. The matrix forms $L_{X_i}$ and $L_{Y_i}$ of the two interaction screws $H_{X_i}$ and $H_{Y_i}$ , expressed in $F$ and in $O$, are given by (37) and may be written:

$$\begin{array}{ll} L_{X_i}^T = [ & -1/z_i & 0 & X_i/z_i & X_iY_i & -(1+X_i^2) & Y_i & ] \\ L_{Y_i}^T = [ & 0 & -1/z_i & Y_i/z_i & 1+Y_i^2 & -X_iY_i & -X_i & ] \end{array} \quad (40)$$

Various sensor signals may be generated from image points, as described in [3].

### 4.2.2 Straight lines

A straight line in $E$ may be represented as the cross-section of two planes:

$$h(\underline{x}, \underline{p}) = \begin{cases} a_1x + b_1y + c_1z + d_1 = 0 \\ a_2x + b_2y + c_2z + d_2 = 0 \end{cases} \quad (41)$$

If we exclude the degenerated case where the projection centre belongs to the straight line ($d_1 = d_2 = 0$), the equation of the projected line in the image plane is:

$$AX + BY + C = 0 \text{ with } \begin{cases} A = a_1d_2 - a_2d_1 \\ B = b_1d_2 - b_2d_1 \\ C = c_1d_2 - c_2d_1 \end{cases} \quad (42)$$

Since the parametrization $(A, B, C)$ of 2D straight lines is not minimal, another one should be preferred. The most used, $\underline{P} = (a, b)$, is inadequate because two charts non compatible ($Y = aX + b$, $X = aY + b$) have to be used. We therefore choose $\underline{P} = (\rho, \theta)$ and the equation of a straight line $\mathcal{D}$ is then:

$$g(\underline{X}, \underline{P}) = X \cos \theta + Y \sin \theta - \rho = 0 \quad (43)$$

- **Remark:** The ambiguity on the representation $(\rho, \theta)$ may be easily overcome.

It may then be shown that the related interaction screws are:

$$L_\theta^T = [\; \lambda_\theta \cos\theta \quad \lambda_\theta \sin\theta \quad -\lambda_\theta\rho \quad -\rho\cos\theta \quad -\rho\sin\theta \quad -1 \;]$$
$$L_\rho^T = [\; \lambda_\rho \cos\theta \quad \lambda_\rho \sin\theta \quad -\lambda_\rho\rho \quad (1+\rho^2)\sin\theta \quad -(1+\rho^2)\cos\theta \quad 0 \;]$$

(44)

$$\text{with} \quad \lambda_\theta = (a_2 b_1 - a_1 b_2)/\sqrt{A^2 + B^2}$$
$$\text{and} \quad \lambda_\rho = [(c_2 a_1 - c_1 a_2)\cos\theta + (c_2 b_1 - c_1 b_2)\sin\theta]/\sqrt{A^2 + B^2}$$

Other more complex cases (circles, spheres...) are examined in [1].

## 4.3  Vision-based Task Design

### 4.3.1  On the Model of the Interaction Matrix

It has been seen that, when data provided by a mobile camera are used as sensor signals, the associated interaction matrices $L^T$ take the form $L^T(\underline{p}, \underline{P})$ where $\underline{P} = \underline{P}(\bar{r}, t)$ may be measured in the image, and where $\underline{p} = \underline{p}(\bar{r}, t)$ represents the 3D information coming from the considered primitives. Since this last information is a-priori unknown, it is necessary to choose a model $\hat{L}$ of $L$. The task vector (21) has therefore to be derived by choosing $C = W\hat{L}$ or $C = W\hat{L}^{T\theta}$. Properties (14) and (16) will thus be satisfied if we may ensure that, respectively, $R\left(\hat{L}\right) = R(L)$ and $CL^TW^T > 0$. Recall that the interest of satisfying these properties lies in the possibility of simply choosing the identity matrix as a model of $\frac{\partial e}{\partial \bar{r}}$. Several possibilities exist:

- $\hat{L} = L(\hat{p}, \underline{P})$ when $\underline{p}$ may be concurrently estimated.

- $\hat{L} = L(\underline{p}^*, \underline{P})$ where $\underline{p}^*$ is the value of $\underline{p}$ at $s = s^*$ which realizes $e_1 = 0$. Some further assumptions are then needed. We will come back to this point later.

- $\hat{L} = L(\hat{\underline{p}}^*, \underline{P})$ where $\hat{\underline{p}}^*$ is an estimate of $\underline{p}^*$ when no 3D information is available.

The above choices need the matrix $C$ to be updated at the same rate as the control loop. This may be difficult, for example when $C$ is chosen equal to $W\hat{L}^{T\theta}$, because of the computing time required by the computation of generalized inverses. It is also sometimes necessary to anticipate the possible crossing of an isolated singularity. A simpler solution consists in using a *constant* model $\hat{L}$, determined within the task design step. It may then be chosen $\hat{L} = L(\underline{p}^*, \underline{P}^*)$, also denoted as $L_{|s=s^*}$, which is the value of the interaction matrix at the position corresponding to the selected feature, $s = s^*$.

The positivity condition (22) is then often only satisfied in the neighbourhood of the desired position $s = s^*$, whatever the choice of $C$. Fortunately, it should be emphasized that this condition is only *sufficient*, and, in practical experiments, the convergence of the control law was always obtained even from initial conditions far away from the goal position. This choice of $\hat{L}$ requires the knowledge of $\underline{p}^*$, which is equivalent to making assumptions about the geometry of the 3D scene. Such assumptions may often be done when the task is being defined, and seem then not too strong: for example, if the task consists in positioning the camera in front of a door, it may be assumed that there is a door in the scene, and that characteristic signals of this door (for example its 4 corners) may be extracted. In addition, if it is desired to place the camera at a given range of the door, its dimensions should be known in order to determine the goal feature in the image.

However, in some cases, the form assumption is the only condition required for the image feature determination (in the previous case this means that the obtained range would be indifferent to the user). Then, if $L^T(\underline{p}^*, \underline{P}^*)$, where the value of $\underline{p}^*$ is unknown, may be written as $L^T(\underline{p}^*, \underline{P}^*) = B^T(\underline{P}^*)\, D(\underline{p}^*, \underline{P}^*)$ where $D$ is a $n \times n$ positive matrix, and where $B$ and $L$ are of same rank, we may choose $\hat{L} = B(\underline{P}^*)$ when the positivity condition is satisfied around $s^*$, for example when $D$ is diagonal or when $B$ is of full rank.

Finally, when the previous conditions are not fulfilled, or if no scene knowledge is available (for example when it is wished to track an unknown object with a goal image feature extracted from an initial image of the object), it may be chosen $\hat{L} = L(\hat{\underline{p}}^*, \underline{P}^*)$ where $\hat{\underline{p}}^*$ is an estimate of $\underline{p}^*$, not necessarily very accurate, but ensuring the asymptotic stability of the controlled system in some neighbourhood of $s = s^*$. Ensuring the positivity condition to be satisfied is then difficult, even when $s = s^*$, since the value $L_{|s=s^*}$ remains unknown.

In the experiments we have conducted, form and dimensional assumptions were done. Therefore, $\hat{L} = L_{|s=s^*}$. Furthermore, the matrix $C$ was always chosen equal to $W\hat{L}^{T^g}$, where $W$ is such that $R\left(W^T\right) = R(\hat{L})$, because of better obtained decoupling properties than with the choice $C = W\hat{L}$.

# 5    Results and Concluding Remarks

Several examples, obtained in simulation or with an experimental testbed, are reported in [3] and [1]. We only give here a simple illustration of the proposed approach.

Let us consider a task aimed to position a camera with respect to a 'road', which is symbolized by three parallel straight lines in a plane (lateral and central white bands). The goal position is such that the camera lies at a height $y^*$ at the middle of the right lane and that the camera axis $\vec{z}$ coincides with its direction and its axis $\vec{y}$ is vertical.

By using equations (41) and (43), fonctions $h(\underline{x}, \underline{p})$ and $g(\underline{X}, \underline{P})$ associated to the three lines are immediatly obtained:

$$h_1(\underline{x},\underline{p}^*) : \begin{cases} y + y^* = 0 \\ x + l/4 = 0 \end{cases} \Rightarrow \begin{cases} \theta_1^* = \arctan(-l/4y^*) \\ \rho_1^* = 0 \end{cases} \tag{45}$$

$$h_2(\underline{x},\underline{p}^*) : \begin{cases} y + y^* = 0 \\ x - l/4 = 0 \end{cases} \Rightarrow \begin{cases} \theta_2^* = \arctan(l/4y^*) \\ \rho_2^* = 0 \end{cases} \tag{46}$$

$$h_3(\underline{x},\underline{p}^*) : \begin{cases} y + y^* = 0 \\ x - 3l/4 = 0 \end{cases} \Rightarrow \begin{cases} \theta_3^* = \arctan(3l/4y^*) \\ \rho_3^* = 0 \end{cases} \tag{47}$$

The sensor signals to be selected for describing this task are the parameters which represent the three lines: $s = (\theta_1, \rho_1, \theta_2, \rho_2, \theta_3, \rho_3)$. Therefore: $s^* = (\theta_1^*, 0, \theta_2^*, 0, \theta_3^*, 0)$. Furthermore, the interaction matrix associated with $s^*$ may be easily derived from (44):

$$L_{|s=s^*}^T = \begin{pmatrix} -\cos^2\theta_1^*/y^* & -\cos\theta_1^*\sin\theta_1^*/y^* & 0 & 0 & 0 & -1 \\ 0 & 0 & 0 & \sin\theta_1^* & -\cos\theta_1^* & 0 \\ -\cos^2\theta_2^*/y^* & -\cos\theta_2^*\sin\theta_2^*/y^* & 0 & 0 & 0 & -1 \\ 0 & 0 & 0 & \sin\theta_2^* & -\cos\theta_2^* & 0 \\ -\cos^2\theta_3^*/y^* & -\cos\theta_3^*\sin\theta_3^*/y^* & 0 & 0 & 0 & -1 \\ 0 & 0 & 0 & \sin\theta_3^* & -\cos\theta_3^* & 0 \end{pmatrix} \tag{48}$$

$L_{|s=s^*}^T$ is always of rank 5, and Ker $L_{|s=s^*}^T = (0\,0\,1\,0\,0\,0)^T$.

Let us now apply the previous approach to the derivation of $e$. The following $5 \times 6$ matrix may be chosen as a matrix $W$:

$$W = \begin{pmatrix} 1 & 0 & 0 & 0 & 0 & 0 \\ 0 & 1 & 0 & 0 & 0 & 0 \\ 0 & 0 & 0 & 1 & 0 & 0 \\ 0 & 0 & 0 & 0 & 1 & 0 \\ 0 & 0 & 0 & 0 & 0 & 1 \end{pmatrix} \tag{49}$$

The combination matrix $C$ is chosen equal to $WL_{|s=s^*}^{T_s}$ and, by using (21), the following task vector $e$ is obtained:

$$e = W^+WL_{|s=s^*}^{T_s}(s(\bar{r},t) - s^*) + \alpha\left(I_6 - W^+W\right)g_s^{rT} \tag{50}$$

The secondary task may consist in specifying a time trajectory along $\vec{z}$, for example a constant velocity $v$. The associated secondary cost to be minimized is $h_s = \frac{1}{2}(z(t) - z_0 - vt)^2$ with $z(0) = z_0$. Therefore $g_s^r = (0\ 0\ (z(t) - z_0 - vt)\ 0\ 0\ 0)$. Note that tasks $e_1$ and $e_2 = z(t) - z_0 - vt$ are then compatible and independent since:

$$e = \begin{pmatrix} 1 & 0 & 0 & 0 & 0 & 0 \\ 0 & 1 & 0 & 0 & 0 & 0 \\ 0 & 0 & 0 & 0 & 0 & 0 \\ 0 & 0 & 0 & 1 & 0 & 0 \\ 0 & 0 & 0 & 0 & 1 & 0 \\ 0 & 0 & 0 & 0 & 0 & 1 \end{pmatrix} L_{|s=s^*}^{T_s}(s(\bar{r},t) - s^*) + \alpha\begin{pmatrix} 0 \\ 0 \\ z(t) - z_0 - vt \\ 0 \\ 0 \\ 0 \end{pmatrix} \tag{51}$$

Figure 3 gives an example of the obtained behaviour. Left and right top windows show respectively initial and final positions of the camera (symbolized by a pyramid) with respect to the target. Middle windows represent the associated images. On bottom windows, the time variation of $\|s(\bar{r}, t) - s^*\|$ and of the components of $T_c$ are respectively plotted. Finally, Figure 4 presents a sequence of real images and the obtained plots corresponding to the positioning of a camera handled by a six jointed robot with respect to a 4-point target.

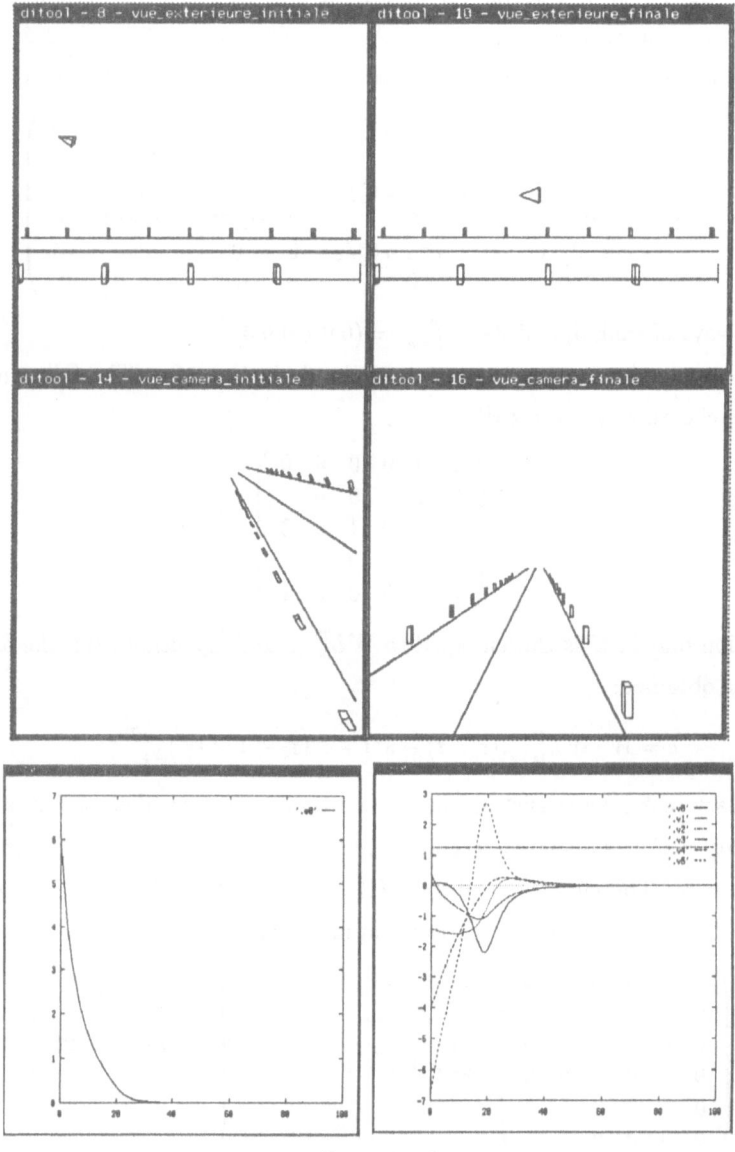

Figure 3: Road following

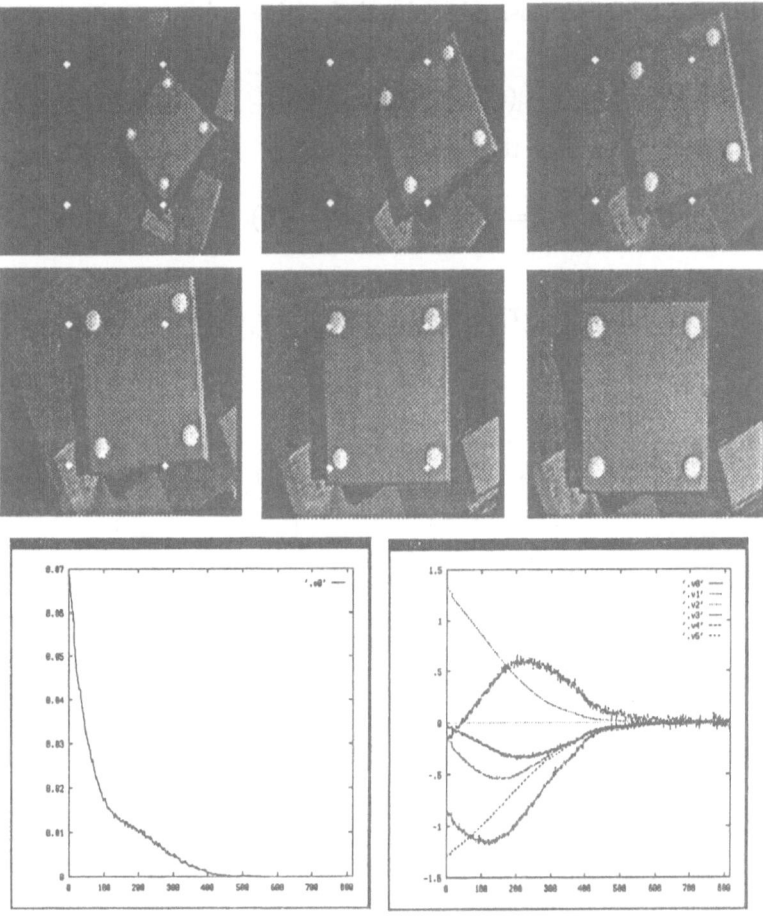

Figure 4: Positioning task

One possible development of this work lies in the use of an adaptive approach of the control scheme. Indeed, we may consider that intrinsic system parameters (inertia, kinematics, camera parameters...) which are not liable to large variations, may be computed or estimated off-line. On the contrary, uncertainties on the environment, which have a strong influence on the control behaviour, have to be considered carefully. In the present case, an example of uncertainty is the own object velocity, which contributes to the term $\frac{\partial e}{\partial t}$. Setting it equal to zero, as done here, may lead to tracking errors all the more large that the gains are small. Since the system is very sensitive to this parameter, an on-line estimation of the task vector velocity or of the object velocity within an indirect adaptive control approach would be useful.

Indeed, let us consider the vision-based-task function (with a constant combination matrix):

$$e(q,t) = C \left( s(q,t) - s^* \right) \tag{52}$$

Let $\mathcal{T}$ the object velocity, we then have:

$$\dot{e} = CL^T(q,t) \left( J(q)\,\dot{q} - \mathcal{T} \right) \tag{53}$$

and

$$\ddot{e} = CL^T(q,t)\,J(q)\,\ddot{q} + f(q,\dot{q},\mathcal{T},\dot{\mathcal{T}},t) \tag{54}$$

with

$$f(q,\dot{q},\mathcal{T},\dot{\mathcal{T}},t) = CL^T(q,t) \left( \begin{bmatrix} \vdots \\ \dot{q}^T \frac{\partial J_i^T}{\partial q}\,\dot{q} \\ \vdots \end{bmatrix} - \dot{\mathcal{T}} \right) + C \left[ \begin{matrix} \vdots \\ (J(q)\dot{q} - \mathcal{T})^T \frac{\partial L_i}{\partial \bar{r}}\,(J(q)\dot{q} - \mathcal{T}) \\ \vdots \end{matrix} \right]$$

where $J_i$ and $L_i (i = 1, \cdots, n)$ are the i-th row of $J(q)$ and $L(q,t)$.

We can use the control scheme (11) with a model $\hat{L}$ of the interaction matrix (see section 4.3.1) and with:

$$\widehat{\frac{\partial e}{\partial q}} = C\hat{L}^T J(q) \quad , \quad \widehat{\frac{\partial e}{\partial t}} = -C\hat{L}^T \hat{\mathcal{T}} \tag{55}$$

and

$$\hat{f} = C\hat{L}^T \left( \begin{bmatrix} \vdots \\ \dot{q}^T \frac{\partial J_i^T}{\partial q}\,\dot{q} \\ \vdots \end{bmatrix} - \hat{\dot{\mathcal{T}}} \right) + C \left[ \begin{matrix} \vdots \\ (J(q)\dot{q} - \hat{\mathcal{T}})^T \frac{\partial \hat{L}_i}{\partial \bar{r}}\,(J(q)\dot{q} - \hat{\mathcal{T}}) \\ \vdots \end{matrix} \right] \tag{56}$$

where $\hat{\mathcal{T}}$ and $\hat{\dot{\mathcal{T}}}$ which appears in $\hat{f}$ may be obtain by estimation algorithms. A simple case, where $\mathcal{T}$ is assumed constant, is treated in [1].

# References

[1] F. Chaumette: *La relation vision-commande: théorie et application à des tâches robotiques*, Thèse de l'Université de Rennes I, Juillet 1990.

[2] P. I. Corke, R. P. Paul: *Video-Rate Visual Servoing for Robots*, First International Symposium on Experimental Robotics, Montreal, Canada, June 1989.

[3] B. Espiau, F. Chaumette, P.Rives: *Une nouvelle approche de la relation vision-commande en robotique*, INRIA Research report no 1172, March 1990; also submitted to IEEE Trans. on Robotics and Automation.

[4] B. Espiau, J.P. Merlet, C. Samson: *Force Feedback Control and Non-contact Sensing: a Unified Approach*, 8th CISM-IFTOMM Symposium on Theory and Practice of Robots and Manipulators, 2-6 july 1990, Cracow-Poland.

[5] J. T. Feddema, C. S. G. Lee and O. R. Mitchell: *Automatic selection of image features for visual servoing of a robot manipulator*, Conf. IEEE Robotics and Automation, Scottsdale, Arizona, USA, May 14-19, 1989.

[6] J. T. Feddema, O. R. Mitchell: *Vision-Guided Servoing with Feature-Based Trajectory Generation*, IEEE Transaction on Robotics and Automation, Vol. 5, n.5, October 1989.

[7] M. Kabuka, E. McVey, P. Shironoshita: *An Adaptive Approach to Video Tracking*, IEEE Journal of Robotics and Automation, 4(2):228-236, April 1988.

[8] C. Samson: *Une approche pour la synthèse et l'analyse de la commande des robots manipulateurs*, Rapport de recherche INRIA, n.669, Mai 1987.

[9] C. Samson, B. Espiau, M. Le Borgne: *Robot Redundancy: an Automatic Control Approach*, NATO Advanced Research Workshop on Robots with Redundancy, Salo, Italia, Juin 1988.

[10] C. Samson, B. Espiau: *Application of the Task Function Approach to Sensor-Based-Control of Robot Manipulators*, IFAC, Tallin, USSR, July 1990.

[11] C. Samson, B. Espiau, M. Le Borgne: *Robot Control: the Task Function Approach*, Oxford University Press, 1990.

[12] A. C. Sanderson, L. E. Weiss: *Adaptive Visual Servo Control of Robots*, Reprinted in *Robot Vision*, A. Pugh, Ed. Bedford, UK:IFS Pub. Ltd., 1983.

[13] L. E. Weiss: *Dynamic Visual Servo Control of Robots. An Adaptive Image based Approach*, Technical Report, CMU-RI-TR-84-16; Carnegie Mellon, 1984.

[14] L. E. Weiss, A. C. Sanderson: *Dynamic Sensor-Based Control of Robots with Visual Feedback*, IEEE Journal of Robotics and Automation, Vol. RA-3, n. 5, Oct. 1987.

# ARTIFICIAL IMPEDANCE APPROACH OF THE TRAJECTORY GENERATION AND COLLISION AVOIDANCE FOR SINGLE AND DUAL ARM ROBOTS

by

D.S. Necsulescu
Department of Mechanical Engineering
University of Ottawa
Ottawa, K1N 6N5 Canada

## I.  INTRODUCTION

Among the desired features for robot manipulation for extending the field of future applications, are autonomous trajectory generation and automatic collision avoidance with known or unexpected objects in the work volume.  The artificial mechanical impedance method analyzed in this paper is based on the modulation of the torques in such a way as to produce, between the target and the manipulator endpoint, the dynamic effect of a desired linear or nonlinear mechanical impedance.  The result of using this method is the integration of the controller model with the nonlinear dynamics model of the robot into a single mechanical model which is used to obtain the desired mechanical properties.  The Artificial Potential Field control schemes are formulated as continuum mechanical problems and so require equations describing the geometric shapes of objects in the robot work volume in continuum mechanical form [1],[2],[9].  The actual application of these schemes is based on lumped parameter control laws which suggest that it is perhaps more appropriate to consider a lumped parameters formu- lation of the dynamics in robot Cartesian space and Artificial Mechanical Impedances [3],[4].  These control schemes were developed for single arm robots.

Dual-arm robot motion control requires a higher level of controller complexity compared to the single-arm robot motion control because of the need to continuously coordinate the motion of the two arms.

Various combinations of position and/or force control of the two-arm motion have been reported in the literature, [11] to [14].

The literature review shows that the dual-arm motion coordination results to-date cover only a limited variety of types of payloads and do not include such payloads as a string/blanket or a vibrating structure. Also, collision avoidance for dual-arm robots has not been analyzed in detail.

In this paper, artificial impedance method is used for the trajectory generation of a jointed manipulator in free motion and for the trajectory correction in order to avoid collisions with obstacles. For obstacle avoidance the artificial impedance approach is used in conjunction with a 'coastal navigation' scheme which provides a means of generating trajectories in non-convex spaces utilizing proximity-sensor data. The technique has the potential of permitting operation within an unknown workspace and of providing trajectory correction to avoid collisions with unexpected or moving objects. Simulation results illustrate this technique.

Dual arm robot motion is analyzed for two distince phases, the approaching phase, when each arm is moving independently toward the desired positions, and the coordination phase, when the two arms are transporting together an object to a desired position. The motion in the approaching phase will be coordinated by introducing artificial attractive impedances between each arm and the desired positions (to generate attractive forces between the arms end points and the targets) and artificial repulsive impedances between the arms and obstacles (to generate artificial repulsive forces and avoid the collisions of the moving arms) in the case of the detection of obstacles on the arms trajectories. The motion in the second phase, when the two arms are holding a rigid object, represents the motion of a closed loop mechanism. Artificial impedance approach to motion coordination, in this case, is based on designating one arm as a leader arm and the other as a follower arm.

The paper presents the analytical model of the impedance based control of a dual arm robot, as well as simulations of the operation of the impedance based controller in approaching and coordination phases of the motion of the two arms. The objects will be assumed in simulations as rigid bodies, elastic bodies and as strings.

In the last part of the paper, implementation aspects of the artificial impedance approach are analyzed.

## II. ARTIFICIAL IMPEDANCE APPROACH OF THE INTERCEPTION OF A MOVING TARGET--TRANSLATIONAL MOTION CASE

### Motion control of a frictionless rigid body [5].

The Artificial Impedance approach can be conceptually illustrated for a simple position control problem. We assume a mass m, (positioned at x) separated from a moving target (positioned at $x_d$) by an artificial elastic spring with stiffness K in parallel with an artificial damper with coefficient B. A term $M\ddot{x}$, where M is a virtual mass, is also included.

The equation of contact motion of the mass is given by

$$f = m\ddot{x} + f_{ext} \tag{1}$$

where $f_{ext}$ is the contact force.

An artificial impedance between x and $x_d$ would be described in this case by

$$- f_{ext} = M\ddot{x} + B(\dot{x}-\dot{x_d}) + K(x-x_d) \tag{2}$$

When the motion of the mass m is impeded by an obstacle continuously in contact with the mass the motion of the obstacle is imposed on the motion of the mass, $x(t)=x_1(t)$ where $x_1(t)$ is the coordinate of the obstacle. The motion of the obstacle is assumed faster than the motion of the target so that eventual interception is possible. The contact force between the mass with the obstacle produces $f_{ext}$ and the states $x(t)$ and $\dot{x}(t)$ are measured. We obtain from Eq. (2), the computed acceleration

$$\ddot{x}^{(c)} = M^{-1}B(\dot{x_d}-\dot{x}) + M^{-1}K(x_d-x) - M^{-1} f_{ext} \tag{3}$$

Substituting $\ddot{x}$ in (1) by $\ddot{x}^{(c)}$ we obtain

$$f = m\left\{\left[M^{-1}B(\dot{x_d}-\dot{x}) + M^{-1}K(x_d-x)\right] -M^{-1} f_{ext}\right\}+f_{ext} \tag{4}$$

The control law (4) permits the calculation of the force command f for given measurements of $x$, $\dot{x}$ and $f_{ext}$ and known values for $x_d$ and $\dot{x_d}$. For M=m, the coefficient of $f_{ext}$, $1-mM^{-1}=0$, and the contact force does not influence position control command any more.

## III. THE MODEL AND SIMULATION RESULTS FOR A SINGLE ARM ROBOT

The model [5],[7]

Assuming that the measurement of Cartesian contact forces 6x1 vector $F_{ext}$ between the end point of a 6DOF jointed robot and the environment is available, the dynamic equation in terms of 6x1 vector $\theta$ of joint angles gives the following 6x1 vector $\tau$ of joint torques

$$\tau = M(\theta)\ddot{\theta} + V(\theta,\dot{\theta}) + J^T(\theta)F_{ext}. \tag{5}$$

The term $M(\theta)$ is the configuration-dependent matrix of inertias.

Here $V(\theta,\dot{\theta})$ includes Coriolis, centrifugal, gravitational and frictional terms and $J(\theta)$ is the robot Jacobian matrix.

Specific to the Artificial Impedance approach is the choice of a desired artificial impedance between the end point, positioned at $X$(a 6x1 vector) and the target positioned at $X_d$ (a 6x1 vector), producing the following desired dynamics

$$-F_{ext} = M\ddot{X} + B(\dot{X}-\dot{X}_d) + K(X-X_d). \tag{6}$$

M, B and K are 6x6 diagonal matrices of the desired artificial impedance in Cartesian space. From robot kinematics, the following relationships between state variables in joint space and in Cartesian endpoint space can be written [6]

$$X = KIN(\theta) \tag{7}$$

$$\dot{X} = J(\theta)\dot{\theta} \tag{8}$$

$$\ddot{X} = J(\theta)\ddot{\theta} + \dot{J}(\theta)\dot{\theta} . \tag{9}$$

where the Jacobian 6x6 matrix $J(\theta)$ relates joint velocities $\dot{\theta}$ to base frame defined endpoint velocities $\dot{X}$. Using Eqs. (7) to (9) in Eq. (6) the computed acceleration is given by

$$\ddot{\theta}^{(c)} = -J^{-1}(\theta)M^{-1}F_{ext}-J^{-1}(\theta)\dot{J}(\theta)\dot{\theta}+J^{-1}(\theta)M^{-1}B[\dot{X}_d-J(\theta)\dot{\theta}]$$

$$+ J^{-1}(\theta)M^{-1}K[X_d - KIN(\theta)] . \tag{10}$$

Substituting $\ddot{\theta}$ in eq. (5) by $\ddot{\theta}^{(c)}$ we obtain

$$\tau = M(\theta)J^{-1}(\theta)\{M^{-1}K[X_d-KIN(\theta)]+M^{-1}B[\dot{X}_d-J(\theta)\dot{\theta}]-\dot{J}(\theta)\dot{\theta}-M^{-1}F_{ext}\}$$

$$+ V(\theta,\dot{\theta}) + J^T(\theta)F_{ext} \tag{11}$$

Using the notations $M_x$ for the Cartesian mass matrix and $V_x$ for the velocity dependent and gravitational terms,[6], we obtain

$$M_x(\theta) = J^{-T}(\theta)M(\theta)J^{-1}(\theta) \tag{12}$$

and

$$V_x(\theta,\dot{\theta}) = J^{-T}(\theta)[V(\theta,\dot{\theta})-M(\theta)J^{-1}(\theta)\dot{J}(\theta)\dot{\theta}] \tag{13}$$

Using the notations (12) and (13) in (11) we obtain

$$\tau = J^T(\theta)\left\{M_x(\theta)\left\{ M^{-1}K[X_d-KIN(\theta)]+M^{-1}B[\dot{X}_d-J(\theta)\dot{\theta}]-M^{-1}F_{ext}\right\} + V_x(\theta,\dot{\theta})+F_{ext} \right\} \tag{14}$$

For $M=M_x(\theta)$, the coefficient of $F_{ext}$ is $I-M_xM^{-1}=0$ and $F_{ext}$ does not influence the position control commands.

We can denote the computed Cartesian acceleration by

$$\ddot{x}^{(c)} = M^{-1}K[X_d-KIN(\theta)]+M^{-1}B[\dot{X}_d-J(\theta)\dot{\theta}]-M^{-1}F_{ext} \tag{15}$$

Using the relationship between the Cartesian force 6x1 vector $F$ and $\tau$, [6], $F=J^{-T}(\theta)\tau$, we obtain the Cartesian space dynamics equation

$$F = M_x(\theta)\ddot{x}^{(c)}+V_x(\theta,\dot{\theta})+F_{ext} \tag{16}$$

For $F$ and $\ddot{x}^{(c)}$ defined with regard to the robot base frame, $M_x(\theta)$ results a diagonal matrix.

Collision avoidance features can be incorporated in the control scheme, based on eqs. (15) and (16), shown in Fig. 1 by posing artificial repulsive impedances between the robot arm and the obstacles [5].

Difficulties associated with the application of the artificial impedance approach to robot trajectory generation.

The resultant of the action of the attractive and repulsive impedances is zero for $x=x_d$ whenever the artificial impedances are defined to simulate the effect of physical impedances applied to the same Cartesian decoupled robot arm [4]. The repulsive artificial impedances were previously reduced to virtual springs [4,9]. Multiple obstacles, in particular configurations, can lead to poorly damped or unstable robot arms motion. The choice of spring-damper type of artificial repulsive impedances creates the necessary damping associated with each artificial impedance.

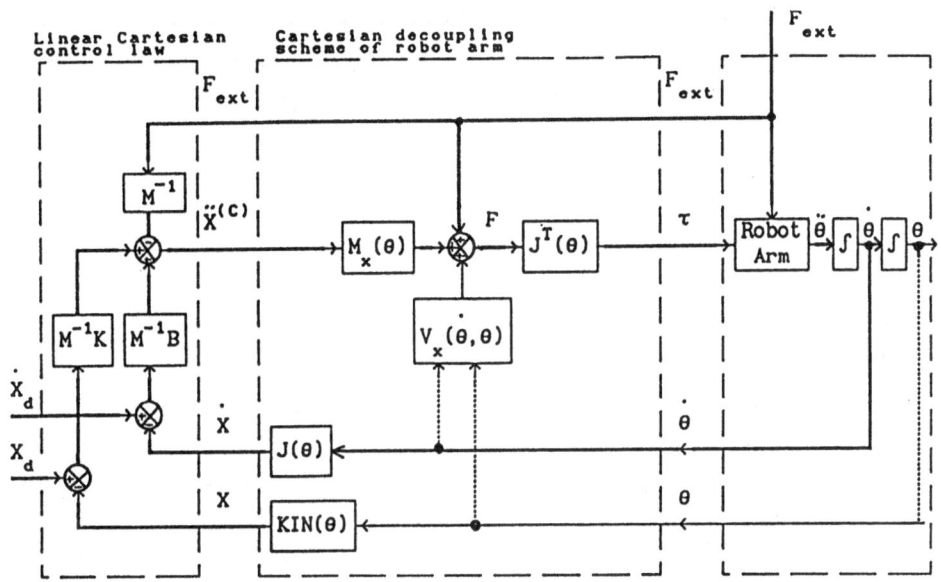

<u>Figure 1:</u>

Another important difficulty in this case is the possibility
of local static equilibrium points in the robot Cartesian space.
This possibility is obvious in case of obstacles which create
potential field concavities in the path of the end effector toward
the desired position.  Further modifications of the choice of
Cartesian space artificial impedances can lead to solutions for
escaping the local potential traps by using a coastal/surface
navigation scheme.  The schemes consist of Cartesian tracking of
virtual attractive points moving along a safe path.  In the case
of a planar manipulator, coastal navigation scheme consists in
creating a virtual attractive point $X_v$ which will move continuously on a path orthogonal to the vector starting at end effector
position and ending in the point $X_o$ detected by proximity sensors
on the obstacle as the closest to the end effector.  As long as
the line of sight between the end-effector position $X$ and the
desired position $X_d$ is obstructed, the desired position is replaced
by a virtual attractive point $X_v$ and the end effector will move
toward $X_v$.The resulting perpendicularity of the vectors $\overline{XX_v}$ and
$\overline{XX_o}$ leads to a trajectory generation on a path along the coast at
a safe distance from the obstacle.  Further heuristics can be used
in both planar and spatial motion in order to find a feasible path.

In spatial motion, the surface navigation scheme defines the virtual attractive point in a plane defined by $X$, $X_d$ and $X_o$. These schemes can be further improved based on minimum energy requirements.

Another difficulty analyzed in this paper refers to the verification of non-violating the actuators' saturation limits. In the case of the artificial impedance approach applied to a Cartesian decoupled robot a solution to the violation is a linear rescaling of the Cartesian force $F$ defined by equation (16) using the instantaneous linear dependence

$$F = J^{-T}(\theta)\tau \tag{17}$$

In case the resulting torque command, say $\tau_i$, violates one of the joint actuation saturation limit $\tau_{i,max}$, i.e. $\tau_i > \tau_{i,max}$, the linear rescaling results in a feasible torque command $\tau_f$ defined by

$$(\frac{\tau_{i,max}}{\tau_i})F = J^{-T}(\theta)(\frac{\tau_{i,max}}{\tau_i})\tau = J^{-T}\tau_f \tag{18}$$

where $\tau_{i,max}/\tau_i$ is a scalar recalculated at every computation step.

## Simulations of a 2 DOF Manipulator Using Impedance Control of the Free Motion [5].

The simulations are for a 2 DOF jointed manipulator. The desired impedance parameters are $M = I_2$, $B = 4I_2$, $K = 4I_2$, where $I_2$ is the 2x2 identity matrix. The manipulator is frictionless and operates in a vertical plane with gravitational forces included.

The paths from an initial position $I$ to a destination $D$ for the three control schemes are compared in Fig. 2. Path (a) results from the PD position controller with $K_p=K_v=4$. The path is constrained to start at $I$ and end at $D$, but is otherwise established arbitrarily by the controller. This scheme requires an inverse kinematic computation. Path (b) is the result for the case of a joint decoupling scheme. The result indicates that the performance of the artificial impedance based controller applied to a joint decoupled-linearized manipulator is poor; however, good performance of the artificial impedance approach is illustrated by path (c) which results for an impedance controlled manipulator in which a Cartesian decoupling scheme is used. The straight line path is obtained without an inverse kinematics computation or a trajectory planning scheme.

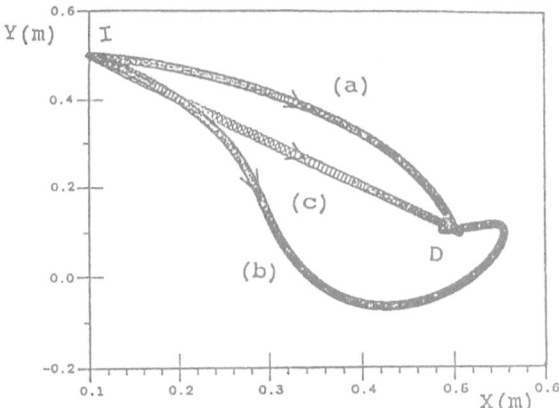

**Figure 2:**

Multiple obstacles avoidance is illustrated in Fig. 3. Figures 3a and 3b represent simulations of the same situation except for the repulsive stiffness which is two times higher in Fig. 3b ($K_0$=200) versus Fig. 3a (K=100). In Fig. 3a the path goes between the two obstacles. Higher repulsive stiffness diverts the path around both $O_1$ and $O_2$ as shown in Fig. 3b. Also shown in Fig. 3a is the straight line path between I and D, which is obtained when no obstacles exist. In Figs. 3c and 3d the same data are used except for a five times higher stiffness in the case of Fig. 3d ($K_0$=500) versus Fig. 3c ($K_0$=100). The path in Fig. 3c results from a first correction from the repulsive force associated with $O_1$ followed by that produced by the repulsive force associated with $O_2$ and so on until it turns toward a path leading towards D. In Fig. 3d the reflections between $O_1$ and $O_2$ are much stronger, actually preventing the manipulator from proceding past the obstacles to D. The repulsive force is produced here by a nonlinear spring and the motion is unstable. This result suggests that repulsive forces used in collision avoidance should contain a damping term to avoid instability in some cases. These simulations show, in general, that the choice of the structure and the gains used for repulsive forces are important and require a design technique.

The simulations illustrated in Fig. 2 and 3 assume that there is always enough actuator torque to produce the command torque required by the controller. Separate simulations were performed in which a Cartesian rescaling scheme based on Eq. (18) is used to correct the command torque when the actuator limits are reached [5].

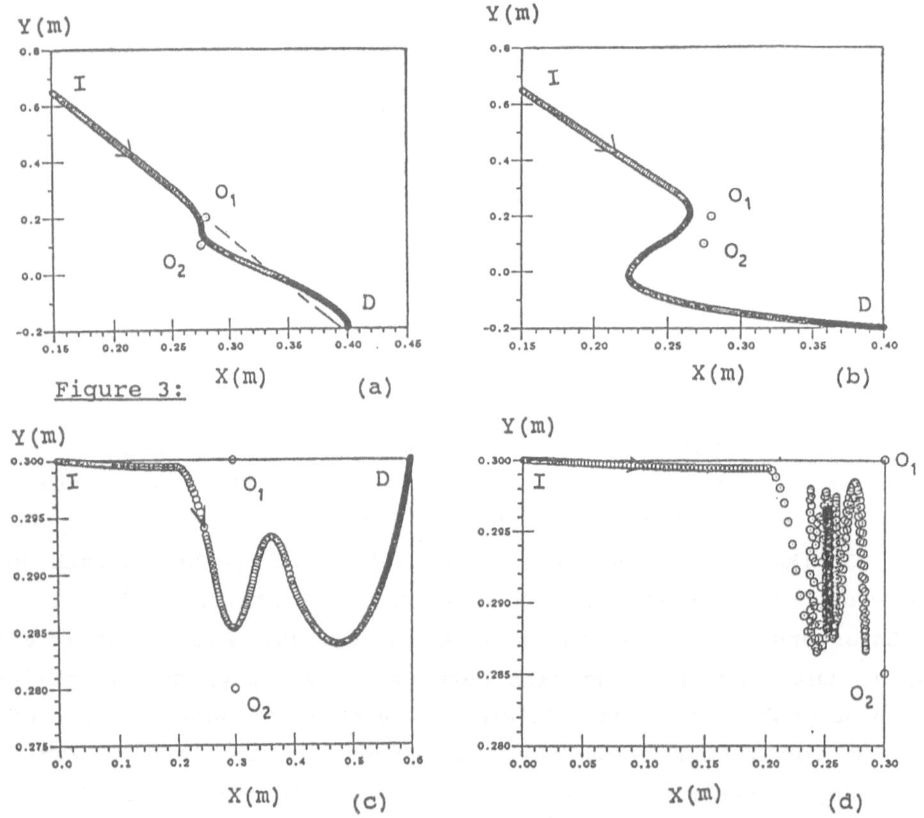

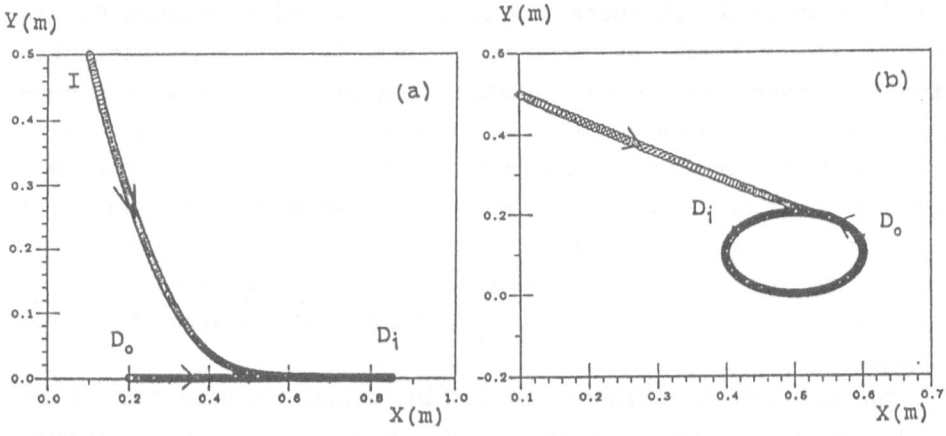

Figure 3:

The interception of moving targets is illustrated in Fig. 4.

Figure 4:

In Fig. 4a, a target is moving in a straight line starting at $D_0$ and is intercepted by the manipulator at $D_i$. A target moving in a circular trajectory with an angular speed 0.3 rad/sec is intercepted rapidly (Fig. 4b).

## IV.   MODEL AND SIMULATIONS OF A DUAL ARM ROBOT [15].

In this paper a dual arm robot handling a flexible rod and a string is assumed.  The model of the dynamics of each arm is an extension of the model developed by Luh and Zheng [11].

The generalized vector of input forces to the joints of the n DOF leader and the n DOF follower arms are

$$T^l - [J_b(\theta^l)]^T F^l - [J_a(\theta^l)]^T N^l - D(\theta^l)\ddot{\theta}^l - f(\dot{\theta}^l, \theta^l) - g(\theta^l) = 0 \qquad (19)$$

$$T^f - J_b((\theta^f)^T F^f - [J_a(\theta^f]^T N^f - D(\theta^f)\theta^f - f(\theta^f, \theta^f) - g(\theta^f) = 0 \qquad (20)$$

where

| | |
|---|---|
| $J(\theta)$ | $= [J_b^T(\theta), J_a^T(\theta)]^T$ |
| $J(\theta)$ | = the Jacobian matrix |
| $\theta$ | = n-dimensional vector of joint angular displacements |
| $F, N$ | = the vector forces and moments, respectively, applied at the origin of the end effector coordinates |
| $D(\theta)$ | = n*n inertia matrix |
| $f(\dot{\theta}, \theta)$ | = n-dimensional vector of Coriolis and centrifugal terms |
| $g(\theta)$ | = n-dimensional vector of gravity terms and, |
| $\ell, f$ | = superscripts for the leader and the follower respectively. |

The vectors  $F$  and  $N$  result from specific load dynamics model.  The leader and follower end effectors are assumed linked by artificial mechanical impedances to the target position.  The control scheme is similar to the control scheme for single arms, given in Fig. 1.

A dual-arm planar horizontal robot with two degrees-of-freedom arms is simulated (see Fig. 5) [15].  The handled object is assumed massless, with an unstressed length of 0.2 m.  The parameters of the two arms are $L_1{}^l = 0.5m$, $L_2{}^l = 0.4$ m, $m_1{}^l = 0.4$ kg, $m_2{}^l = 0.4$ kg, $L^f = 0.5$ m, $L_2{}^f = 0.4$ m, $m_1{}^f = 0.4$ kg, $m_2{}^f = 0.4$ kg, d=1.0 m.  The artificial impedance parameters are M=1.0 kg, B=4.0 Ns/m and K=4.0N/m.

### Simulations of a Robot Handling a Flexible Rod

The flexible rod is modelled as a linear spring with a damper in parallel, having a spring constant of 1.0 N/m and a damping constant of 1.0 Ns/m. The end effectors are located initially at $R_i^l$ and $R_i^f$, respectively, and move in the approaching phase to $R_{d_1}^l$ and $R_{d_1}^f$ where the flexible rod is grasped (Fig. 6). In the next phase the flexible rod is transported to $R^l{}_{d_2}$ and $R_{d_2}^f$ where the leader reaches the final position. Subsequently, only

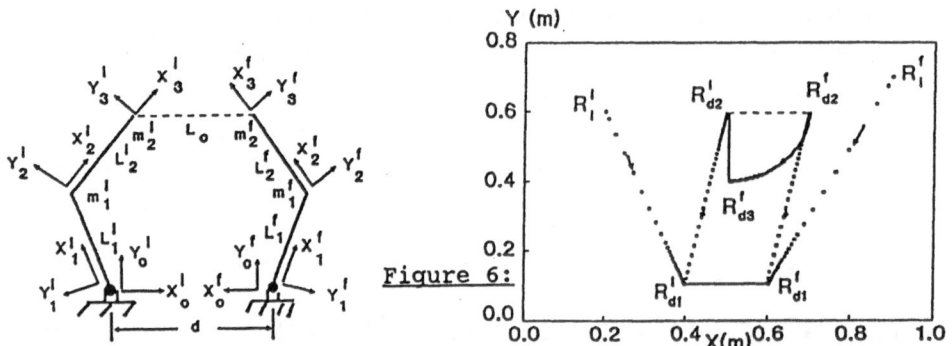

Figure 6:

Figure 5:

the follower moves, rotating the rod until the follower reaches its final position in $R_{d_3}^f$. The length of the flexible rod varies in time within ± 1%. Further simulations in which the sampling time is reduced from 0.005 s to 0.001 s leads to a reduction of the length variation by a factor of 50. These simulations show that the impedance-based controller can coordinate the motion of a two-arm robot in a sequential fashion. Improvements to the control scheme discussed in the next section, eliminate this sequential approach and result in a smooth simultaneous motion of the two arms to the final positions. This control scheme is tested in handling a string.

## Simulations of a Robot Handling a String

The string is modelled, when stretched, as a very compliant spring and, when compressed, as producing no reactive forces on the end effectors. Only the phase in which the object is actually handled is simulated here since the approaching phase poses the same coordination problems for all tasks. The initial positions of the end effectors are $R_1^l$ and $R_1^f$, respectively, where the object is assumed grasped (Fig. 7). The artificial impedances created between the initial positions of the end effectors and their desired positions $R_d^l$ and $R_d^f$, respectively, generate straight line trajectories in the case of a string which when compressed does not produce any reactive forces (Fig. 7). The distance between the two end effectors varies during the motion with a minimum value of 0.14 m (Fig. 8), which leads to a large sag in the string of 0.2 m. An artificial impedance approach overcomes the problem of the lack of reactive forces when compressing a string in length by creating an artificial impedance in parallel with the string. Simulations were performed for an artificial parallel impedance with a spring constant of 1.0 N/m and a damper constant of 2.0 Ns/m. Simulations results in Fig. 9 show that the leader's end effector keeps the straight line trajectory while the follower's end effector trajectory is generated such that the string during the motion has practically no sag, as shown on Fig. 10.

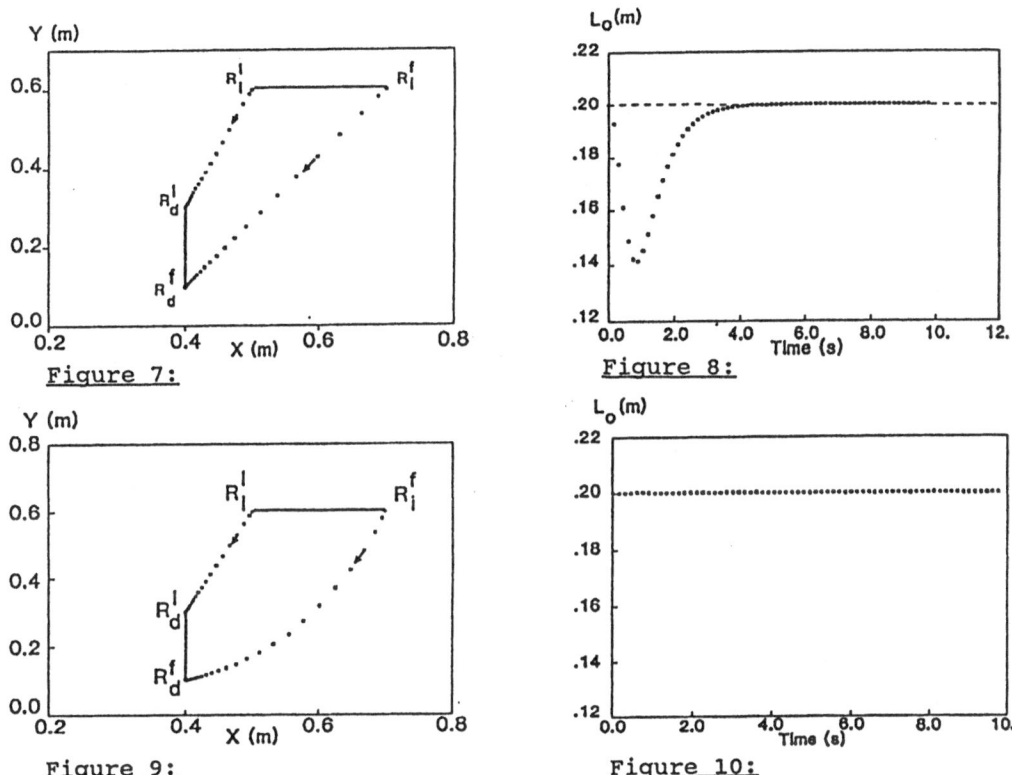

Figure 7:

Figure 8:

Figure 9:

Figure 10:

## V.    IMPLEMENTATION ISSUES

In order to achieve the target impedance described by Eq. (6) in Cartesian space, the overall system dynamics of the robot arm-controller-actuator have to be reduced to independent second order linear passive systems.

The implementation of the Artificial Impedance approach was initially based on using a two-part control [16] (or a partitioned control law [6]) which contains a linear servo control law in Cartesian space and a nonlinear state feedback compensator. In Fig. 1, Eqs. (15) and (16) are represented in a diagram in which the linear servo control output $\dot{X}^{\cdot (c)}$ is applied to a nonlinear state feedback compensator which consists of $M_x(\theta)\ddot{X}^{\cdot (c)}$

$+ V_x(\theta,\theta)+F_{ext}$. For $F_{ext} = 0$ this type of Cartesian space control has been known earlier as Resolved-Acceleration Control [8].

The decoupling method for nonlinear systems has been formulated by Freund and has been illustrated for a joint decoupling case [10].

Artificial Impedance approach and its continuum mechanics counterpart Artificial Potential Field approach, are special cases of Cartesian control in which, not only the errors are controlled in Cartesian space, but the robot arm is reduced for the linear controller to a decoupled Cartesian system of independent inertial systems. The implementation issues of a Cartesian two-part control law refer here mainly to difficulties in achieving the Cartesian decoupling of a highly nonlinear robot arm. The parameters of the dynamic equations, for example the load parameters, are often imprecisely known and an analysis of the uncertainty effect on the Cartesian controlled robot arm motion is needed. This situation led to various solutions proposed for improving robustness of Cartesian control law, in general, and of Artificial Impedance approach, in particular.

Several solutions were proposed, in general, for robust robot control. Some solutions are applicable to the Artificial Impedance approach as for example Model Reference Adaptive Control [19], [20] and Second Method of Lyapunov-based designs. Model Reference Adaptive Control benefits from the robustness of the approach, but only asymptotically the robot arm would behave like the desired Cartesian space impedances.

Second Method of Lyapunov-based design has been used for obtaining sliding mode controllers for robots [22]. The resulting discontinuous control law, inherent to the application of the second method of Lyapunov, can be approximated by a boundary layer-based continuous control law. Maintaining sliding mode requires sufficient torque output, large actuator bandwidths and fast closed loop components other than the robot arm.

Second Method of Lyapunov can be applied directly for obtaining a corrective term added to a linear control law in order to compensate for the errors resulting from inexact cancellation of the nonlinear dynamic terms by nonlinear state feedback compensator. A discontinuous corrective term would result and a continuous approximation solution is proposed for joint de-coupling case [26, pp. 227-236]. An interesting alternative in this case, to the use of Lyapunov functions, is the use of Hamiltonian functions. An application of a nonlinear version of MRAC to a two DOF robot arm modelled by Hamiltonian dynamics using a single rigid link reference model led to a nonlinear-discontinuous control law [25].

The implementation of a Cartesian two-part control law represented in Fig. 1 requires extensive computations of the highly

nonlinear terms $M_x(\theta)$ and $V_x(\theta,\dot{\theta})$ besides the kinematics compu-tations present in all Cartesian control schemes. Compared to Cartesian decoupling schemes, joint decoupling schemes are com-putationally less intensive. The proposed disturbance observer scheme proposed in [23] is an example of computationally efficient joint decoupling scheme.

In what follows is proposed a scheme based on a three-part control scheme which can be obtained from Eqs. (11) and (15)

$$\ddot{\theta}^{(c)} = J^{-1}(\theta)\{\ddot{x}^{(c)} - \dot{J}(\theta)\dot{\theta}\} \tag{21}$$

$$\tau = M(\theta)\ddot{\theta}^{(c)} + V(\theta,\dot{\theta}) + J^T(\theta)F_{ext} \tag{22}$$

These equations are represented in Fig. 11 where the three parts of the control scheme are identified. Eq. (22) corresponds to a joint decoupling scheme, while Eq. (21) contains the Cartesian decoupling scheme of a joint decoupled manipulator. It can be observed that the two-step control of Fig. 1 has been replaced in Fig. 11, by a three-step control: (a) linear Cartesian control law, (b) Cartesian decoupling of a joint decoupled arm and (c) joint decoupling. In this scheme, step (b) requires only manipulations of the Jacobian matrix, i.e. only kinematics computations. In the case of no contact force, $F_{ext}=0$, this scheme is the same as Fig. 5 in [8] for the Resolved-Acceleration Control for which the PD gains can be interpreted as $M^{-1}K$ and $M^{-1}B$. Artificial Impedance approach brings extra compensation for the

contact force and a physical interpretation of the PD gains.   In
fact, Resolved Acceleration control did not produce the development

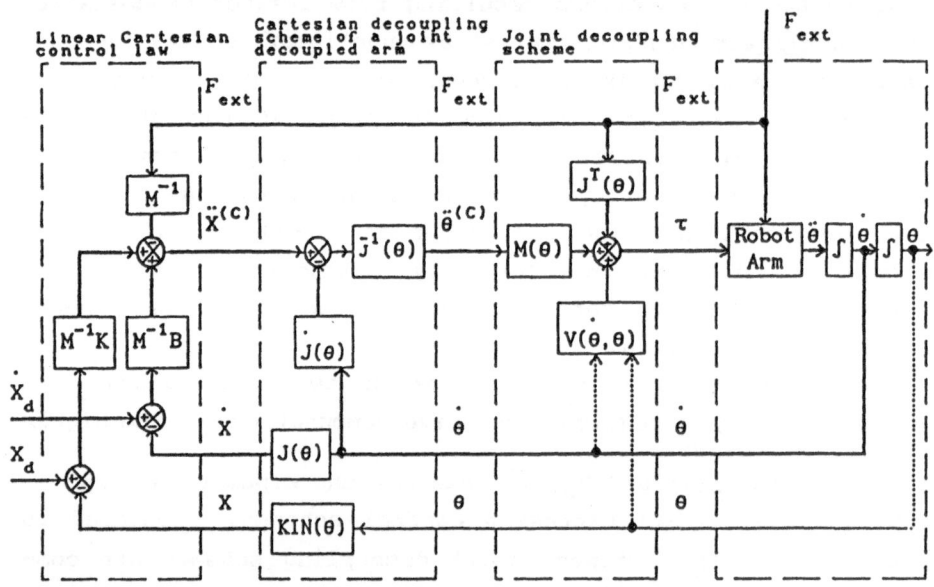

Figure 11:

of collision avoidance schemes probably because of not incor-
porating the Cartesian control of actuators in an overall dynamic
model in which the actuator effects are replaced by passive
artificial impedances in Cartesian space.


## VI.   CONCLUSIONS

Artificial Impedance approach is an important candidate for
manipulation control of complex tasks.   Collision avoidance,
actuator limits and moving targets interception are features that
can be naturally implemented in an impedance controller without
need for complete geometric description of the work volume and of
the intruding obstacles.

Further research in the implementation of the artificial
impedance approach holds the promise of an autonomous manipulator
which generates and corrects on-line the trajectory when obstacles
or actuator limits are encountered.

The coordinated motion of a dual-arm robot can also be
controlled by applying the Artificial Impedance approach.   The
simulations were performed to illustrate the manipulation of non-
rigid bodies such as a flexible rod and a string.   The results show
that this control approach is a candidate for solving some of the

difficulties of coordinated dual-arm robot operation both in the approaching phase and in the object handling phase, in particular, for the case of handling a string with sag constraints and other non-rigid objects. Further simulations and experiments are required at this stage to show the practicality of the Artificial Impedance approach to dual-arm trajectory generation. The implementation of the Artificial M-B-K Impedance approach is proposed in the form of a three-part control scheme: (a) a linear Cartesian control law with position gain $M^{-1}K$ and velocity gain $M^{-1}B$, (b) a Cartesian decoupling scheme of a joint decoupled manipulator using only Jacobian manipulations, and (c) a non-linear feedback joint decoupling compensator. In the case of non-contact motion, this scheme is identical to Resolved Acceleration control. The importance of an Artificial Impedance control scheme results from the physical interpretation of the linear controller gains and the possibility of generating and correcting robot trajectory in case of detecting obstacles by posing artificial impedances in the robot work volume.

**ACKNOWLEDGEMENTS**

The work presented in sections II to IV benefited from the cooperation with W.B. Graham of Canadian Space Agency and R. Jassemi-Zargani, Ph.D. student at the University of Ottawa, and contains computer simulation results obtained by the latter. This work is documented in greater detail in references [5],[7] and [15].

**REFERENCES**

[1]  Khatib, O., Commande dynamique dans l'espace opérationnel des robots manipulateurs en présence d'obstacles, Thèse docteur ingénieur, ENSAE, Toulouse, 1980.

[2]  Takegaki, M., Arimoto, S., A New Feedback Method for Dynamic Control of Manipulators, *Journal of Dynamic Systems, Measurement and Control*, Vol. 102, June 1981, pp. 119-125.

[3]  Hogan, N., Impedance Control: An Approach to Manipulation, *Journal of Dynamic Systems, Measurement and Control*, Vol. 107, March 1985, pp. 1-24.

[4]  Hogan, N., Stable Execution of Contact Tasks Using Impedance Control, *The Proceedings of the IEEE International Conference on Robotics and Automation*, 1987, pp. 1047-1054.

[5]  Necsulescu, D.S., Jassemi-Zargani, J., Graham, W.B., Impedance Control for Robotic Manipulation, *The Proceedings of the Second Workshop on Military Robotics Applications*, Royal Military College, Kingston, Ont., Aug. 1989.

[6]  Craig, J.J., *Introduction to Robotics*, Addison-Wesley, 1986.

[7]  Necsulescu, D.S., Jassemi-Zargani, R., and Graham, W.B., The Methods of Artificial Potential Field and Impedance Control in Robotics, *CASI Symposium on Space Station*, Ottawa, Canada, 8-9 Nov., 1989.

[8]  Luh, J.Y.S., Walker, M.W., and Paul, R.P.C., Resolved Acceleration of Mechanical Manipulators, *IEEE Trans. of AC*, No. 3, 1980, pp. 236-241.

[9]  Khatib, O., A Unified Approach for Motion and Force Control of Robot Manipulators: The Operational Space Formulation, *IEEE Journal of Robotics and Automation*, No. 1, 1987, pp. 43-53.

[10] Freund, E., Fast Nonlinear Control with Arbitrary Pole ^R Placement for Industrial Robots and Manipulators, *The Int. Journal of Robotics*, No. 1, 1982, pp. 67-78.

[11] Luh, J.Y. and Zheng, Y.F., Constrained Relations Between Two Coordinated Industrial Robots for Motion Control, *Int. Journal of Robotics Research*, 1987, No. 3, pp. 60-70.

[12] Kazerooni, H. and Tsai, T.I., Compliance Control and Unstructured Modelling of Cooperating Robots, *IEEE Int. Conf. on Robotics and Automation*, 1988, pp. 510-515.

[13] Uchiyama, M. and Dauchez, P., A Symmetric Hybrid Position/Force Control Scheme for the Coordination of Two Robots, *IEEE Int. Conf. on Robotics and Automation*, 1988, pp. 350-356.

[14] Kopf, C.D. and Yabuta, T., Experimental Comparison of Master/Slave and Hybrid Two Arm Position/Force Control, *IEEE Int. Conf. on Robotics and Automation*, 1988, pp. 1633-1637.

[15] Necsulescu, D.S., Jassemi-Zargani, R. and Graham, W.B., Trajectory Generation for Dual-Arm Robots Using Artificial Impedance Approach, *Can. Conf. on El. and Comp. Eng.*, Ottawa, Sept. 3-6, 1990, pp. 50.1.1-50.1.4.

[16] Wolovich, W.A., Robotics: Basic Analysis and Design, Holt, Rinehart and Winston, 1987.

[17] An, C.H., Atkeson, G.G., Griffiths, J.D., and Hollerbach, J.M., Experimental Evaluation of Feedforward and Computed Torque Control, *IEEE Trans. on Robotics and Automation*, 1989, No. 3, pp. 368-373.

[18] Spong, M., and Vidyasagar, M., Robust Linear Compensator Design for Nonlinear Robotic Control, *IEEE Journal on Robotics and Automation*, No. 4, 1987, pp. 345-351.

[19] Tomizuka, M.R., Horowitz, R. and Landau, Y.D., On the Use of Model Reference Adaptive Control Techniques for Mechanical Manipulators, *2nd IASTED Symp. on Identification, Control and Robotics*, Davos, March 1982.

[20] Nicosia, S. and Tomei, P., Model Reference Adaptive Control Algorithms for Industrial Robots, *Automatica*, No. 5, 1984, pp. 635-644.

[21] Craig, J.J., Adaptive Control of Mechanical Manipulators, Addison-Wesley, 1988.

[22] Asada, H., and Slotine, J.J.E., Robot Analysis and Control, J. Wiley, 1985.

[23] Nakao, M., Ohnishi, K. and Miyachi, K., A Robust Decentralized Joint Control Based on Interference Estimation, *IEEE Conf. on Robotics and Automation*, 1987, pp. 376-331.

[24] Komada, S. and Ohnishi, K., Force Feedback Control of Robot Manipulator by the Acceleration Tracing Orientation Method, *IEEE Trans. on Industrial Electronics*, No. 1, 1990, pp. 6-12.

[25] Flashner, H., and Skowronski, J.M., Model Tracking Control of Hamiltonian Systems, *J. Dyn. Syst., Measurement and Control*, Dec. 1989, pp. 656-660.

[26] Spong, M., Vidyasagar, M. Robot Dynamics and Control, J.W., 1989.

# End-Effector Trajectory Tracking in Flexible Arms: Comparison of Approaches Based on Regulation Theory

A. De Luca, L. Lanari, G. Ulivi

Dipartimento di Informatica e Sistemistica
Università degli Studi di Roma "La Sapienza"
Via Eudossiana 18, 00184 Roma, Italy

## Abstract

*Accurate tracking of end-effector trajectories is one of the most demanding tasks for robot arms with flexible links. This problem is tackled here using regulation theory, considering the nonlinearities of the general dynamic model. The control design is presented in detail, including output trajectory generation, associated reference state computation, and different feedforward/feedback realizations of the regulation concept. Extensive simulations on a simple but representative case study validate the analysis and allow to compare the various approaches.*

## 1. Introduction

Lightweight flexible structures are receiving increasing interest in robotic applications as they require smaller actuators to obtain higher performance, e.g. in terms of motion speed [1]. Typically, the most significative but demanding task to be executed is the tracking of desired trajectories for the robot end-effector. Starting from the experience gained in vibration suppression for large space structures [2], a common control approach is to superimpose active modal damping to standard techniques for controlling the rigid body motion [3,4]. For point-to-point tasks, a clever application of such composed strategies may lead to acceptable results. However, when considering motion along a trajectory, these methods cannot avoid deviations of the arm tip from the nominal path during the slew. On the other hand, use of inversion techniques for exact reproduction of end-effector trajectories induces in general instability in the closed-loop behavior [5]. When the tip position is the controlled output, the system input-output mapping is *non-minimum phase*, a concept defined both in the linear and nonlinear setting [6,7]. As a consequence, direct inversion is unfeasible due to the forced cancellation of unstable zero-dynamics.

For this class of systems, a convenient way to achieve asymptotic output tracking while preserving internal stability is to follow a *regulation* approach. End-effector motion control via nonlinear regulation has been successfully demonstrated in [8,9] for a one-link flexible arm. The rationale of this control scheme goes beyond cancellation of nonlinearities, yet producing quite good tracking results. Suppose to apply in open-loop a bounded input torque at the joint level: the flexible arm, being a passive mechanical device, will display bounded deformations and, accordingly, the end-effector will move along a certain trajectory. At this stage, no form of instability may ever occurr. If the desired output trajectory coincides with (or is part of) this motion, the robot arm must experience the same deformation history, even under closed-loop control. Therefore, the controller itself has to generate this evolution as a reference for the system state, starting from the specified output trajectory. Moreover, any feedback action should force the internal state of the system towards this 'natural' time-varying deflection, so that the end-effector will eventually move as desired. The regulator control structure is tailored to this qualitative description: an exosystem generates both the desired output trajectory and the associated state evolution, while a stabilizing linear feedback guarantees attractivity of this state-space reference trajectory.

The initial conditions of the arm play an important role in understanding the type of end-effector tracking behavior (exact vs. asymptotic) with respect to an arbitrary motion. If the full state is assigned at the initial time, i.e. when the trajectory starts, and does not match the necessary one, then a transient output error will exist, although the stabilizing action of the regulator will let it decay to zero. On the other hand, when input torques are allowed to be applied to the system also ahead of time, it is possible to bring the flexible arm into the proper deformation state at the initial time. From there on, the desired trajectory will be exactly reproduced at the system output. This interpretation points out the relations with other strategies, sometimes referred as *non-causal* solutions to the inversion problem, which are typical of frequency domain approaches [10] and of iterative learning schemes [11]. Off-line optimal control laws also yield similar input profiles [12].

In [8], nonlinear regulation theory was applied for the first time to the tracking of sinusoidal end-effector motions of a flexible arm. Some viable alternatives in the regulator design have been presented in [9]. Taking advantage of a preliminary input-output inversion control law, regulation can be performed using several combinations of feedforward/feedback terms. In particular, *direct*, *indirect* and *mixed* designs were introduced, including different amounts of nonlinear state feedback. For illustration purposes, developments in [8,9] were carried out for a one-link flexible arm, using a simple nonlinear dynamic model with flexibility concentrated in an elastic spring located along the link.

In this paper, the regulator approach is further investigated for the common non-linear model of flexible robotic arms, thus including also the multi-link case, with the aim of tracking general end-effector trajectories. Further insights are provided into the actual computation of the controller terms, showing in particular some properties of an efficient approximation technique. Next, the single steps involved in the regulator synthesis will be outlined using the same one-link flexible arm as in [8,9]. In this case study, the direct, indirect and mixed nonlinear regulator designs are compared in terms of practical tracking performance. Reference trajectories are chosen so to obtain smooth interpolated motions. With this respect, cubic and fifth-order polynomials are tested by simulation and numerical results indicate that very limited transient errors can be obtained, even when starting the arm from the standard rest condition. For completeness, also tracking results obtained with a linear regulator are given. The reported analysis provides the basis for dealing with more complex robotic applications, well beyond the case study presented here.

## 2. Nonlinear Regulation of Flexible Robot Arms

The results on output tracking via nonlinear regulation using static state feedback are briefly recalled, referring to the seminal paper [13] for details and technical assumptions. Alternate design procedures are presented for the class of nonlinear invertible systems. These theoretical findings are then reformulated for a general dynamic model of flexible robot arms, illustrating the main computational steps involved and their significance.

### 2.1 Output Regulation of Nonlinear Systems

Consider an $n$-dimensional nonlinear system

$$\dot{\mathbf{x}} = \mathbf{f}(\mathbf{x}) + \mathbf{g}(\mathbf{x})\mathbf{u}, \qquad \mathbf{y} = \mathbf{h}(\mathbf{x}), \tag{1}$$

with $m$ inputs and outputs, and assume that its linear approximation at $\mathbf{x} = \mathbf{0}$ is stabilizable by means of a linear state feedback $\mathbf{u} = \mathbf{F}\mathbf{x}$. A reference trajectory $\mathbf{y}_d(t)$ is supposed to be generated by an autonomous $r$-dimensional dynamic system (i.e. an exosystem)

$$\dot{\mathbf{w}} = \mathbf{s}(\mathbf{w}), \qquad \mathbf{y}_d = \mathbf{q}(\mathbf{w}). \tag{2}$$

Vector functions $\mathbf{f}$ and $\mathbf{h}$ are assumed to be zero at $\mathbf{x} = \mathbf{0}$, as well as $\mathbf{s}$ and $\mathbf{q}$ at $\mathbf{w} = \mathbf{0}$. Although the control design is carried out for a fixed exosystem, note that a whole class of trajectories can be generated by varying the initial conditions $\mathbf{w}(0)$.

The problem of tracking a sufficiently well behaved reference trajectory $\mathbf{y}_d(t)$ on a limited time horizon, while preserving stability in the closed-loop system, can be solved finding a state feedback law of the form

$$\mathbf{u} = \gamma(\mathbf{w}) + \mathbf{F}\big(\mathbf{x} - \pi(\mathbf{w})\big), \tag{3}$$

where the smooth vector functions $\gamma(\mathbf{w})$ and $\pi(\mathbf{w})$ are zero at $\mathbf{w} = \mathbf{0}$ and satisfy the following equations:

$$\frac{\partial \pi}{\partial \mathbf{w}}\mathbf{s}(\mathbf{w}) = \mathbf{f}(\pi(\mathbf{w})) + \mathbf{g}(\pi(\mathbf{w}))\gamma(\mathbf{w}), \tag{4a}$$

$$\mathbf{q}(\mathbf{w}) = \mathbf{h}(\pi(\mathbf{w})). \tag{4b}$$

The mapping $\pi(\mathbf{w})$ characterizes the desired state evolution associated to the output trajectory $\mathbf{y}_d$, and thus it will also be denoted as $\mathbf{x}_d$. The feedforward action $\gamma(\mathbf{w})$ in the regulator law (3) is needed to keep the state evolving in time exactly as $\mathbf{x}_d(t)$, once an initial state error has been reduced to zero. The linear feedback action (i.e. $\mathbf{F}$) guarantees this attraction to $\mathbf{x}_d$, at least locally.

The set of $n$ partial differential equations (4a) implicitly constrains the search for $\pi(\mathbf{w})$ to actual trajectories of the forced system. The $m$ algebraic relations (4b) simply express the fact that the output behaves as desired, or $\mathbf{y} = \mathbf{y}_d$, when the state is the required one. In particular, if the initial state matches the desired one, $\mathbf{x}(0) = \mathbf{x}_d(0) = \pi(\mathbf{w}(0))$, then exact output reproduction will result. Otherwise, only asymptotic tracking is obtained. Moreover, boundedness of the exosystem state $\mathbf{w}$, and thus of the desired output trajectory $\mathbf{y}_d$, implies in general that of the solution $\pi(\mathbf{w})$ to (4), i.e. of the state trajectory $\mathbf{x}_d$. Note that, if a suitable approximation $\hat{\pi}(\mathbf{w})$ to

$\pi(\mathbf{w})$ is used in the regulator realization, the closed-loop stability will still be guaranteed while the resulting output error is easily computed as $\mathbf{y}_d(t) - \mathbf{h}(\widehat{\pi}(\mathbf{w}(t)))$.

The overall block diagram of this *direct* nonlinear regulator is reported in Fig. 1. The term 'nonlinear' refers here to the nature of the controlled system (1), implying that both $\pi(\mathbf{w})$ and $\gamma(\mathbf{w})$ are in general nonlinear functions of the exosystem state.

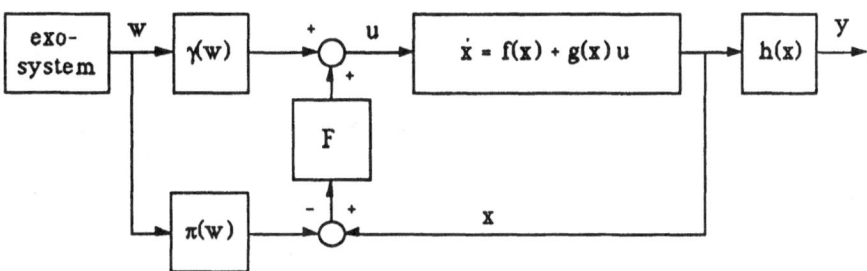

Fig. 1 – Nonlinear regulator: direct design

Indeed, replacing the system model (1) with a linear one

$$\mathbf{f}(\mathbf{x}) = \mathbf{A}\mathbf{x}, \qquad \mathbf{g}(\mathbf{x}) = \mathbf{B}, \qquad \mathbf{h}(\mathbf{x}) = \mathbf{H}\mathbf{x}, \tag{5}$$

and using a linear exosystem

$$\mathbf{s}(\mathbf{w}) = \mathbf{S}\mathbf{w}, \qquad \mathbf{q}(\mathbf{w}) = \mathbf{Q}\mathbf{w}, \tag{6}$$

yields linear forms $\pi(\mathbf{w}) = \Pi\mathbf{w}$, $\gamma(\mathbf{w}) = \Gamma\mathbf{w}$ as solutions of (4), which in turn collapse into the standard matrix equations

$$\Pi\mathbf{S} = \mathbf{A}\Pi + \mathbf{B}\Gamma, \tag{7a}$$
$$\mathbf{Q} = \mathbf{H}\Pi. \tag{7b}$$

The resulting linear regulator is

$$\mathbf{u} = \Gamma\mathbf{w} + \mathbf{F}(\mathbf{x} - \Pi\mathbf{w}). \tag{8}$$

Note that (5) could represent the linear approximation of (1) at $\mathbf{x} = 0$.

In [8], two other schemes that realize the same nonlinear regulation concept were introduced, assuming that system (1) is invertible [14]. In the case of a flexible robot arm, it is known [5] that input-output inversion can be achieved by means of a purely static state feedback of the form

$$\mathbf{u} = \alpha(\mathbf{x}) + \beta(\mathbf{x})\mathbf{v}, \qquad \text{with } \beta(\mathbf{x}) \text{ nonsingular.} \tag{9}$$

When this nonlinear feedback is applied, a proper change of coordinates $\widetilde{\mathbf{x}} = \Psi(\mathbf{x})$ will display the linearity in the resulting input-output behavior. The state-space trasformation may equivalently be performed before using (9). Completing a direct synthesis of the regulator on the new input **v**, the input fed into the original plant will be a truly nonlinear state feedback control law,

$$\mathbf{u} = \alpha(\mathbf{x}) + \beta(\mathbf{x})[\tilde{\gamma}(\mathbf{w}) + \tilde{\mathbf{F}}(\tilde{\mathbf{x}} - \tilde{\pi}(\mathbf{w}))], \tag{10}$$

where a tilda denotes quantities computed after the application of the inversion law (9). The overall *indirect* regulator is shown in Fig. 2.

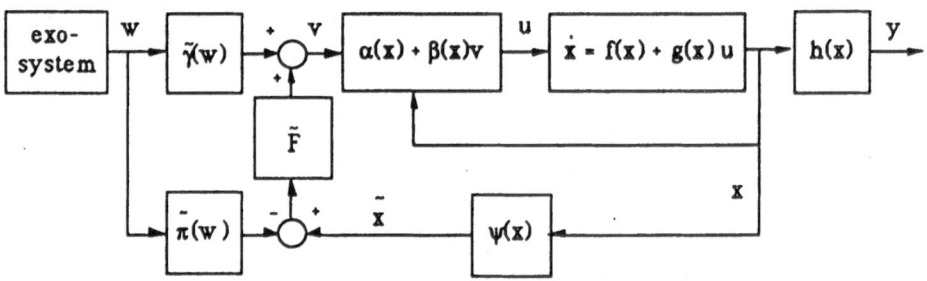

Fig. 2 – Nonlinear regulator: indirect design

In this two-stage approach, inversion control typically cancels 'hard' nonlinearities present in the original system, based on the *measured* state. The composition of (1) with (9) may result in an unstable closed-loop system, which however inherits the original stabilizability property. Thus, the following regulation stage will mainly take care of instabilities.

A third regulation scheme is obtained by computing nonlinear terms in (9) along the nominal state trajectory $\pi(\mathbf{w})$, instead of at the current state **x**. The control input becomes

$$\mathbf{u} = \alpha(\pi(\mathbf{w})) + \beta(\pi(\mathbf{w}))(\tilde{\gamma}(\mathbf{w}) + \mathbf{F}(\mathbf{x} - \pi(\mathbf{w}))), \tag{11}$$

and the resulting block diagram of this *mixed* design is shown in Fig. 3. Here, stabilization (e.g. through pole placement) is obtained using a time-varying matrix $\beta(\pi(\mathbf{w}))\mathbf{F}$. This gain modulation is expected to give better results than the constant linear feedback of the direct design. Note also that any combination of measured or nominal states could be used in evaluating single components within $\alpha$ or $\beta$.

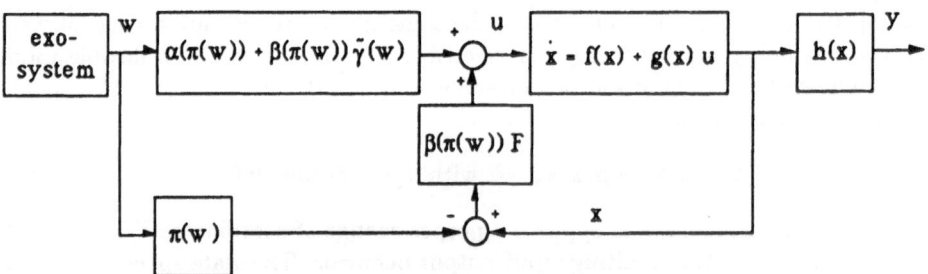

Fig. 3 – Nonlinear regulator: mixed design

## 2.2 End-Effector Regulation of Flexible Robot Arms

In the following, the regulator equations (4) will be specified for the set of second-order nonlinear differential equations, describing a controlled mechanical system. In particular, consider a robot arm with an open kinematic chain structure, rotational joints and flexible links. The dynamic model is of the form

$$
\begin{bmatrix} B_{11}(\theta,\delta) & B_{12}(\theta,\delta) \\ B_{12}^T(\theta,\delta) & B_{22}(\theta,\delta) \end{bmatrix} \begin{bmatrix} \ddot{\theta} \\ \ddot{\delta} \end{bmatrix} + \begin{bmatrix} n_1(\theta,\delta,\dot{\theta},\dot{\delta}) \\ n_2(\theta,\delta,\dot{\theta},\dot{\delta}) \end{bmatrix} + \begin{bmatrix} 0 \\ K\delta \end{bmatrix} + \begin{bmatrix} D_1\dot{\theta} \\ D_2\dot{\delta} \end{bmatrix} = \begin{bmatrix} u \\ 0 \end{bmatrix}, \tag{12}
$$

where $\theta$ are the $N$ rigid joint variables, and $\delta$ are the $N_e$ generalized coordinates associated to deformations. $B_{ij}$ are blocks of the positive definite inertia matrix, partitioned according to the rigid and flexible components. Similarly, $n_i$ contain the Coriolis, centrifugal and gravity terms. The diagonal matrix $D_1$ represents joint viscous friction, while the positive definite, symmetric (and typically diagonal) matrices $K$ and $D_2$ are, respectively, the modal stiffness and damping of the arm links. The model structure (12) holds for any finite dimensional approximation of distributed flexibility, as long as the assumed modes of deformation satisfy the geometric clamped boundary conditions at the base of each link [15]. Accordingly, the input torques $u$ appear only in the first set of equations. With reference to (1), here $n = 2(N + N_e)$, $m = N$.

For the sake of simplicity, consider the simpler but significative case of deflections limited, for each link, to the plane of rigid motion. Instead of taking cartesian-space quantities, the output can be defined as

$$
y = \theta + C\delta, \tag{13}
$$

i.e. each component $y_i$ is the rigid joint variable $\theta_i$ modified by a linear combination of the variables $\delta_{ij}$ associated to link $i$. Under the hypothesis of small link deformation, $y$ is one-to-one related to position and orientation of the end-effector through the standard direct kinematics of the arm.

Using (13), the robot dynamics (12) can be rewritten in the new coordinates $(y, \delta)$:

$$
B_{11}(y - C\delta, \delta)\ddot{y} + [B_{12}(y - C\delta, \delta) - B_{11}(y - C\delta, \delta)C]\ddot{\delta}
$$
$$
+ n_1(y - C\delta, \delta, \dot{y} - C\dot{\delta}, \dot{\delta}) + D_1(\dot{y} - C\dot{\delta}) = u, \tag{14a}
$$
$$
B_{12}^T(y - C\delta, \delta)\ddot{y} + [B_{22}(y - C\delta, \delta) - B_{12}^T(y - C\delta, \delta)C]\ddot{\delta}
$$
$$
+ n_2(y - C\delta, \delta, \dot{y} - C\dot{\delta}, \dot{\delta}) + K\delta + D_2\dot{\delta} = 0. \tag{14b}
$$

The reference state evolution $x_d$ used in the regulator (3) is equivalently expressed in terms of the new coordinates $y_d$ and $\delta_d$, and of their derivatives $\dot{y}_d$ and $\dot{\delta}_d$. By this choice, the problem of finding a solution to (4) reduces to determining only the $N_e$ functions $\delta_d = \pi_\delta(w)$, being $\dot{\delta}_d = \pi_{\dot{\delta}}(w) = [\partial\pi_\delta/\partial w]\, s(w)$ a direct consequence.

Assume that the desired trajectory is generated by a linear exosystem in observable canonical form, so that

$$
w(t) = \{y_{d,i}, \dot{y}_{d,i}, \ddot{y}_{d,i}, \ldots, y_{d,i}^{r_i-1};\ i = 1, \ldots, N\} \triangleq Y_d(t) \tag{15}
$$

can be taken as its state. The reduced solution to (4) is rewritten as $\pi_\delta(\mathbf{Y}_d)$, making explicit that the time evolution of the arm deformation will be a function only of the desired end-effector trajectory and of its time derivatives. Thus, the function $\pi_\delta$ should satisfy equation (14b), evaluated along the nominal output evolution

$$\mathbf{B}_{12}^T(\mathbf{y}_d, \pi_\delta)\ddot{\mathbf{y}}_d + \left[\mathbf{B}_{22}(\mathbf{y}_d, \pi_\delta) - \mathbf{B}_{12}^T(\mathbf{y}_d, \pi_\delta)\mathbf{C}\right]\ddot{\pi}_\delta + \mathbf{n}_2(\mathbf{y}_d, \pi_\delta, \dot{\mathbf{y}}_d, \dot{\pi}_\delta) + \mathbf{K}\pi_\delta + \mathbf{D}_2\dot{\pi}_\delta = 0, \tag{16}$$

where arguments have been indicated in a compact form. It should be stressed that this equation is independent from the applied torque. Being (16) a nonlinear time-varying differential equation, it is hardly impossible to determine a bounded solution in closed-form. Indeed, this would be equivalent to finding proper initial conditions for $\pi_\delta$ and $\dot{\pi}_\delta$ at time $t = 0$ such that forward integration of (16) yields a bounded evolution. A more feasible approach is to approximate the solution of (16), using elements which are bounded functions of their arguments, e.g. making use of polynomials in $\mathbf{Y}_d$. For a second-order expansion, this would lead to an approximation of the form

$$\delta_d = \pi_\delta(\mathbf{Y}_d) = \hat{\pi}_\delta(\mathbf{Y}_d) + o(\|\mathbf{Y}_d\|^3), \tag{17}$$

with

$$\hat{\pi}_\delta(\mathbf{Y}_d) = \Pi_1 \mathbf{Y}_d + \sum_{i=1}^{N_e}(\mathbf{Y}_d^T \Pi_{2,i} \mathbf{Y}_d)\mathbf{e}_i, \tag{18}$$

where $\mathbf{e}_i$ is the $i$th column of the identity matrix, and $\Pi_1$, $\Pi_{2,i}$ are constant coefficient matrices. As long as each component $y_{d,i}(t)$ of the desired trajectory is bounded together with its derivatives up to the $(r_i - 1)$ order, the approximation $\hat{\pi}_\delta(\mathbf{Y}_d)$ is necessarily a bounded function of time. The constant coefficients in (18) are determined by plugging this into (16), expanding nonlinear terms up to the second order, and applying the polynomial identity principle. This procedure can be iteratively applied for increasing orders of the polynomial approximation, starting with the linear one. Note that coefficients computed in the $k$th order expansion are kept also in the following one, resulting in large computational savings: at each step, a linear system of equations has to be solved for the unknown coefficients.

Once this solution is obtained, with any desired order of accuracy, backsubstitution of the reference deformation $\delta_d$, of the desired output trajectory $\mathbf{y}_d$, and of their time derivatives into (14a) will give the nominal feedforward term in the regulation law (3). In fact, $\mathbf{u} = \gamma(\mathbf{Y}_d)$ because $\mathbf{x} = \mathbf{x}_d$ is being assumed. The only approximation involved in this process is the one in (17), while all subsequent steps are consistent. In fact, the obtained feedforward term will keep the whole state evolving along a suitable close approximation to the nominal state reference $\mathbf{x}_d()$.

The above computational procedure corresponds to a direct design of the nonlinear regulator. For the indirect design, the control input $\mathbf{u} = \alpha(\mathbf{x}) + \beta(\mathbf{x})\tilde{\gamma}(\mathbf{w})$ can be recovered again from (14). Under the assumption that $\mathbf{B}_{22} - \mathbf{B}_{12}^T\mathbf{C}$ is nonsingular, acceleration $\ddot{\delta}$ is isolated from (14b) and introduced in (14a). Next, by evaluating $\ddot{\mathbf{y}}$ everywhere as $\ddot{\mathbf{y}}_d$, (14a) will give a nonlinear control law with feedback from the tranformed state $\tilde{\mathbf{x}} = (\mathbf{y}, \delta, \dot{\mathbf{y}}, \dot{\delta})$. The latter is immediately rewritten in terms of the original state $\mathbf{x} = (\theta, \delta, \dot{\theta}, \dot{\delta})$, through (13). Indeed, a further stabilization around $\mathbf{x}_d$ (or

$\tilde{x}_d$) has to take place. The mixed regulation design is accomplished in a similar way. In this case, the feedback gain will be modulated by part of the inertia matrix.

### 3. Case Study: A One-Link Flexible Arm

*3.1 Dynamic Model*

A one-link flexible planar robot arm will be used as a case study for the end-effector tracking problem. For a single link, linear dynamic models are usually assumed when limiting the analysis to small deflections. However, in case of fast motion and/or in presence of heavy carried loads, nonlinear effects arise also in this case due to the larger deformations coming into play. A simple modeling technique divides the flexible link into rigid segments that are connected by elastic springs, where link deformation is concentrated. Following the Lagrangian approach, a nonlinear dynamic model can be obtained in the standard form (12). Explicit expressions that are parametrized in the number of segments have been derived in [8], so that model order can be varied easily to achieve the prescribed accuracy. The following treatment will be limited to the case of two equal segments of uniform mass, moving on the horizontal plane.

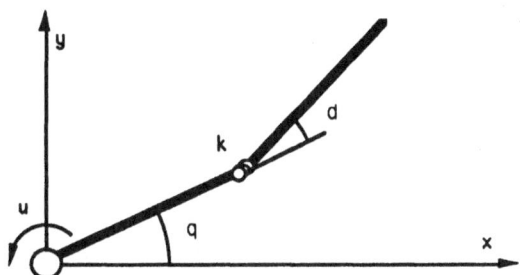

Fig. 4 – A simple one-link flexible arm

Let $m$ and $\ell$ denote the total link mass and length, and $k$ the spring elasticity. With reference to Fig. 2, $\theta$ is the angular position of the joint and $\delta$ is the flexible variable, so that $N = N_e = 1$. The dynamic equations are

$$\begin{bmatrix} b_{11}(\delta) & b_{12}(\delta) \\ b_{12}(\delta) & b_{22} \end{bmatrix} \begin{bmatrix} \ddot{\theta} \\ \ddot{\delta} \end{bmatrix} + \begin{bmatrix} n_1(\delta, \dot{\theta}, \dot{\delta}) + d_1 \dot{\theta} \\ n_2(\delta, \dot{\theta}) + k\delta + d_2 \dot{\delta} \end{bmatrix} = \begin{bmatrix} 1 \\ 0 \end{bmatrix} u, \qquad (19)$$

with the elements of the inertia matrix $\mathbf{B}(\delta)$ given by

$$b_{11}(\delta) = a + 2c\cos\delta, \qquad b_{12}(\delta) = b + c\cos\delta, \qquad b_{22} = b,$$

and Coriolis and centrifugal terms

$$n_1(\delta, \dot{\theta}, \dot{\delta}) = -c(\dot{\delta}^2 + 2\dot{\theta}\dot{\delta})\sin\delta, \qquad n_2(\delta, \dot{\theta}) = c\dot{\theta}^2 \sin\delta,$$

where $a = 5m\ell^2/24$, $b = m\ell^2/24$, $c = m\ell^2/16$. State equations can be obtained by setting $\mathbf{x} = (\theta, \delta, \dot{\theta}, \dot{\delta}) \in \mathbb{R}^4$.

The linearized expression of the end-effector angular position as seen from the base

$$y = \theta + \frac{1}{2}\delta, \tag{20}$$

will be taken as controlled output for the system. The above finite-dimensional model, although of reduced-order, displays the same basic control properties of more accurate and complex distributed models. In particular, (20) is a non-minimum phase output.

### 3.2 Direct Regulator Design

In order to obtain output tracking of end-effector trajectories for the considered one-link flexible arm, the direct nonlinear regulator design will be followed. Stabilizability of the linear approximation of (19) around the origin $x = 0$ is easily verified.

An exosystem will be considered, capable of generating polynomial trajectories up to the *fifth order*. In particular, the following linear system of order $r = 6$ in canonical form,

$$\dot{w} = \begin{bmatrix} 0 & 1 & 0 & 0 & 0 & 0 \\ 0 & 0 & 1 & 0 & 0 & 0 \\ 0 & 0 & 0 & 1 & 0 & 0 \\ 0 & 0 & 0 & 0 & 1 & 0 \\ 0 & 0 & 0 & 0 & 0 & 1 \\ 0 & 0 & 0 & 0 & 0 & 0 \end{bmatrix} \begin{bmatrix} w_1 \\ w_2 \\ w_3 \\ w_4 \\ w_5 \\ w_6 \end{bmatrix} = Sw, \qquad y_d = w_1 = Qw, \tag{21}$$

properly initialized at $w(0) = (a_0, a_1, 2a_2, 6a_3, 24a_4, 120a_5)$, generates the reference output

$$y_d(t) = a_5 t^5 + a_4 t^4 + a_3 t^3 + a_2 t^2 + a_1 t + a_0. \tag{22}$$

The chain structure of integrators (21) can be extended to the $n$th order for generating, as $y_d$, polynomial trajectories of degree $n - 1$. Note that, if an infinite time horizon is considered, polynomial trajectories will become unbounded. However, the proper initialization of these exosystems and the limited time span considered in typical robotic applications overcomes this critical point. In the quintic case, different combinations of boundary conditions could be imposed on initial/final position, velocity, acceleration, and jerk. The most practical choice is to set a finite time $t_f$, with specified initial and final position, velocity and acceleration. For

$$\begin{aligned} y_d(0) &= y_0 & \dot{y}_d(0) &= y_0' & \ddot{y}_d(0) &= y_0'' \\ y_d(t_f) &= y_f & \dot{y}_d(t_f) &= y_f' & \ddot{y}_d(t_f) &= y_f'' \end{aligned} \tag{23}$$

coefficients in (22) take on the values:

$$\begin{bmatrix} a_5 \\ a_4 \\ a_3 \end{bmatrix} = \frac{1}{2t_f^5} \begin{bmatrix} 12(y_f - y_0) - 6t_f(y_f' + y_0') + t_f^2(y_f'' - y_0'') \\ -30t_f(y_f - y_0) + 2t_f^2(7y_f' + 8y_0') - t_f^3(3y_f'' - 2y_0'') \\ 20t_f^2(y_f - y_0) - 4t_f^3(2y_f' + 3y_0') + t_f^4(y_f'' - 3y_0'') \end{bmatrix}, \qquad \begin{bmatrix} a_2 \\ a_1 \\ a_0 \end{bmatrix} = \begin{bmatrix} y_0'' \\ y_0' \\ y_0 \end{bmatrix}. \tag{24}$$

In order to obtain the reference state trajectory, the general procedure outlined through (14) and (16) will be followed, using the second-order differential representation (18) of the robot system. In particular, (16) becomes

$$(b + c\cos\delta)\ddot{y}_d + (\frac{b - c\cos\delta}{2})\ddot{\delta} + c(\dot{y}_d - \frac{\dot{\delta}}{2})^2 \sin\delta + k\delta + d_2\dot{\delta} = 0, \qquad (25)$$

which should be solved for $\delta(t) = \pi_\delta(Y_d(t))$. A first-order approximation of this solution will be determined by choosing

$$\delta_d = \Pi_1 Y_d = p_0 y_d + p_1 \dot{y}_d + p_2 \ddot{y}_d + p_3 \dddot{y}_d + p_4 y_d^{(4)} + p_5 y_d^{(5)},$$
$$\dot{\delta}_d = \Pi_1 \dot{Y}_d = p_0 \dot{y}_d + p_1 \ddot{y}_d + p_2 \dddot{y}_d + p_3 y_d^{(4)} + p_4 y_d^{(5)}, \qquad (26)$$
$$\ddot{\delta}_d = \Pi_1 \ddot{Y}_d = p_0 \ddot{y}_d + p_1 \dddot{y}_d + p_2 y_d^{(4)} + p_3 y_d^{(5)}.$$

Introducing these formulas for $\delta$, $\dot{\delta}$, $\ddot{\delta}$ into the linearized version of (25) around $\delta = \dot{\delta} = 0$,

$$(b + c)\ddot{y}_d + \frac{b - c}{2}\ddot{\delta} + k\delta + d_2\dot{\delta} = 0, \qquad (27)$$

and solving for the coefficients of $y_d$ and of its derivatives, yields the following explicit expressions:

$$p_0 = 0, \qquad p_1 = 0, \qquad p_2 = -\frac{b + c}{k}, \qquad p_3 = \frac{d_2(b + c)}{k^2},$$
$$p_4 = \frac{(b + c)((c - b)k + 2d_2^2)}{2k^3}, \qquad p_5 = \frac{d_2(b + c)((c - b)k + d_2^2)}{k^4}. \qquad (28)$$

In [8], this kind of approximation was found to leave some steady-state error in the case of a sinusoidal trajectory so that more terms had to be included, at least up to the third order. The second-order approximation (18) simplifies to

$$\delta_d = \Pi_1 Y_d + Y_d^T \Pi_2 Y_d. \qquad (29)$$

Substituting (29) into (25), expanding nonlinear functions and retaining terms up to the second-order, linear terms in the components of $Y_d$ are matched by the same values (28), while identities on quadratic terms are satisfied by $\Pi_2 \equiv 0$. For the third-order approximation

$$\delta_d = \Pi_1 Y_d + Y_d^T \Pi_2 Y_d + \sum_{\substack{0 \leq i \leq 5 \\ k \geq j \geq i}} p_{ijk} y_d^{(i)} y_d^{(j)} y_d^{(k)}, \qquad (30)$$

(25) should be expanded accordingly as

$$(b + c - \frac{c}{2}\delta^2)\ddot{y}_d + (\frac{b - c}{2} + \frac{c}{4}\delta^2)\ddot{\delta} + c(\dot{y}_d - \frac{\dot{\delta}}{2})^2\delta + k\delta + d_2\dot{\delta} = 0. \qquad (31)$$

Again, using (30) into (31) leads to a linear system of equations for the $p_{ijk}$ unknowns weighting cubic terms in the elements of $Y_d$. Arising terms of order 4 or more are neglected at this stage. Out of the 56 cubic coefficients, only 34 are non zero: in particular, all coefficients $p_{0jk}$ vanish since deformation cannot depend on $y$ (viz. $y_d$)

— a cyclic coordinate in the dynamic model (19); furthermore, $p_{111} = 0$ because there are no cubic terms in the velocity $\dot{y}_d$. The rest of the coefficients can be computed numerically, although for this example explicit symbolic expressions in terms of the model parameters were also derived by using $Mathematica^{TM}$. Note that most of the cubic coefficients in (30) turn out to be very small numbers, but they multiply high-order derivatives. Thus, simplifications can be carried out only depending on the time scaling of the desired trajectory (i.e. on $t_f$).

The feedforward term of the direct regulator is obtained using the reference deformation $\delta_d = \delta_d(\mathbf{Y}_d)$ given by (30) in the first equation of the model (19). This provides

$$u = \gamma(\mathbf{Y}_d) = \left(a + 2c\cos\delta_d\right)\ddot{y}_d + \left(b - \frac{a}{2}\right)\ddot{\delta}_d - 2c\dot{y}_d\dot{\delta}_d(\mathbf{Y}_d)\sin\delta_d + d_1\left(\dot{y}_d - \frac{\dot{\delta}_d}{2}\right). \quad (32)$$

The feedback action is based on a stabilizing matrix $\mathbf{F}$ (here, a row vector) penalizing the error on the full state $\mathbf{x}$. As a result, the regulator will be

$$u = \gamma(\mathbf{Y}_d) + \mathbf{F}\begin{bmatrix} \theta - (y_d - \delta_d(\mathbf{Y}_d)/2) \\ \delta - \delta_d(\mathbf{Y}_d) \\ \dot{\theta} - (\dot{y}_d - \dot{\delta}_d(\mathbf{Y}_d)/2) \\ \delta - \delta_d(\mathbf{Y}_d) \end{bmatrix}. \quad (33)$$

*3.3 Indirect Regulator Design*

Since the relative degree of output (20) is two, the synthesis of an inversion-based control is accomplished by deriving twice the output and setting $\ddot{y} = v$. Solving for $u$ yields

$$u = n_1(\delta, \dot{\theta}, \dot{\delta}) + d_1\dot{\theta} + \frac{b_{11}(\delta) - 2b_{12}(\delta)}{2b_{22} - b_{12}(\delta)}(n_2(\delta, \dot{\theta}) + k\delta + d_2\dot{\delta}) + \frac{2\det B(\delta)}{2b_{22} - b_{12}(\delta)}v$$

$$= -c(\dot{\delta}^2 + 2\dot{\theta}\dot{\delta})\sin\delta + d_1\dot{\theta} + \frac{(a - 2b)(c\dot{\theta}^2\sin\delta + k\delta + d_2\dot{\delta}) + 2(ab - b^2 - c^2\cos^2\delta)v}{b - c\cos\delta}$$
$$(34)$$

which is in the standard form $u = \alpha(\mathbf{x}) + \beta(\mathbf{x})v$. The input-output linearizing coordinates in the system after inversion are $\tilde{\mathbf{x}} = (y, \dot{y}, \delta, \dot{\delta})$. In view of (20), this implies only a linear transformation in the state space. The closed-loop equations can be written as

$$\ddot{y} = v,$$
$$\ddot{\delta} = \frac{2(n_2(\dot{y}, \delta, \dot{\delta}) + k\delta + d_2\dot{\delta})}{b_{12}(\delta) - 2b_{22}} + \frac{2b_{12}(\delta)}{b_{12}(\delta) - 2b_{22}}v, \quad (35)$$

and it is easy to see that this system is unstable in the first approximation. In particular, this instability is reflected in the system zero-dynamics, which is obtained [8] by imposing $y(t) \equiv 0$ in (35):

$$\ddot{\delta} = -\frac{(c/2)\dot{\delta}^2\sin\delta + 2(k\delta + d_2\dot{\delta})}{b - c\cos\delta}. \quad (36)$$

This confirms that tracking of a desired output trajectory $y_d(t)$ cannot be achieved by simply stabilizing the (linear) input-output behavior in (35), e.g. using

$$v = \ddot{y}_d + \tilde{F}_1(y - y_d) + \tilde{F}_2(\dot{y} - \dot{y}_d), \qquad \tilde{F}_1, \tilde{F}_2 < 0, \tag{37}$$

as specified in a pure inversion-based approach [16].

As a matter of fact, (34) with (37) will force the state of the system to become unbounded. This is always true, except when the initial arm deformation is exactly in that particular state specified by a bounded solution of the regulator equations, i.e. when $\delta(0) = \delta_d(Y_d(0))$ and $\dot{\delta}(0) = \dot{\delta}_d(Y_d(0))$. This instability will be overruled by the linear feedback part in the regulation synthesis of $v$. The indirect design takes advantage of the simplified structure of system (35). In particular, the reference behavior for the first two states and the feedforward term are in this case

$$\tilde{\pi}_1(Y_d) = y_d, \qquad \tilde{\pi}_2(Y_d) = \dot{y}_d, \qquad \tilde{\gamma}(Y_d) = \ddot{y}_d, \tag{38}$$

i.e. the output reference position, velocity, and acceleration, as expected. On the other hand, references for the deflection variables are the same ones computed for the direct design. The resulting $v$ will be of the form

$$\begin{aligned} v &= \ddot{y}_d + \tilde{F}_1(y - y_d) + \tilde{F}_2(\dot{y} - \dot{y}_d) + \tilde{F}_3(\delta - \delta_d(Y_d)) + \tilde{F}_4(\dot{\delta} - \dot{\delta}_d(Y_d)) \\ &= \ddot{y}_d + \tilde{F}(\tilde{x} - \tilde{x}_d), \end{aligned} \tag{39}$$

which should be compared with (37), as a clear distinction between inversion and regulation. The actual input torque applied to the flexible robot arm is obtained combining (39) with (34).

## 3.4 Mixed Regulator Design

A mixed regulator can be derived from (34), evaluating nonlinear terms at their desired values but including also the linear stabilization part (39) into $v$. After some manipulation, the control input becomes

$$u = \gamma(Y_d) + \frac{2(ab - b^2 - c^2 \cos^2 \delta_d(Y_d))}{b - c \cos \delta_d(Y_d)} \, \tilde{F}(\tilde{x} - \tilde{x}_d), \tag{40}$$

where $\gamma(Y_d)$ is the same as in (32). This expression points out that, in case of matched initial conditions for the full state, the direct regulator (33), the indirect regulator (34) with (37) or (39), and the mixed one (40) all collapse into a unique feedforward law that assigns the same steady-state behavior. However, when initial state matching is impractical, the above control laws will produce different transient errors.

## 4. Simulation Results

The nonlinear regulator approach has been tested by simulation using as parameters for the one-link flexible arm:

$$\ell = 1 \text{ m}, \quad m = 0.2 \text{ kg}, \quad k = 5 \text{ Nm/rad}, \quad d_1 = d_2 = 0.01 \text{ Nm sec/rad}. \tag{41}$$

Simulations were run using Matlab™, with a fourth order Runge-Kutta integration method. For the quintic polynomial trajectory, the following data were used:

$$y_0 = 0°, \quad y_f = 90°, \quad y'_0 = y'_f = y''_0 = y''_f = 0, \quad t_f = 1 \text{ sec.} \tag{42}$$

After $t_f$, the reference state is forced to zero, except for $y_d = \theta_d = 90°$. This corresponds to reset instantaneously the exosystem state, a critical operation that will cause disturbance on tracking. The feedback gain matrix $\bar{F}$ of the indirect design was chosen by assigning poles at $-20 \pm i30$ and $-30 \pm i25$ to the linearized system. In all other cases, gains were determined in a consistent way so to allow a significative comparison.

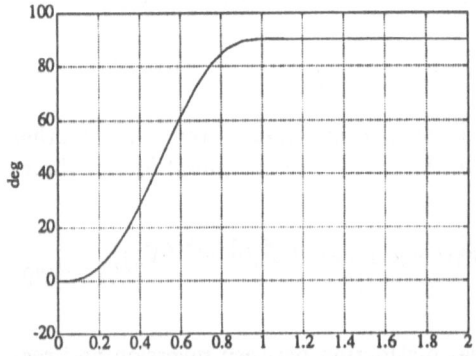 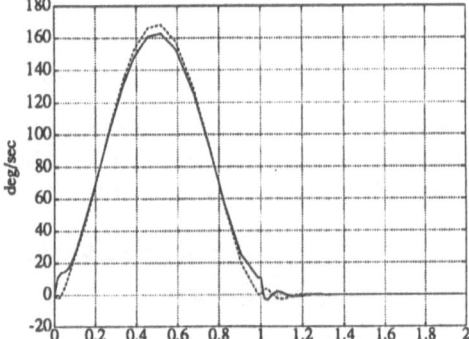

Fig. 5 – Output for a quintic polynomial      Fig. 6 – Joint and tip velocity

Figures 5–8 show the obtained tracking results for the quintic polynomial case using the indirect regulator law (34) with (39). The desired and the obtained output are practically coincident, even in the presence of a peak velocity reaching more than 160°/sec at the trajectory midpoint. As shown in Fig. 6, the velocity of the joint (continuous line) and that of the tip (dashed) are slightly different in magnitude. The non-minimum phase nature of the system is displayed in the reverse initial motion of the tip. A more detailed view of the tracking error $e = y - y_d$ is given in Fig. 7. The maximum error value is about 0.15° and vanishes in 0.2 sec: from there on, exact tracking follows. The presence of a transient error at $t_f = 1$ sec is a consequence of the exosystem reset which produces a mismatch with the current state of the flexible arm. The input torque $u$ in Fig. 8 is similar to the one necessary for an equivalent rigid arm — a cubic as the desired acceleration profile. Differences arise at the beginning and at the end of the motion, in response to state errors. Note that no excess torque is required for the flexible case. When applying the direct regulator design (33), which uses pure linear feedback and feedforward based on the nonlinear model, no relevant differences were found.

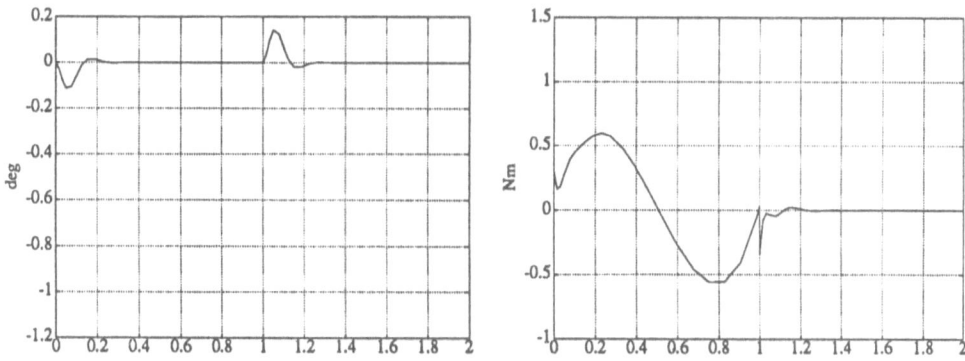

Fig. 7 – Error for a quintic polynomial     Fig. 8 – Torque for a quintic polynomial

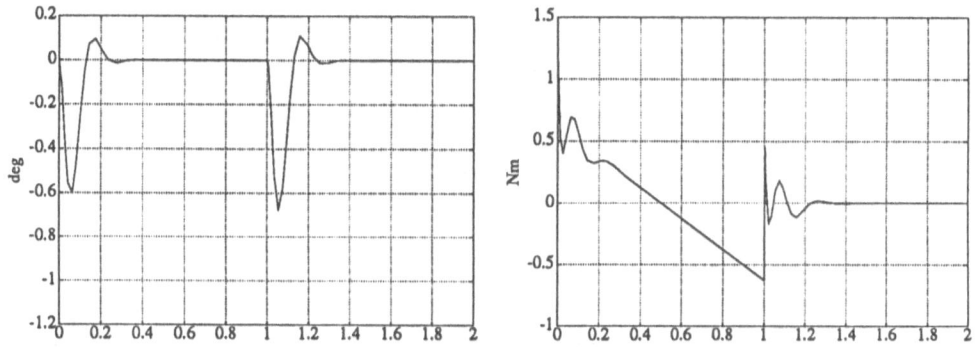

Fig. 9 – Error for a cubic polynomial     Fig. 10 – Torque for a cubic polynomial

In order to assess the effect of trajectory smoothness, the same net motion was accomplished using a cubic polynomial with zero initial and final velocity. Tracking error and required torque are shown respectively in Fig. 9 and 10. The maximum error is about six times higher, but still very small. Non-zero torques at the initial and final instants correspond to step changes in the desired acceleration, with an added peak due to transients. Moreover, the quintic polynomial trajectory roughly halves the deflection $\delta$ and its time derivative (Fig. 11) with respect to the cubic case. Note that, for this very smooth trajectory, the maximum deflection is about 2.5°.

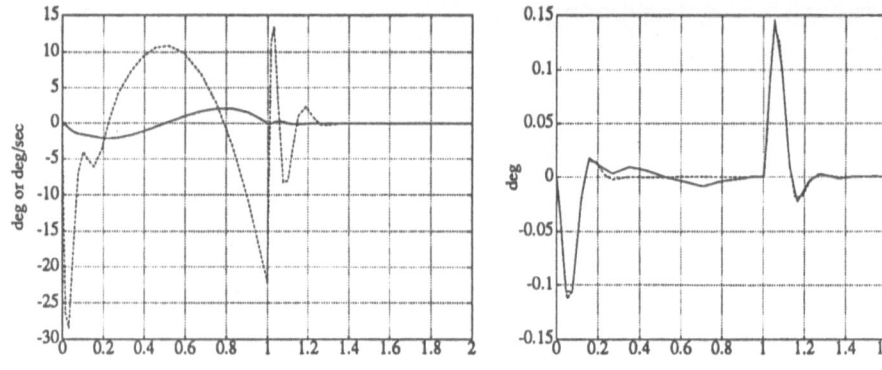

Fig. 11 – Flexible variables

Fig. 12 – Linear vs. nonlinear
regulation error

It is also interesting to compare the performance of the nonlinear regulator with respect to a linear one, i.e. using (8) designed on the linearized system. Figure 12 shows a very similar error behavior in the linear (continuous line) and in the nonlinear (dashed) case. This is not unexpected because of the very small nonlinearities coming into play. However, while peak transient errors are identical, the remaining behavior during motion (between 0.2 and 1 sec) is qualitatively different, giving perfect tracking only for the nonlinear regulator. This becomes also more apparent when considering the case of fast sinusoids, as reported in [8].

The performance of the mixed regulator design (40) is also quite satisfactory, as shown in Figs. 13–14 for the quintic polynomial. An initial position error of about 9° was assumed in this case, so to emphasize closed-loop convergence of state trajectories towards the reference one. In particular, Fig. 14 indicates a rapid and well damped transient according to the chosen pole location of the linear feedback. The small error appearing after $t_f$ is the same as in Fig. 7, shown with a different scale.

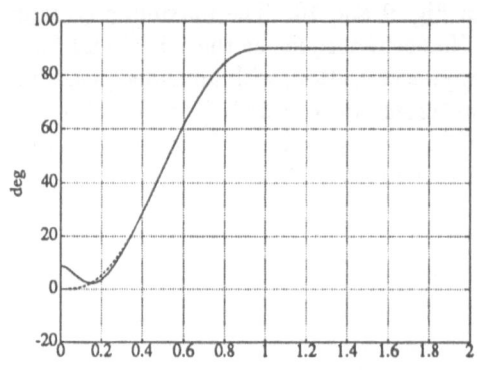

Fig. 13 – Output with mixed
regulator design

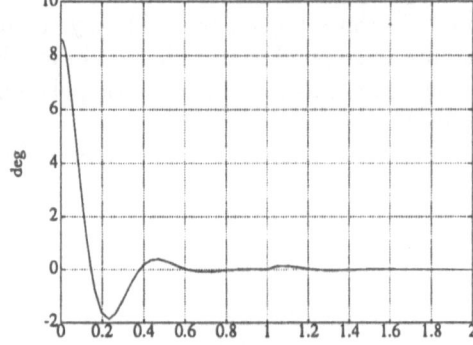

Fig. 14 – Error with mixed
regulator design

## 5. Conclusions

Nonlinear control theory offers several techniques to deal with the problem of tracking trajectories for robot arms with flexible links. In particular, the regulation approach is well suited in the case of end-effector tracking, as it is a natural framework for solving the closed-loop instability phenomena arising with standard inversion techniques. It has been pointed out that, for flexible manipulators, solving the regulator equations accounts in determining the nominal deformation associated with the desired output motion. This allows to set up an efficient computational procedure for deriving the regulation control law.

Different possible realizations of nonlinear regulators have been presented, highlighting tracking capabilities and ease of implementation. Simulation results have shown that tracking can be quite accurate when smooth reference trajectories are selected. Work is under way for extending the application of this control strategy to multimode/multi-link flexible robot arms. The proposed feedforward/feedback schemes will be evaluated in the experimental test bed available in our Robotics Laboratory [17].

Finally, it is worth to remark that the conditions under which a nonlinear regulator can be successfully designed using static state-feedback are the same which allow regulation using only output measurements, provided that a reasonable observability hypothesis is satisfied and that dynamics is included in the controller. Conversely, the stabilizability of the nonlinear system, which is needed for regulation, does not necessarily ask for full state feedback. For instance, feedback from the flexible variables may be avoided. Further investigation is being devoted to these aspects.

## Acknowledgements

This paper is based on work supported by the *Ministero dell'Università e della Ricerca Scientifica e Tecnologica* under 40% funds (Scientific Director: Prof. F. Nicolò).

## References

[1] W.J. Book, "Modeling, design, and control of flexible manipulators arms: A tutorial review," *29th IEEE Conf. on Decision and Control* (Honolulu, HI, Dec. 5–7, 1990), pp. 500–506.

[2] M.J. Balas, "Fedback control of flexible systems", *IEEE Trans. on Automatic Control*, vol. AC-23, no. 4, pp. 673–679, 1978.

[3] B. Siciliano and W.J. Book, "A singular perturbation approach to control of lightweight flexible manipulators," *Int. J. of Robotics Research*, vol. 7, no. 4, pp. 79–90, 1988.

[4] S.N. Singh and A.A. Schy, "Control of elastic robotic systems by nonlinear inversion and modal damping," *ASME J. of Dynamic Systems, Measurements, and Control* vol. 108, no. 5, pp. 180–189, 1986.

[5] A. De Luca, P. Lucibello, and G. Ulivi, "Inversion techniques for trajectory control of flexible robot arms," *J. of Robotic Systems*, vol. 6, no. 4, pp. 325–344, 1989.

[6] R.H. Cannon, Jr. and E. Schmitz, "Initial experiments on the end-point control of a flexible one-link robot," *Int. J. of Robotics Research*, vol. 3, no. 3, pp. 62–75, 1984.

[7] C. Byrnes and A. Isidori, "Local stabilization of critically minimum phase nonlinear systems," *Systems and Control Lett.*, vol. 11, no. 1, pp. 9–17, 1988.

[8] A. De Luca, L. Lanari, and G. Ulivi, "Output regulation of a flexible robot arm," *9th INRIA Int. Conf. on Analysis and Optimization of Systems* (Antibes, F, Jun. 12–15, 1990), pp. 833–842.

[9] A. De Luca, L. Lanari, and G. Ulivi, "Nonlinear regulation of end-effector motion for a flexible robot arm," *Joint Conf. on New Trends in Systems Theory* (Genova, I, Jul. 9–11, 1990).

[10] E. Bayo, "A finite-element approach to control the end-point motion of a single-link flexible robot," *J. of Robotic Systems*, vol. 4, no. 1, pp. 63–75, 1985.

[11] M. Poloni and G. Ulivi, "Iterative trajectory tracking for flexible arms with approximate models," *5th Int. Conf. on Advanced Robotics (ICAR'91)* (Pisa, I, Jun. 20–22, 1991).

[12] T. Kokkinis and M. Sahraian, "Flexible robot arm control: An optimal solution for the inverse dynamics," *IEEE Work. on Intelligent Motion Control* (Istanbul, TR, Aug. 20–22, 1990), pp. 529–534.

[13] A. Isidori and C. Byrnes, "Output regulation of nonlinear systems," *IEEE Trans. on Automatic Control*, vol. AC-35, no. 2, pp. 131–140, 1990.

[14] A. Isidori, *Nonlinear Control Systems*, 2nd Edition, Springer Verlag, Berlin, 1989.

[15] L. Meirovitch, *Analytical Methods in Vibrations*, Macmillan, New York, 1967.

[16] R.M. Hirschorn, "Output tracking in multivariable nonlinear systems," *IEEE Trans. on Automatic Control*, vol. AC-26, no. 2, pp. 593–595, 1981.

[17] A. De Luca, L. Lanari, P. Lucibello, S. Panzieri, and G. Ulivi, "Control experiments on a two-link robot with a flexible forearm," *29th IEEE Conf. on Decision and Control* (Honolulu, HI, Dec. 5–7, 1990), pp. 520–527.

# An inversion procedure for nonlinear time-varying systems

M.D.Di Benedetto          P.Lucibello
Dipartimento di Informatica e Sistemistica
Università di Roma "La Sapienza"
Via Eudossiana, 18
00184 Roma (Italia)

**Abstract.** *In this paper, a new procedure for inverting nonlinear systems is proposed, which presents some computational advantages when applied to flexible robot arms or, more generally, to mechanical structures. This procedure will be presented in a general setting, for nonlinear time-varying systems. An application to a flexible two-link manipulator illustrates the claimed advantages.*

## 1. Introduction

Different control strategies have recently been proposed in the literature for flexible robots (see e.g. [1, 4, 5, 12, 13]). In particular, for the trajectory tracking problem, inversion techniques are investigated in [6], where the inversion algorithm of [8, 15] is applied to the given system in order to recover the control required for the output to behave as desired, whenever the initial conditions are appropriately chosen. In this paper, a different inversion procedure is proposed, which presents some computational advantages when applied to flexible robot arms or, more generally, to mechanical structures. This procedure will be presented in a general setting, for nonlinear time-varying control systems described by differential equations of the form

$$\dot{x} = f(x, t) + g(x, t)\, u$$

$$y = h(x, t) \tag{1.1}$$

defined for all $t \in J$, where $J$ is an interval on the real axis. The $n \times 1$ vector $x$ is the state (which belongs to some open subset $U$ of $\mathbb{R}^n$), the $m \times 1$ vector $u$ is the control input and the $m \times 1$ vector $y$ is the output. It is assumed that $f$ and the $m$ columns $g_1, g_2, ..., g_m$ of the matrix $g$ are smooth vector fields and that $h$ is a smooth function with respect to $x$ and $t$. Define, at each $(x, t)$

$\in$ UxJ, $G(x, t) = \text{span}_R \{g_1(x, t), g_2(x, t), ..., g_m(x, t)\}$, and suppose that dim $G(x, t) = m$, for all $(x, t)$ $\in$ UxJ.

The inversion of systems of the form (1.1) is of interest when considering the exact tracking problem, which can be stated as follows. Let $y_d(t)$ be the desired trajectory, which is assumed to be a smooth function. Let $x(t, u, x^0)$ denote the trajectory of system (1.1) corresponding to initial state $x^0$ and control u. The *exact tracking problem* consists in finding a control u, an initial time $t^0$ and an initial condition $x^0$ such that

$$y(x(t, u, x^0), t) = y_d(t) \qquad\qquad t^0 \leq t < T, \ [t^0, T] := I \subset J.$$

The inversion algorithm of [14] for linear time-invariant and time-varying systems, was extended to nonlinear time-invariant systems in [8]. Other inversion algorithms have been proposed in [9, 15]. The so-called zero-dynamics algorithm [3, 10] can be used to establish conditions for the exact output reproducibility of $y_d(t) = 0$, when the given system is time-invariant.

In the next section, a different inversion algorithm is proposed, which presents some computational advantages when considering Lagrange equations. In Section 3, an application illustrates the claimed advantages. In the sequel, it is assumed that $h(x, t)$ already represents the difference between the output of the system and the desired trajectory $y_d(t)$.

## 2. The Inversion procedure

The concept of invariant set is well-known in nonlinear systems analysis [7]. In particular, time-varying invariant manifolds have been recently successfully used in investigations concerning reduced order modelling [11]. In this section, precise definitions of invariant and controlled invariant time-varying manifolds are first given. These definitions are then used to derive the inversion procedure.

**Definition 2.1.** A *smooth time-varying manifold* is the image of a mapping $\pi$: MxI $\rightarrow$U, where M is a smooth submanifold of U, if
(i) for each $t \in$ I, the mapping $\pi_t$: M $\rightarrow$ U , $\xi \rightarrow \pi_t(\xi) = \pi(\xi, t)$ is smooth and is a univalent immersion [2, p.70];
(ii) for each $\xi \in$ M, the image of the mapping $\pi_\xi$: I $\rightarrow$ U , $t \rightarrow \pi_\xi(t) = \pi(\xi, t)$ is a smooth trajectory in U. ∎

The symbol $\pi$ will be used for denoting both the mapping and its image when no confusion is possible.

**Definition 2.2.** A smooth time-varying manifold $\pi$ is an *invariant time-varying manifold* for $t \in I \subset J$ for a system of the form

$$\dot{x} = \varphi(x, t) \tag{2.1}$$

where $\varphi$ is a smooth vector field for all $(x, t) \in U \times I$, if $x(t, x^0) \in \pi_t$ for all $t \in I$, whenever $x^0 \in \pi_{t_0}$.

Define the affine subspace of $T_x \mathbb{R}^n$

$$A_x = T_x \pi_{\bar{t}} + \frac{\partial \pi_{\bar{\xi}}}{\partial t}\Big|_{t=\bar{t}}, \qquad x = \pi(\xi, \bar{t})$$

where

$$\frac{\partial \pi_{\bar{\xi}}}{\partial t}\Big|_{t=\bar{t}}$$

is the tangent vector to the trajectory

$$\pi_{\bar{\xi}}(t) \qquad \text{at } t = \bar{t}.$$

This affine subspace is used to characterize an invariant time-varying manifold, as stated in the following lemma.

**Lemma 2.3.** A smooth time-varying manifold $\pi$ is an *invariant time-varying manifold* of the system (2.1) if and only if there exists an interval $I \subset J$ of the real axis such that $\varphi(\pi_t(\xi), t) \in A_{\pi(\xi, t)}$ for all $(\xi, t) \in M \times I$.

*Proof.* (Only If) If $\pi$ is invariant, for each $t \in I$, there exists $\xi \in M$ such that $x(t, x^0) = \pi(\xi, t)$. For the chain-rule, and because $\pi_t$ is a univalent immersion, $\xi$ viewed as a function of time is differentiable. Then

$$\dot{x} = \frac{\partial \pi}{\partial \xi} \dot{\xi} + \frac{\partial \pi}{\partial t}$$

$$= \varphi(x, t) \tag{2.2}$$

(If) Suppose (2.2) holds, for all $(\xi, t) \in M \times I$, with $x = \pi(\xi, t)$. Being $\partial \pi / \partial \xi$ invertible for each $t$, $\varphi$ and $\partial \pi / \partial t$ smooth, the differential equation

$$\dot{\xi} = \left(\frac{\partial \pi}{\partial \xi}\right)^{-1} \left(\varphi(\pi(\xi, t), t) - \frac{\partial \pi}{\partial t}\right)$$

admits smooth solutions on $I$. ∎

With reference to the control system (1.1), the question arises if it is possible to find a control $u$ which renders a given time-varying manifold invariant for the system (1.1) controlled by $u$. The following definition and lemma are devoted to this problem.

**Definition 2.4.** A smooth time-varying manifold $\pi$ is a *controlled invariant time-varying manifold* of (1.1) if there exists a smooth mapping $u^*: \pi_t \times I \to \mathbb{R}^m$ such that $\pi$ is an invariant manifold for

$$\dot{x} = f(x, t) + g(x, t) u^* \quad \blacksquare$$

Let $v_i(x, t)$, $i = 1, ..., n$, be smooth vectors such that, for each fixed $t \in I$, they constitute a basis for the dual space $T^*_x U$ and

$$v_i(x, t) \cdot g_j(x, t) = \delta_{ij} \qquad\qquad i, j = 1, ..., m$$
$$v_i(x, t) \cdot g_j(x, t) = 0 \qquad\qquad i, = m+1, ..., n, \forall j$$

where the dot denotes the scalar product and $\delta_{ij}$ is the Kronecker operator. One has $G^\perp(x, t) = \text{span}\{ v_{m+1}(x, t), ..., v_n(x, t)\}$ while $R(x, t) = \text{span}\{v_1(x, t), ..., v_m(x, t)\}$ is such that $T^*_x U = G^\perp(x, t) \oplus R(x, t)$. Let $w \in T_x U$, and define

$$P_{x, t}: T_x U \to G^\perp(x, t)$$

$$P_{x, t}(w) = \sum_{i=m+1}^{n} \alpha_i v_i , \qquad\qquad \alpha_i = v_i \cdot w$$

Define also

$$\gamma_{x, t}: T_x U \to R(x, t)$$

$$\gamma_{x, t}(w) = \sum_{i=1}^{m} \beta_i v_i , \qquad\qquad \beta_i = v_i \cdot w$$

Set

$$\bar{f}(x, t) = P_{x, t}(f(x, t))$$

**Lemma 2.5.** A smooth time-varying manifold $\pi$ is a controlled invariant time-varying manifold of (1.1) only if

$$\bar{f}(\pi_t(\xi), t) \in P_{\pi_t(\xi), t} (A_{\pi(\xi, t)}) , \quad \text{for all } (\xi, t) \in M \times I. \tag{2.3}$$

Conversely, if (2.3) holds and the map

$$P_{\pi_t(\xi), t} \circ \frac{\partial \pi_t}{\partial \xi} \quad \text{has a smooth inverse, for all } (\xi, t) \in M \times I, \tag{2.4}$$

(where $\partial \pi_t/\partial \xi$ denotes the jacobian of the map $\pi_t$), the smooth time-varying manifold $\pi$ is a controlled invariant time-varying manifold of (1.1).

*Proof.* (Only if) If $\pi$ is controlled invariant, it follows from Lemma 2.3 that

$$f(\pi_t(\xi), t) + g(\pi_t(\xi), t) u^* \in A_{\pi(\xi, t)} \tag{2.5}$$

The application of the operator $P_{x, t}$ yields the result.

(If) Define

$$f^*(\xi, t) = (P_{\pi_t(\xi), t} \circ \frac{\partial \pi_t}{\partial \xi})^{-1} [\bar{f}(\pi_t(\xi), t) - P_{\pi_t(\xi), t} (\frac{\partial \pi}{\partial t})]$$

(2.6)

Then

$$u = \gamma_{x,t}(\frac{\partial \pi_t}{\partial \xi} f^*(\xi, t) + \frac{\partial \pi}{\partial t} - f(\pi_t(\xi), t))$$

(2.7)

is a smooth solution of

$$f(\pi_t(\xi), t) + g(\pi_t(\xi), t) u = \frac{\partial \pi_t}{\partial \xi} f^*(\xi, t) + \frac{\partial \pi}{\partial t}$$

(2.8)

For, apply operators $P_{x,t}$ and $\gamma_{x,t}$ to (2.8) which yields two identities on $G^\perp$ and $R$. ∎

On the basis of the necessary condition expressed in Lemma 2.5, a smooth time-varying manifold $\pi$ can be iteratively constructed. Under some technical assumptions, this manifold is controlled invariant and $h(\pi, t)$ is zero.

**Procedure 2.6.**
Step 1. Let $M^0$ be a smooth submanifold of $U$ and $\pi^0: M^0 \times I^0 \to U$ a smooth time-varying manifold, where $I^0 \subset I$ is an open interval of the real axis, such that

$$h(\pi^0(\xi, t), t) = 0, \quad \text{for all } (\xi, t) \in M^0 \times I^0.$$

Step k. Set

$$L^k \times T^k = \{ (\xi, t) \in M^{k-1} \times I^{k-1} | \bar{f}(\pi_t^{k-1}(\xi), t) \in P_{\pi_t^{k-1}(\xi), t} (A_{\pi^{k-1}(\xi, t)}) \}$$

Let $M^k \subset L^k$ be a smooth submanifold and $I^k \subset T^k$ an open interval of the real axis. Set

$$\pi^k = \pi^{k-1}|_{M^k \times I^k}$$

Stop if dim $M^k$ = dim $M^{k-1}$. ∎

The following result specifies the hypotheses under which the above procedure converges to a controlled invariant time-varying manifold.

**Theorem 2.7.** Suppose that, at each step k of Procedure 2.6, there exist a nonempty smooth submanifold $M^k \subset L^k$ and an open interval $I^k \subset T^k$ of the real axis. Then, Procedure 2.6 converges in a finite number of steps $k^* < n$. Moreover, if condition (2.4) holds on $M^{k^*} \times I^{k^*}$, $\pi^{k^*}$ is a smooth controlled invariant time-varying manifold such that $h(\pi^{k^*}(\xi, t), t) = 0$, for all $(\xi, t) \in M^{k^*} \times I^{k^*}$.

*Proof.* Since $M^k \subset M^{k-1}$, the dimension of $M^k$ is either equal or smaller than the dimension of $M^{k-1}$. Therefore, the procedure converges in a finite number of steps less than the dimension of $M^0$. By definition, for all $(\xi, t) \in M^{k^*} \times I^{k^*}$,

$$A_{\pi^{k^*-1}(\xi, t)} = A_{\pi^{k^*}(\xi, t)}$$

Therefore, for all $(\xi, t) \in M^{k^*} \times I^{k^*}$,

$$\bar{f}(\pi_t^{k^*}(\xi), t) \in P_{\pi_t^{k^*}(\xi), t}(A_{\pi^{k^*}(\xi, t)})$$

Hence, by Lemma 2.5, the smooth time-varying manifold $\pi^{k^*}: M^{k^*} \times I^{k^*} \to U$ is controlled invariant. ∎

## 3. An example

Consider a two link manipulator, moving in a horizontal plane, with a second flexible link modeled only with one deformation mode (see fig.1), assumed to be a parabola.

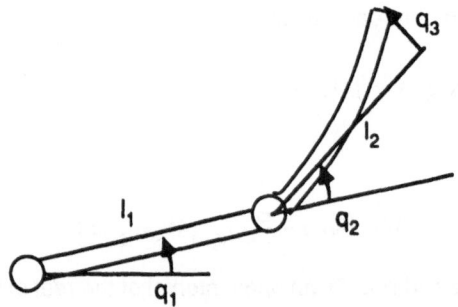

Fig.1   A two link robot arm with a second flexible link

Let $q_1$ be the first joint rotation, $q_2$ the second joint rotation and $q_3$ the flexible link tip displacement with respect to the second link axis. The equations of motion read as

$$b_{11}(q_2)\,\ddot{q}_1 + b_{12}(q_2)\,\ddot{q}_2 + b_{13}(q_2)\,\ddot{q}_3 + n_1(q_2,\dot{q}) = \tau_1$$

$$b_{21}(q_2)\,\ddot{q}_1 + b_{22}\,\ddot{q}_2 + b_{23}\,\ddot{q}_3 + n_2(q_2,\dot{q}) = \tau_2 \qquad (3.1)$$

$$b_{31}(q_2)\,\ddot{q}_1 + b_{32}\,\ddot{q}_2 + b_{33}\,\ddot{q}_3 + n_3(q_2,\dot{q}) = 0$$

where the entries $b_{ij}$ of the symmetrical inertia matrix and the entries $n_i$ of the vector of centripetal, Coriolis and elastic forces are reported in the Appendix of [12] and $\tau_i$ are the applied torques at the joints. In the above model the kinetic energy has been approximated by neglecting those terms depending on $q_3$. The linearized angular end point position is taken as output:

$$z_1 = q_1$$

$$z_2 = q_2 + \frac{q_3}{l_2}$$

where $l_2$ is the second link length.

The equation (3.1) can be put in form (1.1) by inverting the inertia matrix. Note that this is not necessary in order to apply the inversion Procedure 2.6. The second-order structure of the model will be implicitly used in the sequel. The operator P, in these coordinates, is given by

$$P = \begin{bmatrix} b_{31} & b_{32} & b_{33} \end{bmatrix}$$

while the operator $\gamma$ can be written as

$$\gamma = \begin{bmatrix} b_{11} & b_{12} & b_{13} \\ b_{21} & b_{22} & b_{23} \end{bmatrix}$$

The matrix

$$\begin{bmatrix} \gamma \\ P \end{bmatrix}$$

is the inertia matrix and is invertible.

Let $z^*(t)$ be the trajectory to be tracked. The algorithm presented in section 2 can be applied as follows.

Step 1. The output equation can be solved with respect to $q_1$ and $q_2$ thereby obtaining $M^0$ and $\pi^0$. We set

$$M^0 = \{\xi = (q, \dot{q}) \mid q_1 = z_1^*(0), \ \dot{q}_1 = \dot{z}_1^*(0), \ q_2 = z_2^*(0) - \frac{q_3}{l_2}, \ \dot{q}_2 = \dot{z}_2^*(0) - \frac{\dot{q}_3}{l_2}\}$$

and

$$q_1(t) = z_1^*(t)$$

$$q_2(t) = z_2^*(t) - \frac{q_3(t)}{l_2}$$

$$q_3(t) = q_3(t)$$

Step 2. One has

$$\bar{f}(\pi_1^0(\xi), t) = -n_3(z_2^*, q_3, \dot{z}_1^*, \dot{z}_2^*, \dot{q}_3)$$

and $A_{\pi\alpha(\xi, t)}$ is given by

$$\ddot{q}_1(t) = \ddot{z}_1^*(t)$$

$$\ddot{q}_2(t) = \ddot{z}_2^*(t) - \frac{\ddot{q}_3(t)}{l_2} \qquad (3.2)$$

$$\ddot{q}_3(t) = \ddot{q}_3(t)$$

By applying operator P, one has

$$b_{31}^* \ddot{z}_1^* + b_{32}(\ddot{z}_2 - \frac{\ddot{q}_3}{l_2}) + b_{33} \ddot{q}_3$$

where

$$b_{31}^* = b_{31}(z_2^* - \frac{q_3}{l_2})$$

Hence,

$$L^1 \times T^1 = \{(\xi, t) \in M^0 x I^0 |$$

$$(b_{33} - \frac{b_{32}}{l_2}) \ddot{q}_3 = -b_{31}(z_2^* - \frac{q_3}{l_2}) \ddot{z}_1^* - b_{32} \ddot{z}_2^* - n_3(z_2^*, q_3, \dot{z}_1^*, \dot{z}_2^*, \dot{q}_3)\} \quad (3.3)$$

Since the coefficient $b_{33} - b_{32}/l_2$ is nonzero, the above equation can be solved for $d^2q_3/dt^2$. Hence, $M^0 = M^1$ and the algorithm terminates.

The control which keeps the system evolving on $\pi^1 = \pi^0$ is obtained from equation (2.7) which, in our example, corresponds to the first two equations of motion, with $d^2q_1/dt^2$ and $d^2q_2/dt^2$ given by equations (3.2) and $d^2q_3/dt^2$ by equation (3.3).

When following the inversion procedure as given in [12, 17], both the inertia matrix and the decoupling matrix have to be inverted [7], whereas this is not the case in the proposed procedure. When tackling problems with a large number of degrees of freedom, the inversion of the inertia matrix and the subsequent manipulations needed to invert the system may be computationally demanding.

**References**

[1] Bayo, E., Computed torque for the position control of open chain flexible robot, *Proc. IEEE Int Conf. on Robotics and Automation*, (1988).

[2] Boothby, W.M., An introduction to differentiable manifolds and Riemannian geometry, Academic Press, (1975).

[3] Byrnes, C.I., Isidori, A., Local Stabilization of Minimum-phase Nonlinear Systems, *Sys. & Contr. Lett.*, vol.11 (1988), pp. 9-17.

[4] Canudas de Wit, C., Van den Bossche, E., Adaptive control of a flexible arm with explicit estimation of the payload mass and friction, *IFAC Int. Symp. on Theory of Robots*, Vienna, (1986).

[5] Das, A., Singh, S.N., Dual mode control of an elastic arm: nonlinear inversion and stabilization by pole assignment, *Proc. American Control Conference*, (1989).

[6] De Luca, A., Lucibello, P., Ulivi G., Inversion techniques for trajectory control of flexible robot arms, *Int. J. of Robotic Systems*, vol.6, 4 (1989).

[7] Hahn, W., Stability of motion, Springer-Verlag, (1969).

[8] Hirschorn, R.M., Output tracking in multivariable nonlinear systems, *IEEE Trans. Automatic Control*, AC-26 (1981), pp. 595-598.

[9] Hirschorn, R.M., Invertibility of multivariable nonlinear control systems, *IEEE Trans. Automatic Control*, AC-24 (1979), pp. 855-865.

[10] Isidori, A., Nonlinear control systems, Springer-Verlag, 2nd Edition, Communications and Control Engineering Series (1989).

[11] Kokotovic, P.V., Sauer, P.W., Integral manifold as a tool for reducing order modeling in nonlinear systems: a synchronous machine case study, *Proc. 26th IEEE Conf. on Decision and Control*, Los Angeles, CA, (1987), pp.873-878.

[12] Lucibello, P., Nonlinear regulation with internal stability of a two-link flexible robot arm, *Proc. 28th IEEE Conf. Decision and Control*, (1989).

[13] Siciliano, B., Book, W.J., A singular perturbation approach to control of lightweight flexible manipulators, *Int. J. Robotics Res.*, (1988).

[14] Silverman, S.N., Inversion of multivariable linear systems, *IEEE Trans. Automatic Control*, AC-14-3 (1969), pp. 270-276.

[15] Singh, S.N., A modified algorithm for invertibility in nonlinear systems, *IEEE Trans. Automatic Control*, AC-26 (1981), pp. 595-598.

# POSITIONING CONTROL OF FLEXIBLE JOINT ROBOTS*

Patrizio Tomei

Dipartimento di Ingegneria Elettronica, Seconda Università di Roma, *Tor Vergata*, Via O. Raimondo, 00173 Roma, Italy.

*Keywords.* Robots, elastic joint, PD controller.

*Abstract.* The subject of this paper is the problem of point-to-point control of robots whose joints are not rigid. We show that a simple linear feedback (plus a constant compensation) globally stabilizes a robot with elastic joints. Simulation results referred to a one-link elastic joint robot are also presented.

## 1. INTRODUCTION

If accurate models of robot manipulators are needed the elasticity phenomena have to be taken into account, as shown by experimental tests [1]. We refer in this paper to the so called joint elasticity, which is mainly introduced by special gear-boxes so called "harmonic drives". When the the joint elasticity is considered, the dynamic models become more complicated that those of rigid robot [2]. Moreover, some structural properties, such as static state feedback linearization, which were owned by rigid robots, are lost, in general [3]. The design of the corresponding control laws becomes, consequently, a more complex task [2], [4]-[6].

As known [7], a simple PD controller plus gravity compensation is sufficient to globally stabilize a rigid robot. Even though rigid robots are linearizable by state feedback [8], the previous property is based on the property of mechanical systems of being passive. This fact suggests that the same approach may be, in principle, applied also to elastic joint robots, which are mechanical systems as well.

We show in this paper that a simple proportional plus derivative control law globally stabilizes, about a desired position, manipulators having elastic joints.

---

*This work was supported by CNR under contract No. 89.00531.67.

## 2. ROBOT MODEL AND PROPERTIES

We refer to robots having $n + 1$ rigid links, interconnected by $n$ elastic revolute joints. Let $q_1$ represent the $n \times 1$ vector of the position of the links, and let $q_2$ represent the $n \times 1$ vector of the actuator positions. The external torque applied to the $i$th joint is denoted by $u_i$. The kinetic energy of the whole structure is given by

$$T = \frac{1}{2}\dot{q}^T B(q)\dot{q} \tag{2.1}$$

where $q^T = [q_1^T, q_2^T]$ and $B(q)$ is the inertia matrix, which is symmetric positive definite and bounded for any $q$; $B(q)$ is structured as follows [2]

$$B(q) = B(q_1) = \begin{bmatrix} B_1(q_1) & B_2(q_1) \\ B_2^T(q_1) & B_3 \end{bmatrix} \tag{2.2}$$

in which $B_1$ and $B_2$ are $n \times n$ matrices, and $B_3$ is a constant diagonal matrix which depends on the actuator inertias and gear ratios. The potential energy $U$ is given by the sum of two terms. The first one is the gravitational term which takes the form

$$U_1 = U_1(q_1) \tag{2.3}$$

The second one, which is caused by joint elastic torques, is given by

$$U_2 = \frac{1}{2}(q_1 - q_2)^T K_e (q_1 - q_2) \tag{2.4}$$

in which $K_e = \text{diag}[k_1, \ldots, k_n]$, with $k_i$ being the elastic constant of joint $i$. Introducing the matrix

$$K_E = \begin{bmatrix} K_e & -K_e \\ -K_e & K_e \end{bmatrix}$$

(2.4) can be rewritten as

$$U_2 = \frac{1}{2}q^T K_E q \tag{2.5}$$

Viscous frictional forces can be taken into account by introducing the Rayleigh dissipation function [11]

$$\mathcal{F} = \frac{1}{2}\dot{q}^T F \dot{q} \tag{2.6}$$

with $F$ being a symmetric positive semidefinite matrix. According to the Lagrangian formulation, the dynamic equations of the robot motion are

$$\frac{d}{dt}\frac{\partial L}{\partial \dot{q}} - \frac{\partial L}{\partial q} + \frac{\partial \mathcal{F}}{\partial \dot{q}} = m \tag{2.7}$$

where $L(q, \dot{q}) = T(q, \dot{q}) - U(q)$ is the Lagrangian function and $m = [0, \ldots, 0, u_1, \ldots, u_n]^T$. In view of (2.1), (2.3), (2.5) and (2.6), we have

$$B(q_1)\ddot{q} + C(q, \dot{q})\dot{q} + K_E q + F\dot{q} + e(q_1) = m \tag{R}$$

where

$$C(q,\dot{q})\dot{q} = \dot{B}(q_1)\dot{q} - \frac{1}{2}\frac{\partial\dot{q}^T B(q_1)\dot{q}}{\partial q}$$

$$e(q_1) = \frac{\partial U_1(q_1)}{\partial q} = \begin{bmatrix} e_1(q_1) \\ 0 \end{bmatrix} \tag{2.8}$$

in which $e_1 = \dfrac{\partial U_1}{\partial q_1}$.

We assume throughout that matrix $B_2$ is constant. This assumption is true for the quite general robot structure consisting of shoulder, upper arm and forearm, whose motors have rotation axes perpendicular to the axes of the corresponding actuated links.

The dynamic model (R) possesses the following structural properties.

*Property 1.* If the elements of $C(q,\dot{q})$ are defined as

$$C_{ij}(q,\dot{q}) = \frac{1}{2}\left[\dot{q}^T\frac{\partial B_{ij}}{\partial q} + \sum_{k=1}^{2n}\left(\frac{\partial B_{ik}}{\partial q_j} - \frac{\partial B_{jk}}{\partial q_i}\right)\dot{q}_k\right] \qquad i,j = 1,\ldots,2n \tag{2.9}$$

with $q_i$ denoting the $i$th element of vector $q$, then matrix $\dot{B} - 2C$ is skew-symmetric [9].

$\triangle$

*Property 2.* The matrix $C(q,\dot{q})$ has the following form

$$C(q,\dot{q}) = \begin{bmatrix} C_1(q_1,\dot{q}_1) & 0 \\ 0 & 0 \end{bmatrix}$$

with $C_1$ a $n \times n$ matrix. That is a direct consequence of the assumption that $B_2$ is a constant matrix.

$\triangle$

*Property 3.* The structure of matrix $B_2$ is [12]

$$B_2 = \begin{bmatrix} 0 & b_{12} & b_{13} & \cdots & b_{1n} \\ 0 & 0 & b_{23} & \cdots & b_{2n} \\ \vdots & \vdots & \vdots & \vdots & \vdots \\ 0 & 0 & 0 & \cdots & 0 \end{bmatrix}$$

$\triangle$

*Property 4.* Since $e_1(q_1)$ is formed by trigonometric functions of the links positions, a positive real $\alpha$ exists such that

$$\left\|\frac{\partial e_1(q_1)}{\partial q_1}\right\| \leq \alpha, \qquad \forall q_1 \in R^n \tag{2.10}$$

which implies

$$\|e_1(q_1) - e_1(q_{1_0})\| \leq \alpha\|q_1 - q_{1_0}\|, \qquad \forall q_1, q_{1_0} \in R^n \tag{2.11}$$

$\triangle$

## 3. THE PROPOSED CONTROLLER

In this section we show that the dynamic model (R) can be globally stabilized by a very simple control law, consisting of linear feedback from the rotor variables plus a constant compensation term.

Consider the following control law

$$u = -K_p(q_2 - q_{2_0}) - K_d \dot{q}_2 + e_1(q_{1_0}) \tag{3.1}$$

where $K_p$ and $K_d$ are $n \times n$ symmetric positive definite matrices, $q_{1_0}$ is the vector of the desired positions of the links and $q_{2_0}$ is the corresponding vector of the steady-state motor rotor positions,

$$q_{2_0} = q_{1_0} + K_e^{-1} e_1(q_{1_0}) \tag{3.2}$$

Define the matrix

$$K = \begin{bmatrix} K_e & -K_e \\ -K_e & K_e + K_p \end{bmatrix} \tag{3.3}$$

which is symmetric and positive definite, and the variables

$$z_1 = q_1 - q_{1_0}, \qquad z_2 = q_2 - q_{2_0},$$
$$z = [z_1^T, z_2^T]^T, \qquad q_0 = [q_{1_0}^T, q_{2_0}^T]^T,$$

System (R) with the control law (3.1), taking (3.2) into account, becomes

$$B(z_1 + q_{1_0})\ddot{z} + C(z + q_0, \dot{z})\dot{z} + Kz + F\dot{z} + e(z_1 + q_{1_0}) - e(q_{1_0}) = -K_D\dot{z} \tag{3.4}$$

where

$$K_D = \begin{bmatrix} 0 & 0 \\ 0 & K_d \end{bmatrix}. \tag{3.5}$$

System (3.4) can be equivalently obtained by the following Lagrangian and Rayleigh functions

$$L_z = T_z - U_z$$
$$\mathcal{F}_z = \frac{1}{2}\dot{z}^T(K_D + F)\dot{z} \tag{3.6}$$

in which

$$T_z = \frac{1}{2}\dot{z}^T B(z_1 + q_{1_0})\dot{z}$$
$$U_z = \frac{1}{2}z^T Kz + U_1(z_1 + q_{1_0}) - U_1(q_{1_0}) - z^T e(q_{1_0}) \tag{3.7}$$

by means of the Lagrangian equations

$$\frac{d}{dt}\frac{\partial L_z}{\partial \dot{z}} - \frac{\partial L_z}{\partial z} + \frac{\partial \mathcal{F}_z}{\partial \dot{z}} = 0 \tag{3.8}$$

Now, note that

$$U_1(z_1 + q_{1_0}) - U_1(q_{1_0}) - z^T e(q_{1_0}) = U_1(z_1 + q_{1_0}) - U_1(q_{1_0}) - z_1^T e_1(q_{1_0})$$
$$= \int_0^{z_1} \frac{\partial U_1(\xi + q_{1_0})}{\partial \xi} - e_1(q_{1_0}) \, d\xi$$
$$= \int_0^{z_1} e_1(\xi + q_{1_0}) - e_1(q_{1_0}) \, d\xi = \int_0^{z_1} \int_0^{\xi} \frac{\partial e_1(\nu + q_{1_0})}{\partial \nu} \, d\nu \, d\xi$$

which, by virtue of (2.10), implies

$$U_1(z_1 + q_{1_0}) - U_1(q_{1_0}) - z^T e(q_{1_0}) \le \frac{1}{2} \alpha z_1^T z_1$$

Therefore, we can write

$$U_z \ge \frac{1}{2} \|z_1\|^2 K_{e_m} + \frac{1}{2} \|z_2\|^2 (K_{e_m} + K_{P_m}) - K_{e_M} \|z_1\| \|z_2\| - \frac{1}{2} \alpha \|z_1\|^2$$
$$= \frac{1}{2} \begin{bmatrix} \|z_1\| \\ \|z_2\| \end{bmatrix}^T \begin{bmatrix} K_{e_m} - \alpha & -K_{e_M} \\ -K_{e_M} & K_{e_m} + K_{p_m} \end{bmatrix} \begin{bmatrix} \|z_1\| \\ \|z_2\| \end{bmatrix} \qquad (3.9)$$

where $K_{e_m}$ and $K_{e_M}$ are, respectively, the smallest and the largest joint elastic constants, and $K_{p_m}$ is the smallest eigenvalue of $K_p$. From (3.9), we obtain that if the following two conditions hold

$$K_{e_m} > \alpha \qquad (3.10)$$
$$K_{p_m} > \frac{K_{e_M}^2 - K_{e_m}^2 + \alpha K_{e_m}}{K_{e_m} - \alpha} \qquad (3.11)$$

then $U_z$ is a positive definite function of the vector $z$.

Owing to the physical values of the robot parameters, hypothesis (3.10) is generally satisfied. For instance, for a single-link with a rotational flexible joint which can rotate in a vertical plane, we have

$$e_1(q_1) = \frac{1}{2} mgl \sin(q_1)$$

where $m$ is the link mass, $g$ is the gravity constant and $l$ is the link length. Hence,

$$\alpha = \frac{1}{2} mgl \simeq 5ml$$

while the elastic constant is usually 100 times greater than the factor $ml$. Hypothesis (3.11) can be always satisfied by increasing in norm the gain matrix $K_p$.

The stability of the proposed control law (3.1) can be proved by choosing as candidate Lyapunov function the Hamiltonian associated to the Lagrangian function (3.6), that is

$$V(z, \dot{z}) = T_z + U_z$$
$$= \frac{1}{2} [\dot{z}^T B(z_1 + q_{10}) \dot{z} + z^T K z] + U_1(z_1 + q_{10}) - U_1(q_{10}) - z^T e(q_{10})$$

whose time derivative, along (3.4), is given by

$$\dot{V}(z,\dot{z}) = \frac{1}{2}\dot{z}^T \dot{B}(z_1 + q_{10})\dot{z} + \dot{z}^T K z + \frac{\partial U_1}{\partial z_1}\dot{z}_1 - \dot{z}^T e(q_{10})$$

$$+ \dot{z}^T[-C(z+q_0,\dot{z})\dot{z} - Kz - F\dot{z} - e(z_1 + q_{10}) + e(q_{10}) - K_D\dot{z}] \quad (3.12)$$

Recalling Property 1 and (2.8), we have

$$\dot{V}(z,\dot{z}) = -\dot{z}^T F\dot{z} - \dot{z}_2^T K_d \dot{z}_2 \qquad (3.13)$$

Hence, $z = 0, \dot{z} = 0$ is a stable equilibrium point. We now prove that the equilibrium is also globally asymptotically stable. Consider the following set

$$S = \{z, \dot{z} : \quad \dot{V}(z,\dot{z}) = 0\} = \{z, \dot{z} : \quad F\dot{z} = 0, \dot{z}_2 = 0\}$$

The dynamics (3.4) restricted to $S$ is given by

$$B_1(z_1 + q_{1_0})\ddot{z}_1 + C_1(z_1 + q_{10}, \dot{z}_1)\dot{z}_1 + K_e(z_1 - z_2)$$
$$+ e_1(z_1 + q_{1_0}) - e_1(q_{1_0}) = 0 \qquad (3.14)$$
$$B_2^T \ddot{z}_1 + K_e(z_2 - z_1) + K_p z_2 = 0 \qquad (3.15)$$

Differentiating (3.15) along $S$, we have

$$K_e \dot{z}_1 = B_2^T \frac{d}{dt}(\ddot{z}_1)$$

which, due to Property 3, implies

$$z_1 = \text{constant} \qquad (3.16)$$

that substituted in (3.14),(3.15) gives

$$Kz + e(z_1 + q_{10}) - e(q_{10}) = 0 \qquad (3.17)$$

The previous algebraic equation, by virtue of Property 4 and hypotheses (3.10),(3.11), has the unique solution $z = 0$. As a consequence, the only solution of (3.4) which lies entirely in the region $S$ consists of the equilibrium point $z = 0, \dot{z} = 0$. We have thus proved by Lasalle Theorem ([10], p. 108) the global asymptotic stability of that equilibrium point. More specifically , we have proved the following theorem.

*Theorem* 3.1. Consider the system (R) with the control law (3.1). If conditions (3.10) and (3.11) are satisfied, then the equilibrium point $z = 0, \dot{z} = 0$ is globally asymptotically stable.

$$\triangle$$

Theorem 3.1 states that by using only half of the robot state variables it is possible to globally stabilize the closed loop system. In particular, the state variables which are used for the control are the same employed in the rigid robot case. Note that matrix $F$ has been assumed to be positive semidefinite (e.g. it may be null); therefore the stability of the controller is not based on frictional forces.

When the elastic constants and/or the gravity parameters are not exactly known the control law (3.1) cannot be implemented. Let $\hat{K}_e$ and $\hat{e}$ be the available estimates of the elastic constant matrix and of the gravity vector. The implementable control law is

$$u = -K_p(q_2 - \hat{q}_{2_0}) - K_d\dot{q}_2 + \hat{e}_1(q_{1_0}) \qquad (3.18)$$

where

$$\hat{q}_{2_0} = q_{1_0} + \hat{K}_e^{-1}\hat{e}_1(q_{1_0})$$

The following theorem holds [13].

*Theorem 3.2.* Consider system (R) with the control law (3.18). If conditions (3.10) and (3.11) are satisfied, then system (R), (3.18) has only one equilibrium point which is globally asymptotically stable.

*Proof.* See [13].

$$\triangle$$

Theorem 3.2 states that even with not exact knowledge of elastic and gravity parameters the closed loop system (R),(3.18) is still stable, but the equilibrium point may be in general different from the desired one.

## 4. AN EXAMPLE: ONE-LINK ARM.

As an example, we have tested by simulations the performances of the proposed PD controller on a robot consisting of one rigid link with one elastic joint, rotating in a vertical plane. Viscous frictional forces have not been considered. Its dynamic model is represented by

$$J_L\ddot{q}_1 + k(q_1 - q_2) + \frac{1}{2}mgl\sin q_1 = 0$$

$$J_R\ddot{q}_2 + k(q_2 - q_1) = u$$

where $J_L$ and $J_R$ are, respectively, the inertias of the link and of the motor rotor, $m$ is the link mass, $g$ is the gravity constant, $l$ is the link length and $k$ is the elastic constant of the joint. The following values were adopted for the robot parameters (all values are in SI units)

$$m = 1, \qquad l = 1, \qquad k = 100$$
$$J_R = 0.02, \qquad J_L = 0.4.$$

The task was that of positioning the end effector from an initial position (in joint coordinates) of

$$q_1(0) = 0, \qquad q_2(0) = 0$$

to a final desired position of

$$q_{1R} = \pi/4, \qquad q_{2R} = \pi/4 + \frac{1}{k}\frac{mgl}{2}\sin \pi/4.$$

In the first simulation run the following control gains were chosen

$$k_p = 10, \qquad k_d = 3. \qquad (4.1)$$

The results are drawn in Fig. 1. As one can see, these results are acceptable, even though thre is a little overshoot in the transient response. This overshoot can be avoided by increasing the gains, as shown in Fig. 2 which corresponds to the following control parameters

$$k_p = 30, \qquad k_d = 10. \tag{4.2}$$

The same tests were repeated by using the following modified control law

$$u = -k_p(q_1 - q_{1R}) - k_d\dot{q}_2 + e_1(q_{1R}) \tag{4.3}$$

i.e. the feedback of the rotor error has been replaced by feedback of the link error. The control law (4.3) would be better than (3.1) since it does not require the knowledge of the elastic constants. On the contrary, for the control law (3.1) the knowledge of $k$ is needed to compute the rotor reference $q_{2_0}$ (see (3.2)).

In the first simulation the control law (4.3) was adopted with the gains (4.1). The results, that are reported in Fig. 3, show that errors are practically identical to those of Fig. 1.

In the second simulation the gains (4.2) were used. The relative results reported in Fig. 4 show that in this case there are appreciable differences with respect to the results obtained with the control law (3.1) (see Fig. 2). In particular, the responses with the controller (4.3) present a much more oscillatory behavior. However, surprisingly, in both cases the controller (4.3) yields a closed loop stable system. Currently, we are trying to prove theoretically the stability of the control law (4.3) when applied to general $n$-link robots with elastic joints.

## REFERENCES

[1] M.C. Good, L.M. Sweet and K.L. Strobel, "Dynamic models for control system design of integrated robot and drive systems", ASME J. Dynamic Syst., Meas., Contr., Vol. 107, pp. 53-59, 1985.

[2] R. Marino and S. Nicosia, "Singular perturbation techniques in the adaptive control of elastic robots", IFAC Symp. Robot Control, Barcelona, Nov. 1985.

[3] R. Marino and S. Nicosia, "On the feedback control of industrial robots with elastic joints: a singular perturbation approach", R-84.01 Dept. Electrical Eng. 2nd University of Rome, June 1984.

[4] A. De Luca, A. Isidori and F. Nicolo', "Control of robot arm with elastic joints via nonlinear dynamic feedback", 24th Conf. Decision and Control, Ft. Lauderdale, Dec. 1985.

[5] K. Khorasani and P.V. Kokotovic, "Feedback linearization of a flexible manipulator near its rigid body manifold", Systems & Control Letters, Vol. 6, pp. 187-192, 1985.

[6] M.W. Spong, K. Khorasani and P.V. Kokotovic, "An integral manifold approach to the feedback control of flexible joint robots", IEEE J. of Robotics and Automation, Vol. RA-3, pp. 291-300, 1987.

[7] S. Arimoto and F. Miyazaki, "Stability and robustness of PID feedback control for robot manipulators of sensory capability", Robotics Research, eds. M. Brady and R.P. Paul, Cambridge: MIT Press, 1984.

[8] E. Freund, "Fast nonlinear control with arbitrary pole placement for industrial robots and manipulators", The Int. J. of Robotics Research, Vol. 1, pp. 65-78, 1982.

[9] J.E. Slotine and W. Li, "On the adaptive control of robot manipulators", The Int. J. of Robotics Research, Vol. 6, pp. 49-59, 1987.

[10] W. Hahn, "Stability of motion", Springer-Verlag, 1967.

[11] H. Goldstein, "Classical mechanics", Addison-Wesley, 1950.

[12] P. Tomei, "An observer for flexible joint robots", IEEE Trans. Automatic Control, Vol. 35, pp. 739-743, 1990.

[13] P. Tomei, "A simple PD controller for robots with elastic joint", to appear on IEEE Trans. Automatic Control.

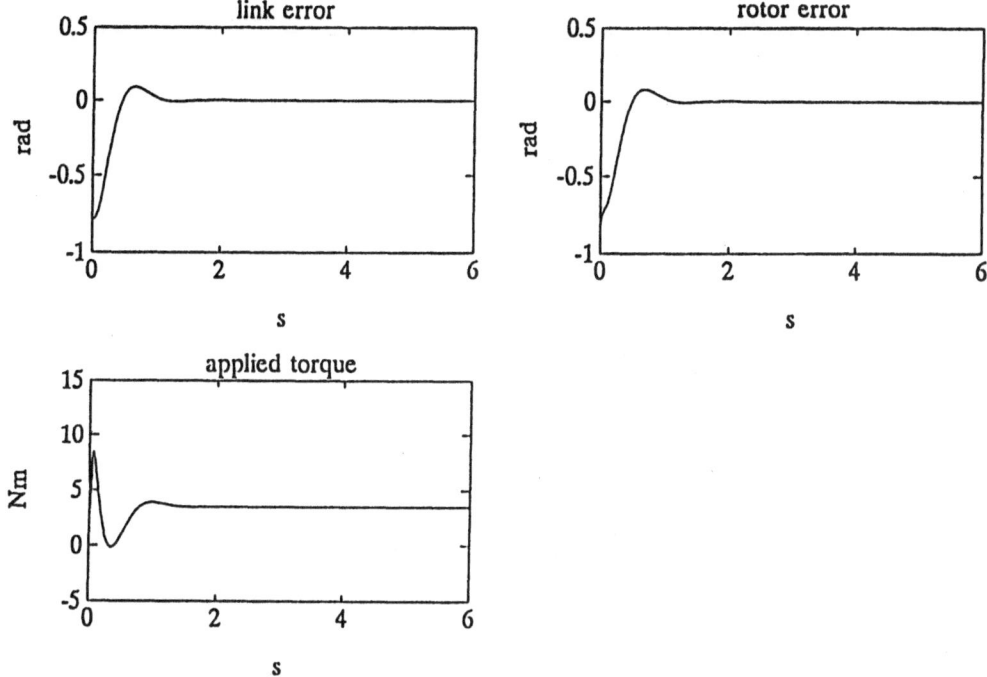

Figure 1. Errors with controller (3.1):
gains (4.1).

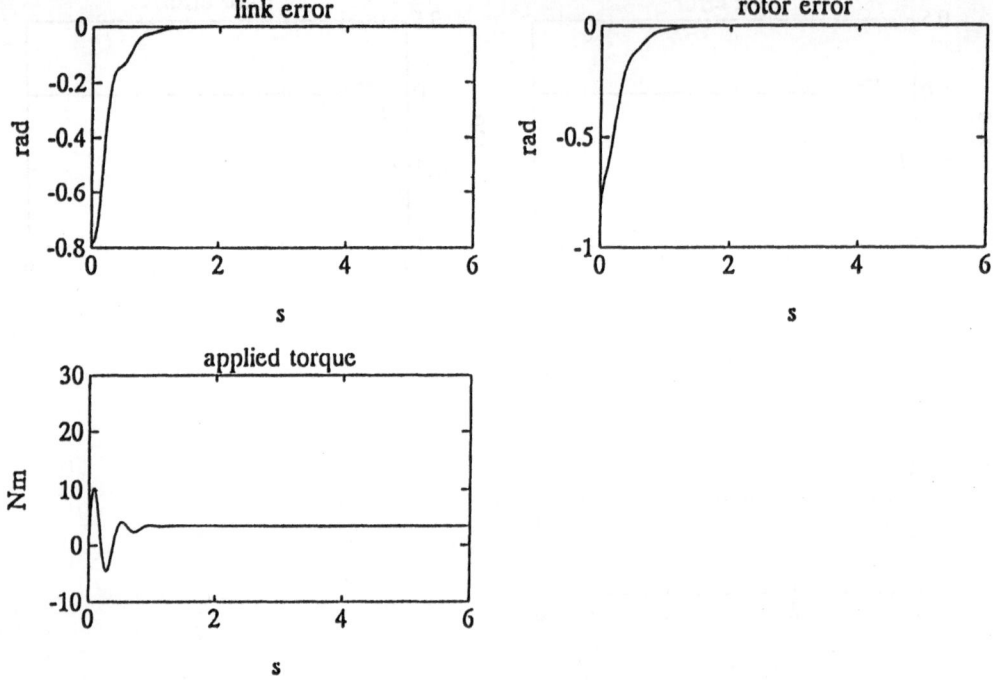

Figure 2. Errors with controller (3.1):
gains (4.2).

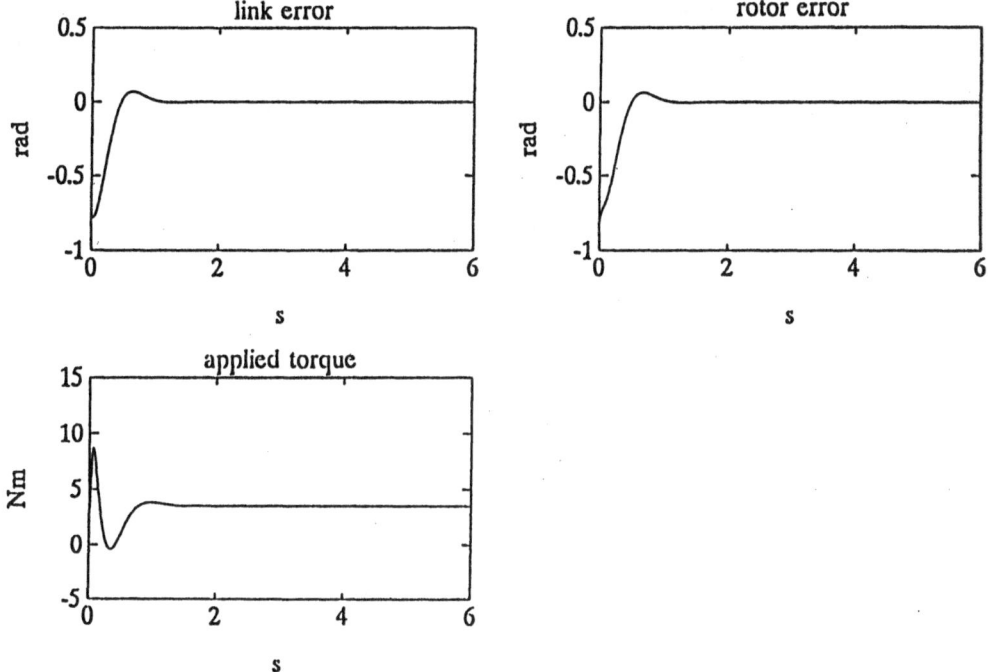

Figure 3. Errors with controller (4.3):
gains(4.1).

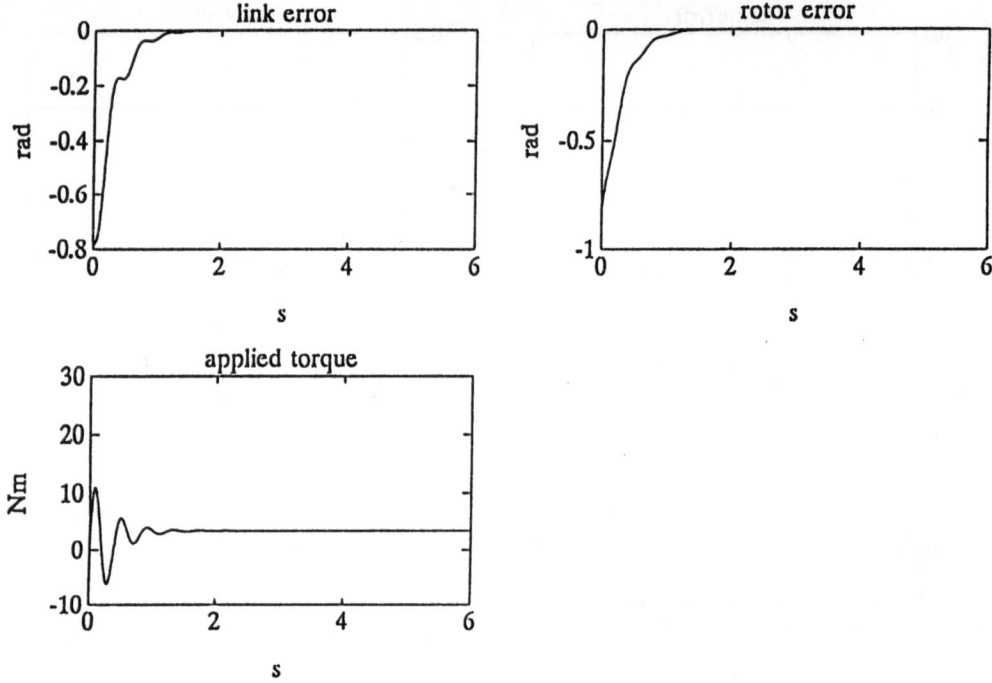

Figure 4. Errors with controller (4.3):
gains (4.2).

# LONG RANGE PREDICTIVE MULTIVARIABLE CONTROL OF A TWO LINKS FLEXIBLE MANIPULATOR

J.M. DION, L. DUGARD, T. NGUYEN THI THANH

Laboratoire d'Automatique de Grenoble (INPG - CNRS)

ENSIEG - BP 46

38402 Saint-Martin d'Hères

The authors are also with the GR "Automatique" - CNRS

**Abstract :**

In this paper, we are interested in the control of a two degree of freedom flexible arm. The objective is to develop a control algorithm that allows to follow some prespecified reference trajectories. Based on a linear two inputs-two outputs model, the control law proposed should cope with the flexibilities and the couplings between the two links for reasonable displacements. To this end, a long range predictive control with a precompensator is used to decouple with stability and to compensate for the vibration modes. Some experimental results on the flexible arm are given.

**Keywords :** Flexible manipulators, multivariable control, long range predictive control.

## I. INTRODUCTION

Control of flexible structures has been a challenge for many years, particularly in aerospace engineering and in robotics. Indeed, the requirements concerning the closed-loop bandwidths impose to take into account the flexibilities of the manipulators. Most experimental studies have focused on single link flexible manipulators. See [1]-[2] for instance. However, relatively little activity has been devoted to the control design of two link (planar) manipulators. Some recent experimental results are given in [3] and [4] for instance.

In this paper, we are interested in the end-point position control for a planar manipulator which has a very flexible link, the other being rigid. With respect to the one degree of freedom flexible arms which exhibit non minimum phase behaviours, several low damped vibration modes, possible non linearities due to friction, multi-link manipulators are characterized by strong couplings and interactions between the links. This makes the control much more difficult.

Based on linear multivariable identified models, long range predictive controllers with fixed parameters are computed and applied to the experimental set-up. The proposed controllers involve a precompensator that allows to decouple the system with stability.

This paper is organized as follows : the experimental set-up is described in section II. Different multivariable ARMA models are given in section III. The control law is presented in section IV. Experimental results are reported in section V, in particular, the influence of the control design parameters and of the underlying models is explained. It is shown that the proposed controller performs quite satisfactorily for reasonable displacements of the end point.

## II - DESCRIPTION OF THE EXPERIMENTAL SET-UP

The experimental set-up is shown in figure 1. This consists of two links, one being very flexible, the other rigid. The flexible beam is made up of two aluminium strips separated by six steel made bridges. Each strip is 60 cm long and 50 mm wide. The first link is clamped on a rigid aluminium hub which is mounted directly on a shaft of a dc-motor. The rigid link mounted at the end of the first link consists of a 20 cm long aluminium cylinder in which lies a dc-motor.

The set-up is very compliant in the horizontal plane to which its motion is confined, but is relatively stiff in vertical bending and torsion.

The two motor axes are locally controlled by numerical PD controllers (IBM PC with an Amadeus card) which are tuned to provide two actuators with a bandwidth of about 15 Hz. In the sequel, the two control inputs ($u_1$ and $u_2$) will be the set-points of the two local loops.

Sensors are installed to measure the hub angles $\theta_1$ and $\theta_2$ (given by potentiometers), and the end tip position of the second link is given by an emitter diode D1 and a photodiode providing with voltages proportional to X, Y displacements).

The inputs of the plant to be controlled $u_1$ and $u_2$ are therefore the voltage commands supplied to the servomotors. The outputs of the plant are either the (x, y) end tip position provided by D1 or the (x, $\theta_2$) position provided by D1 and by the second potentiometer. (See figures 2 and 3a).

The overall control is performed with a HP 1000 - A 900 mini computer.

In the following, different input-output linear models are given according to the considered outputs.

## III - PARAMETRIC PLANT-MODELS

This section is devoted to parametric linear plant-models to be used later for control purposes. As already mentioned, since the plant is truly non linear, linear plant models are only valid for reasonable end-point displacements (velocity and position). By reasonable, we mean that the displacements should be sufficiently small so as the plant model mismatch is relatively small compared to the nominal values of the plant model and sufficiently important so as hard non linearities such as stiction and backlash do not deteriorate too much the behavior of the closed loop system. This will correspond to follow geometrical profiles (about 35 centimeters within 15 seconds).

We present first the class of models and the techniques used to get these models. Then we give some identification results in terms of autoregressive moving average (ARMA) models for both $(x, y)$ or $(x, \theta_2)$ outputs.

### III-1 - *Class of models. Identification methods*

As mentioned, we use the "black box" approach. At this stage, we admit that it is perhaps not the better one, but according to velocity and position objectives, this choice seems justified.

For identification purposes, we use the following parametric model.

$$A(q^{-1}) \, Y(t) = B(q^{-1}) \, U(t) + W(t) \tag{1}$$

where $A(q^{-1})$ and $B(q^{-1})$ are 2 x 2 polynomial matrices

$$A(q^{-1}) = I + A_1 q^{-1} + \ldots + A_{n_a} q^{-n_a}$$

$$B(q^{-1}) = B_1 q^{-1} + \ldots + B_{n_b} q^{-n_b}$$

$$Y(t) = \begin{cases} [x(t), y(t)]^T & \text{for the (x, y) model} \\\\ \left[x(t), \theta_2(t)\right]^T & \text{for the } (x, \theta_2) \text{ model} \end{cases}$$

$$U(t) = [u_1(t), u_2(t)]^T$$

W(t) : external disturbances and modeling errors

$q^{-1}$ : backward shift operator

$$\theta = [A_1, A_2, ..., A_{n_a}, B_1 ... B_{n_b}]$$

Such a model represents the nominal plant input-output dynamics via the transfer matrix $A^{-1}(q^{-1})\, B(q^{-1})$ which is completely defined by the parameter matrix $\theta$.

The identification of the parameter matrix $\theta$ is performed using a simple off line recursive least square algorithm, the inputs being uncorrelated pseudo random binary sequences, more

precisely the parameter estimate $\hat{\theta}(t)$ is given by

$$\hat{\theta}(t) = \hat{\theta}(t-1) + \frac{\varepsilon(t)\, \phi^T(t-1)\, F(t-1)}{1 + \phi^T(t-1)\, F(t-1)\, \phi(t-1)}$$

$$\varepsilon(t) = y(t) - \hat{\theta}(t-1)\, \phi(t-1) \tag{2}$$

$$F(t) = F(t-1) - \frac{F(t-1)\, \phi(t-1)\, \phi^T(t-1)\, F(t-1)}{1 + \phi^T(t-1)\, F(t-1)\, \phi(t-1)}$$

F(t) : identification gain matrix ; F(0) > 0

$$\phi^T(t-1) = [-\, Y^T(t-1), ..., -\, Y^T(t-n_a),\, U^T(t-1), ...,\, U^T(t-n_b)]$$

### III-2 - *Identification results*

Some identification experiments have been performed using an identification software developed by H.N Duong, for different sampling periods and model structures.

The dc-motors are locally controlled so that they can be considered as static gains (bandwidth : 15 Hz). The first vibration modes of the whole system are then only due to the flexibility of the first link. The first two significant vibration modes are approximately 0.65 and 7 Hz, the first being much more energetic (ratio 25 dB). The sampling frequency is accordingly chosen equal to 10 Hz (approximatively 15 times the first vibration mode frequency).

In the sequel, we will use models with $n_a = n_b = 2$ or $n_a = n_b = 3$.

We give here two identified models corresponding to $(x, y)$ see figure 3b, and $(x, \theta_2)$, see figure 3c, outputs respectively. $(x, y)$ model :

$$A_1 = \begin{bmatrix} -1.4759 & 0 \\ 0.1074 & -.6075 \end{bmatrix} \quad A_2 = \begin{bmatrix} .3414 & 0 \\ -.0420 & 0 \end{bmatrix} \quad A_3 = \begin{bmatrix} .3819 & .4013 \\ -.0470 & -.0494 \end{bmatrix}$$

$$(3)$$

$$B_1 = \begin{bmatrix} -.0172 & .0095 \\ .0143 & -.0599 \end{bmatrix} \quad B_2 = \begin{bmatrix} -.0847 & .1442 \\ .0104 & -.0177 \end{bmatrix} \quad B_3 = \begin{bmatrix} -.0441 & -.1437 \\ .0054 & -.0177 \end{bmatrix}$$

Poles :  - .3440
         .9094 ± .3730j
         .6086
         .0006
       - .0006

Zeros :  - 1.7942 ± .3730j
         ± .0018j
         ± .0002j

Static gain :  $G = \begin{bmatrix} -.8024 & .3547 \\ .1309 & -.1938 \end{bmatrix}$

$(x, \theta_2)$ model :

$$A_1 = \begin{bmatrix} -1.5843 & 0 \\ -.0121 & -.4171 \end{bmatrix} \quad A_2 = \begin{bmatrix} .5325 & 0 \\ .0029 & 0 \end{bmatrix} \quad A_3 = \begin{bmatrix} .2404 & -.4200 \\ .0013 & -.0023 \end{bmatrix}$$

$$(4)$$

$$B_1 = \begin{bmatrix} -.0246 & .0096 \\ .0073 & .1926 \end{bmatrix} \quad B_2 = \begin{bmatrix} -.0738 & .1691 \\ -.0004 & .0009 \end{bmatrix} \quad B_3 = \begin{bmatrix} -.0589 & -.2363 \\ -.0003 & -.0013 \end{bmatrix}$$

Poles :    .9104 ± .3744j                        Zeros :  - 2.3767
        - .2460                                            - .8377
          .4267                                            ±  . 0002j
        ± .0002j

Static gain :   $G = \begin{bmatrix} -0.8347 & 0.4457 \\ -0.0000 & 0.3372 \end{bmatrix}$

The main features of these models are :

• a low damped vibration mode (0.62Hz) for both models,
• strong couplings for both models,
• non minimum phase behaviour ("unstable zeros") for both models,

These two models will be used in the sequel to design the control algorithm.


## IV - CONTROL DESIGN

The purpose of the control law is that the arm end-point follows a desired trajectory at a certain speed. In order to achieve this goal, the closed loop system should be decoupled.

The above models are not static state feedback decouplable with stability. This is due to the fact that the rows of the system transfer matrices have no unstable zeros while the transfer matrices have ones. In order to obtain the desired decoupling, we will include a precompensator in the control law. Moreover, due to the good practical properties of the long range predictive controllers, the control law will be designed by minimizing a quadratic cost function on a receding horizon [5].

More precisely, the best precompensator $C(q^{-1})$ adding the minimal number of delays and unstable zeros is given by $B^{-1}(q^{-1}) D(q^{-1})$ where $D(q^{-1})$ is diagonal, with lowest Mac Millan degree and is such that $B^{-1}(q^{-1}) D(q^{-1})$ is proper and stable paper [6]. For simplicity, we will take here :

$$C(q^{-1}) = q B^{-1}(q^{-1}) \det B(q^{-1}) = q \text{ adj } B(q^{-1}) \qquad (5)$$

We consider then the following criterion :

$$J(t) = \sum_{j=1}^{hp} (Y(t+j) - Y_M(t+j))^T R(Y(t+j) - Y_M(t+j)) + \sum_{j=1}^{hc} \Delta U_c^T (t+j-1) Q \Delta U_c(t+j-1) \qquad (6)$$

where hp is the prediction horizon

hc $\leq$ hp is the control horizon $(\Delta U_c(t+j-1) = 0$ for $j >$ hc)

$Y_m(t)$ is the desired trajectory

$\Delta$ is the difference operator $(1 - q^{-1}) I$

R and Q are some positive definite matrices

$U_c(t)$ is the input of the compensated process

$$U(t) = C(q^{-1}) U_c(t) \qquad (7)$$

The control scheme is depicted in figure 4.

The solution $\Delta U_{ct}$ minimizing (6) where $\Delta U^T_{ct} =$

$[\Delta U^T_c(t), \dots , \Delta U^T_c (t + hc - 1) ]$ is given by :

$$\Delta U_{ct} = [G^T \overline{R} G + \overline{Q}]^{-1} G^T \overline{R} [Y_{Mt} - f_{ct}] \qquad (8)$$

where   $- \overline{R} = $ diag $(R, \dots, R)$

$- \overline{Q} = $ diag $(Q, \dots, Q)$

$- Y_{Mt}^T = [Y_M^T (t+1), \dots, Y_M^T (t + hp)]$

$$- G = \begin{bmatrix} G_0 & & \\ G_1 & G_0 & \\ & G_1 & \\ & & \\ G_{hp-1} & G_{hp-hc} & \end{bmatrix} \quad (2\ hp \times 2\ hc)$$

- $f_{ct}$ is a vector built with past control input increments $\Delta U_c (t - i) i > 0$ and with present and past outputs $Y(t - i) i \geq 0$, see [5] for further details.

- the $G_i$'s are the hp first elements of the system impulse response.

The implemented control input U(t) at time t is

$$U(t) = C(q^{-1}) U_c(t)$$

with

$$U_c(t) = U_c(t - 1) + \Delta U_c (t) \tag{9}$$

where $\Delta U_c(t)$ is the first element of $\Delta U_{ct}$.

This control will be effectively applied to the planar flexible arm, using various design parameters.


## V - EXPERIMENTAL RESULTS

In this section, we will give first open loop control responses, then closed loop control responses with a static plant-model and finally closed loop control responses with the proposed control law that exhibits the tracking capability of the control design.

Firstly, experimental results are presented, when using the (x, y) model obtained as mentioned in section III, for the end position given in figure 3b.

The figure 5 shows experimental results with open loop control computed with the static model of the process $(Y = q^{-1} K U)$ i.e. $U(t) = K^{-1} Y_M(t)$ where K is the static gain of the process. The desired trajectory is indicated in dotted line and the true end-point trajectory is given in full line. Five curves are given. It is shown that the effector end-point does not follow in a precise way the desired trajectory. (See the plan x, y). There are strong couplings between x and y and the first natural vibration mode appears clearly for x. The two control inputs $u_1$ and $u_2$ are also given.

The figures 6 shows control results with closed loop with output static feedback computed with the static model. $U(t) = K_1 [Y_M(t) - R_0 Y(t)]$ where $R_0$ and $K_1$ are constant square matrices satisfying $R_0 K K_1 = - \alpha I$ ; where $\alpha$ are the chosen closed loop poles and $I = K K_1 (I + R_0 K K_1)^{-1}$

The behaviour is slightly better than for figure 5, when the non-null closed loop poles are set to 0.2. When the non-null closed loop pole approaches zero, one tends to the open-loop behaviour because $R_0$ tends to zero.

The figures 7 and 8 present more interesting results. The figure 7 shows a control result without precompensator. The prediction horizon is hp = 5, the control horizon is hc = 2, the weighting matrices R = I, Q = 0.3 I, the model orders are $n_a$ = 2 and $n_b$ = 2. The couplings are seen at times 4, 12, 20, 28 seconds when the slope of x changes. There is a drastic improvement in the trajectory following. The figure 8 shows a control result with precompensator (see the control scheme 4). The design parameters are the same except for R = I, Q = 0.005 I. The couplings are attenuated, see the behaviour for Y with respect to Y of figure 7. However, the behaviour in the (x, y) plane is not better than for the figure 7.

The figures 9 and 10 show similar results using a ($n_a$ = $n_b$ = 3) plant model. With the precompensator (figure 10) the tracking is fairly good.

Secondly, experimental results are presented for the $(x, \theta_2)$ model corresponding to the end position given in figure 3c.

In a similar way, the open loop control response is shown in figure 11. The behaviour is fairly bad but better than its counterpart of figure 5.

The figure 12 is the counterpart of figure 6 (closed loop control with static model). The behaviour is slightly better.

The figures 13 and 14 are the counterparts of figures 9, 10. The predictive controller is computed with a ($n_a$ = $n_b$ = 3) plant model. Without precompensator the behaviour is rather good, compared to 9. When the precompensator is used, there is no improvement with respect to 13.

**Remark.** Notice that for figures 3b and 3c the trajectories are quite similar but they do not correspond exactly to the same configuration (arm position). The control inputs are therefore fairly different and the comparisons between them are valid only up to a certain point.

**Concluding remarks.**

We have presented, in this paper, preliminary results concerning long range predictive control of a two degree of freedom flexible arm, based on linear models. The experimental results are encouraging but we have to perform further experiments in order to validate the different models (x, y) and (x, $\theta_2$) in the same configurations. As well, on line estimation of the plant-model parameters should allow to implement an adaptive version of the proposed control scheme when the desired trajectory induces a noticeable change in the parameters. Last, non linear model based controls could be envisaged.

## REFERENCES

[1]  R.H. Cannon, E.Schmitz.
    "Initial Experiments on the End-point Control of a Flexible one Link Robot" . Int. J. of Robotics, vol. 3, n°3, pp. 449-502, 1984.

[2]  M. M'saad, E. Van Den Bossche, A.J. Montano, L. Dugard.
    "A long Range Predictive Adaptive Control : performances enhancements and experimental evaluation". IFAC Workshop on Robustness in Adaptive Control, Newcastle, August 1988.

[3]  S. Yarkovich, A.P. Tzes, I. Lee, K.L. Hillsky.
    "Control and System Identification of a Two-link Flexible Manipulator". IEEE Int. Conf. Robotics and Automation, Cincinnati, May 1990.

[4]  H.G. Lee, S. Kawamura, F. Miyazaki, S. Arimoto
    "External Sensory Control for End-effector of Flexible Multi-link Manipulator". IEEE Int. Conf. Robotics and Automation, Cincinnati , May 1990.

[5]  D.W. Clarke, P.S.Tuffs, C. Mohtadi
    "Generalized Predictive Control", Parts I and II. Automatica, vol 23, n°2, pp. 49-55, 1987.

[6]  J.M. Dion, C. Commault.
    "The Minimal Delay Decoupling Problem Feedback Implementation with Stability". SIAM Journal on Control, n°26, pp. 66-82 1988.

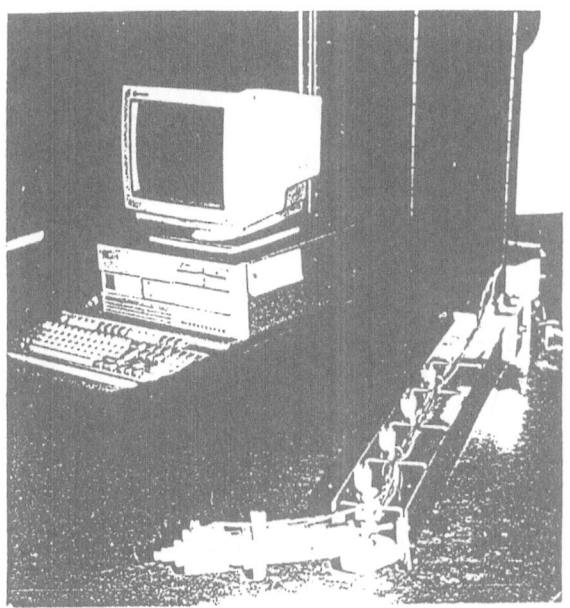

**Fig.1 : Experimental Setup**

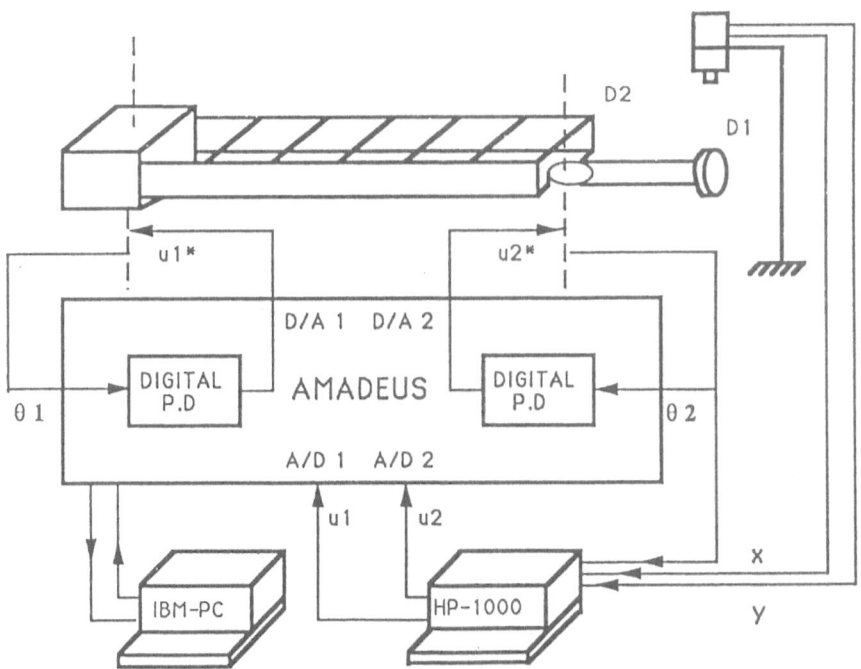

**Fig.2 : Description Of The Process**

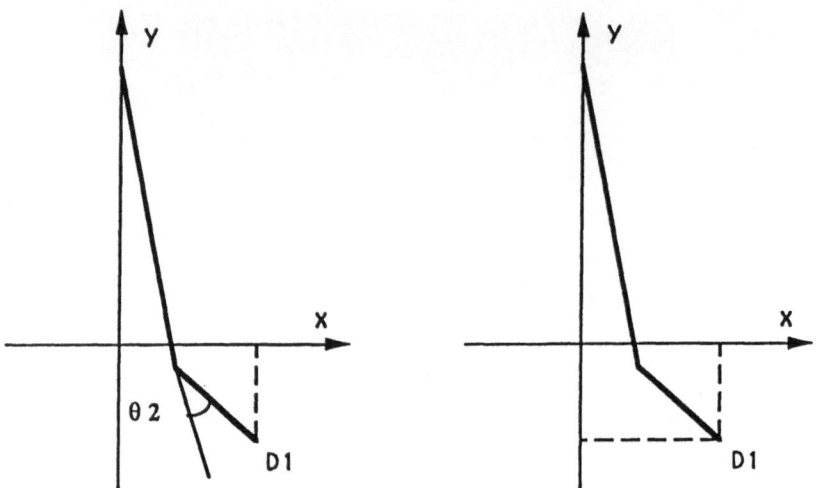

Fig.3a : Process Outputs

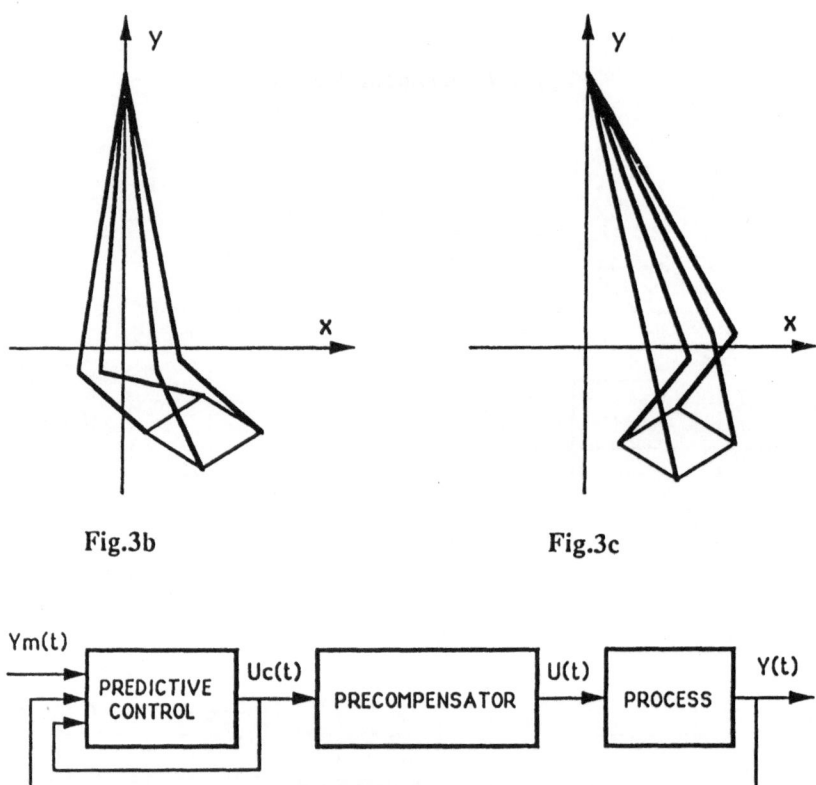

Fig.3b          Fig.3c

Fig.4 : Control Scheme

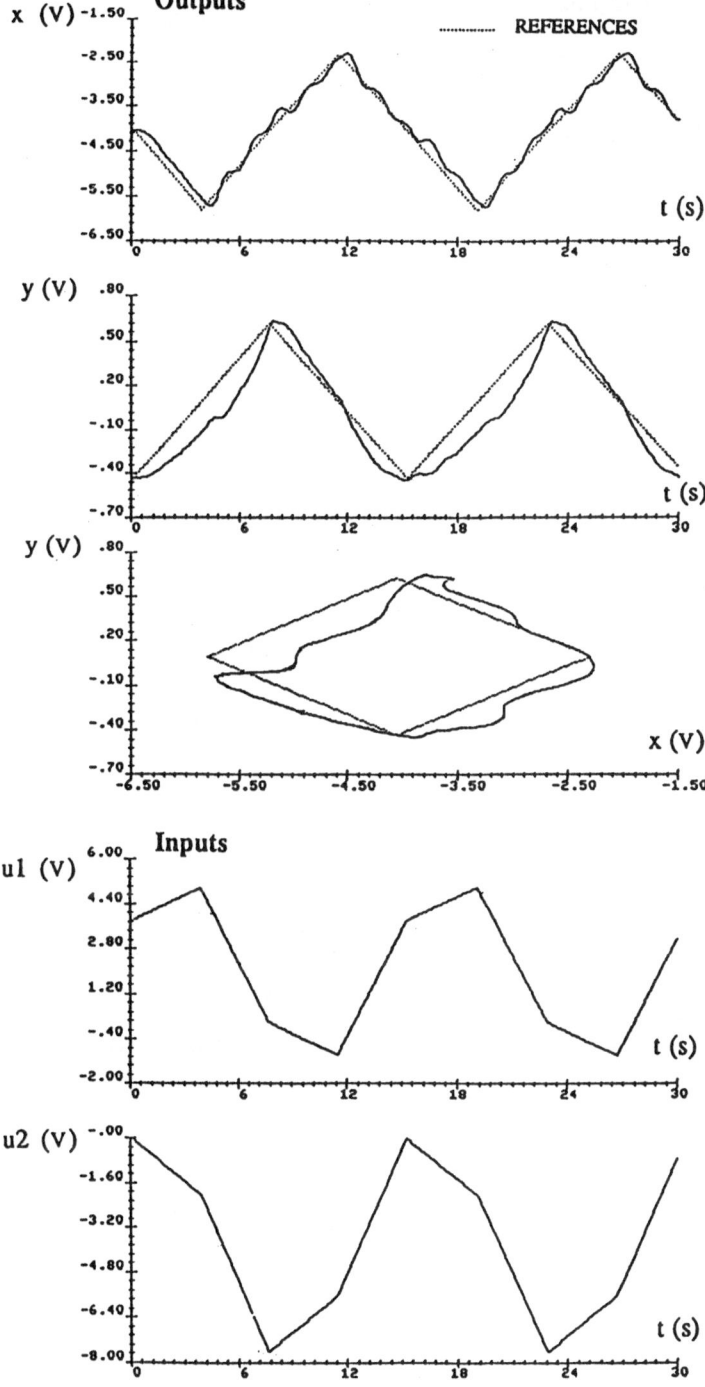

Fig.5 : Open Loop Control with Static Gain

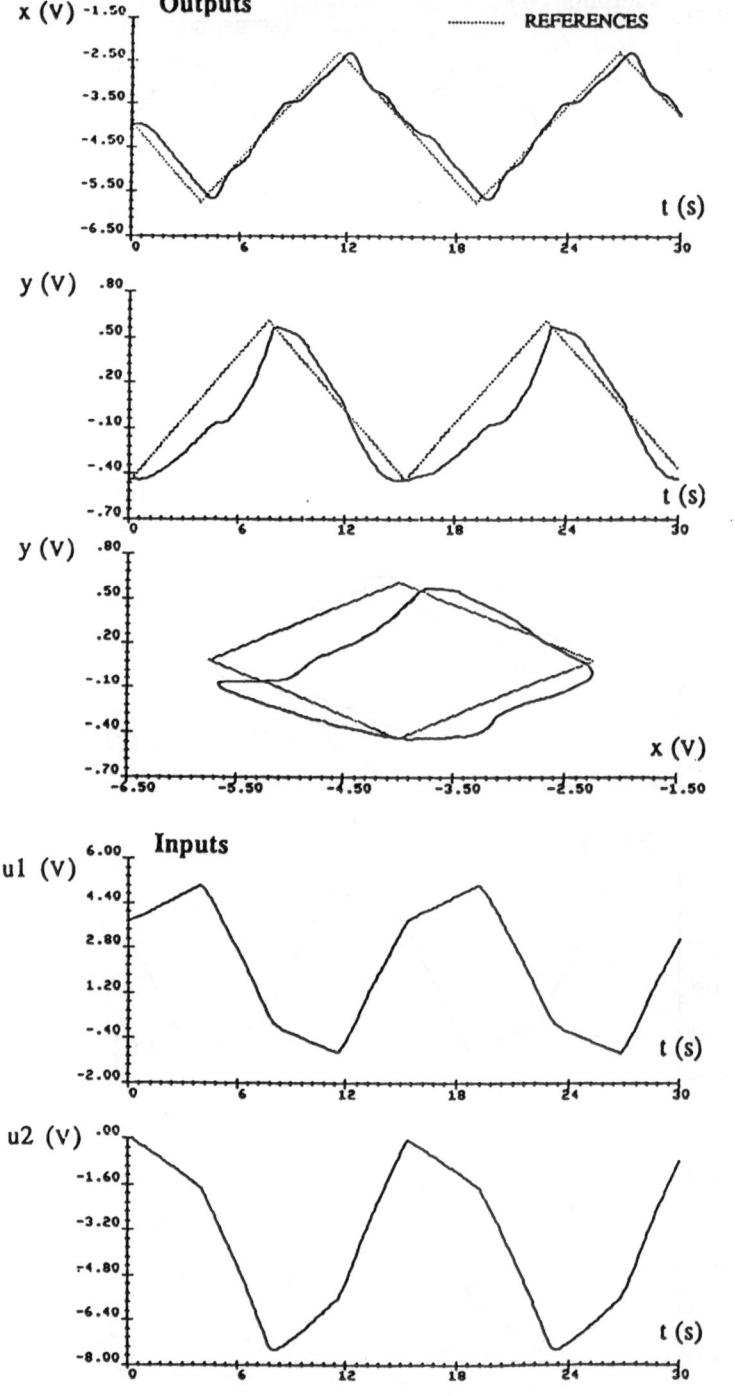

Fig.6 : Closed Loop Control with Static Gain. Pole=0.2

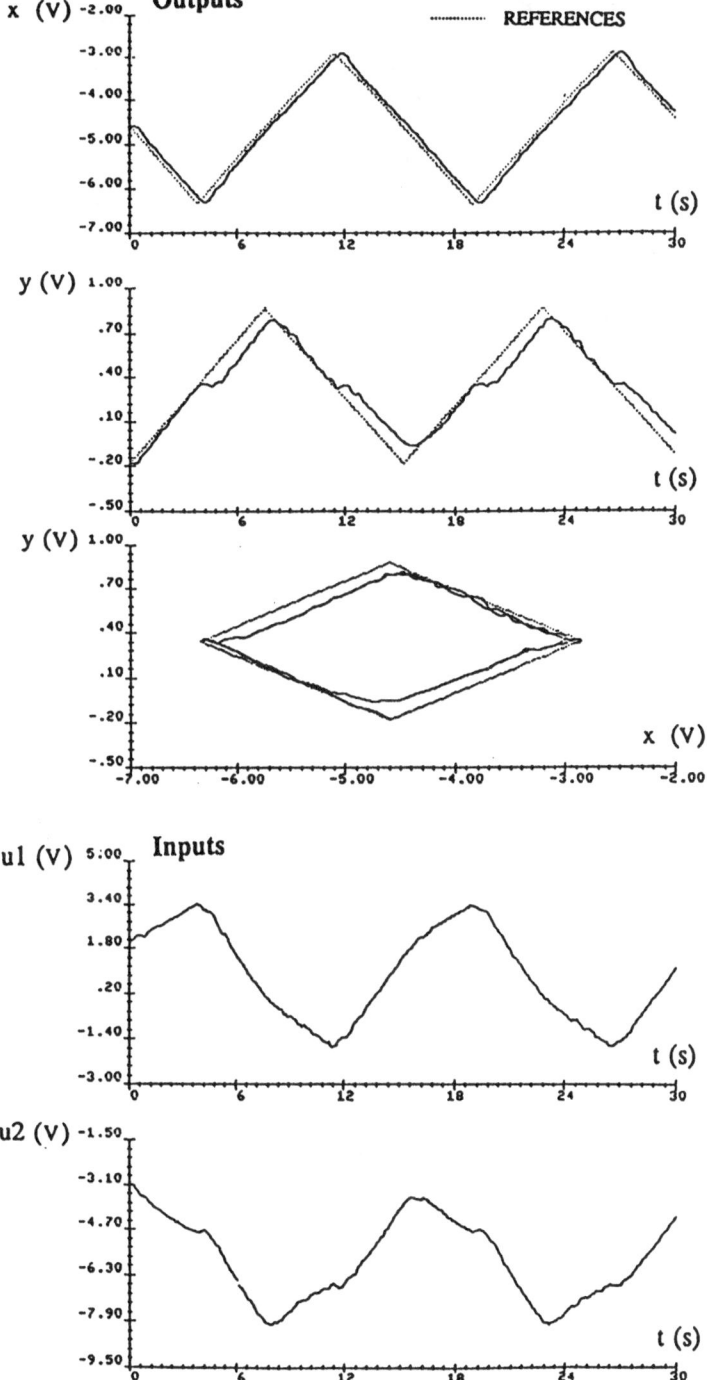

Fig.7 : G.P.C without Precompensator.hp=5;hc=2;Q=0.3.I; na=nb=2

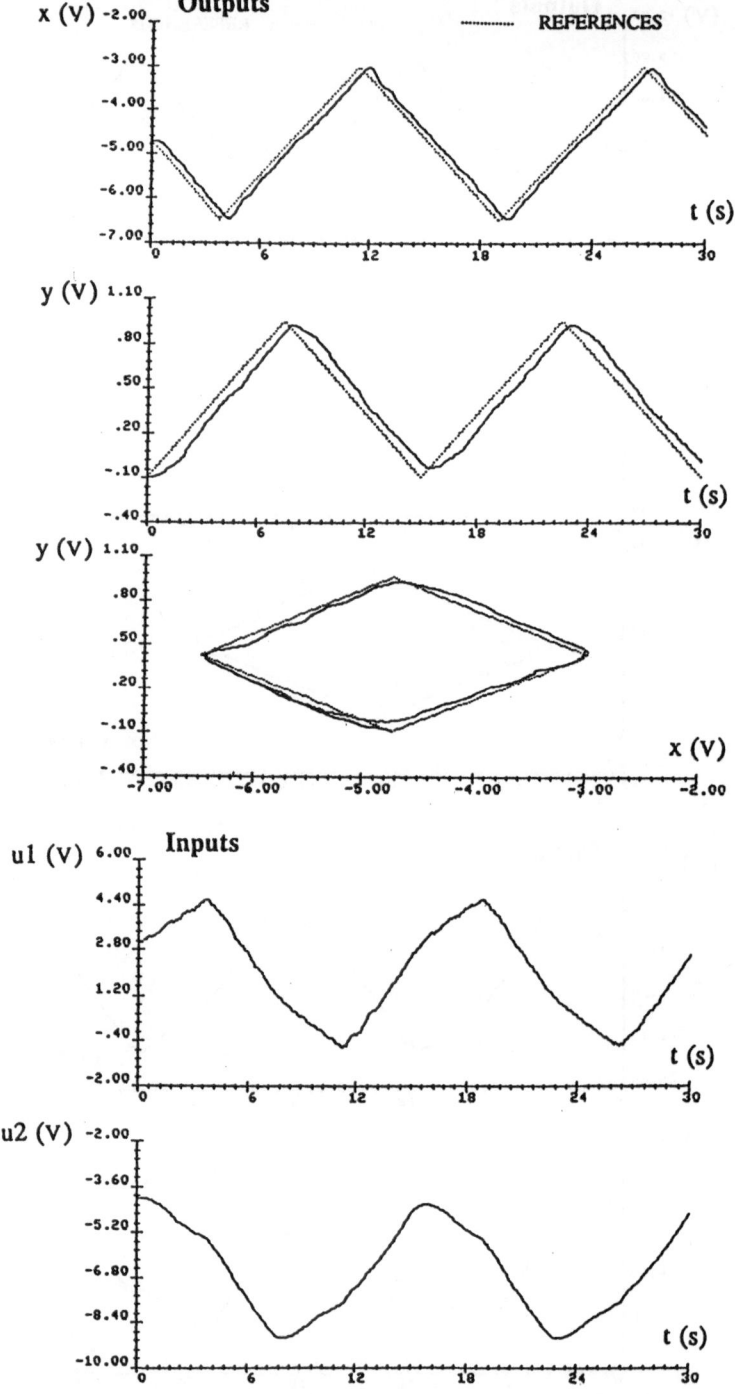

Fig.8 : G.P.C  with  Precompensator. hp=5; hc=2; Q=0.005.I; na=nb=2

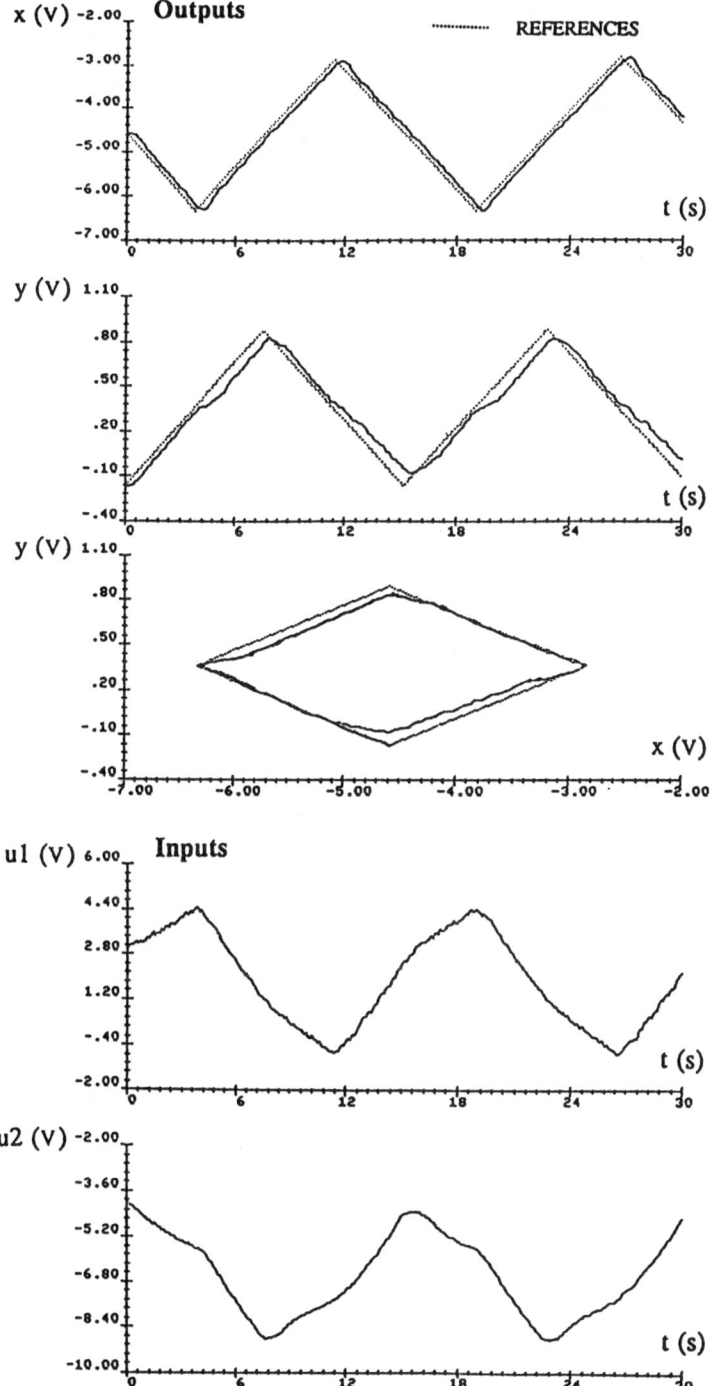

Fig.9 : G.P.C without Precompensator. hp=5; hc=2; Q=0.3.I; na=nb=3

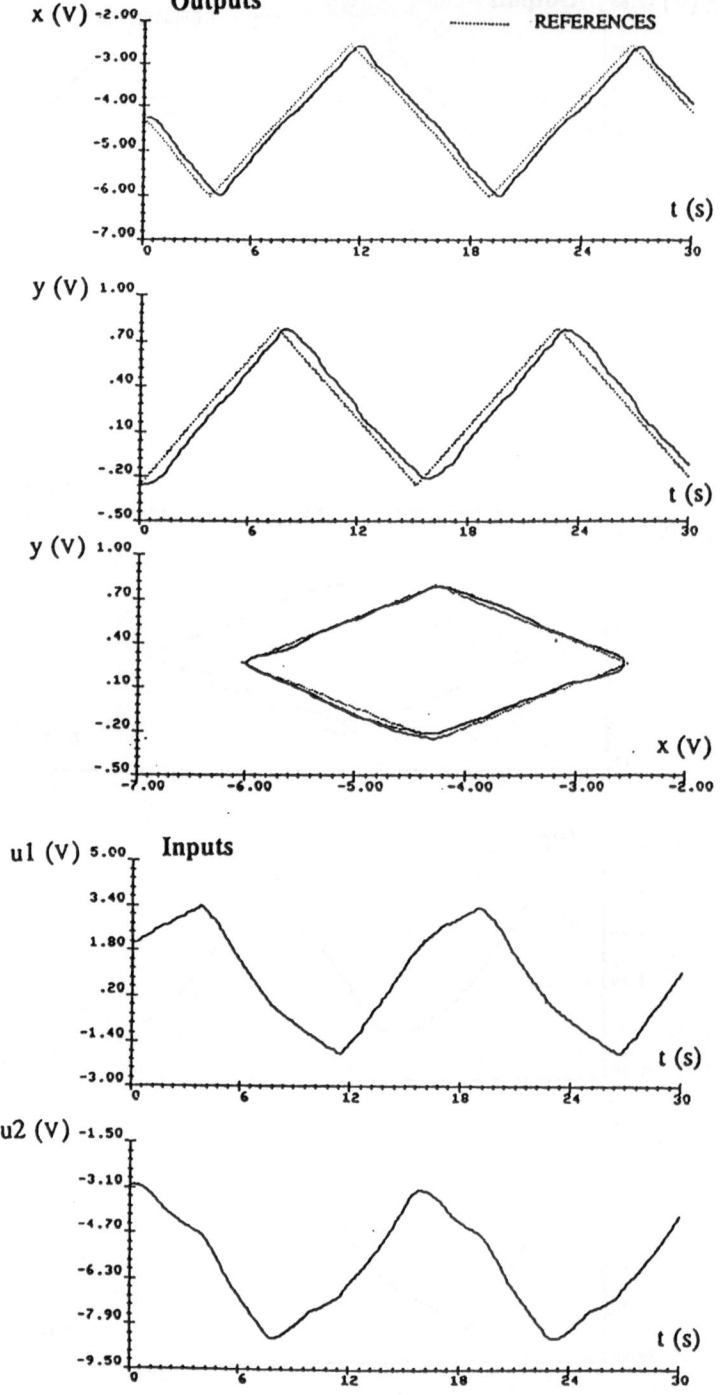

Fig.10 : G.P.C with Precompensator. hp=5; hc=2; Q=0.005.I; na=nb=3

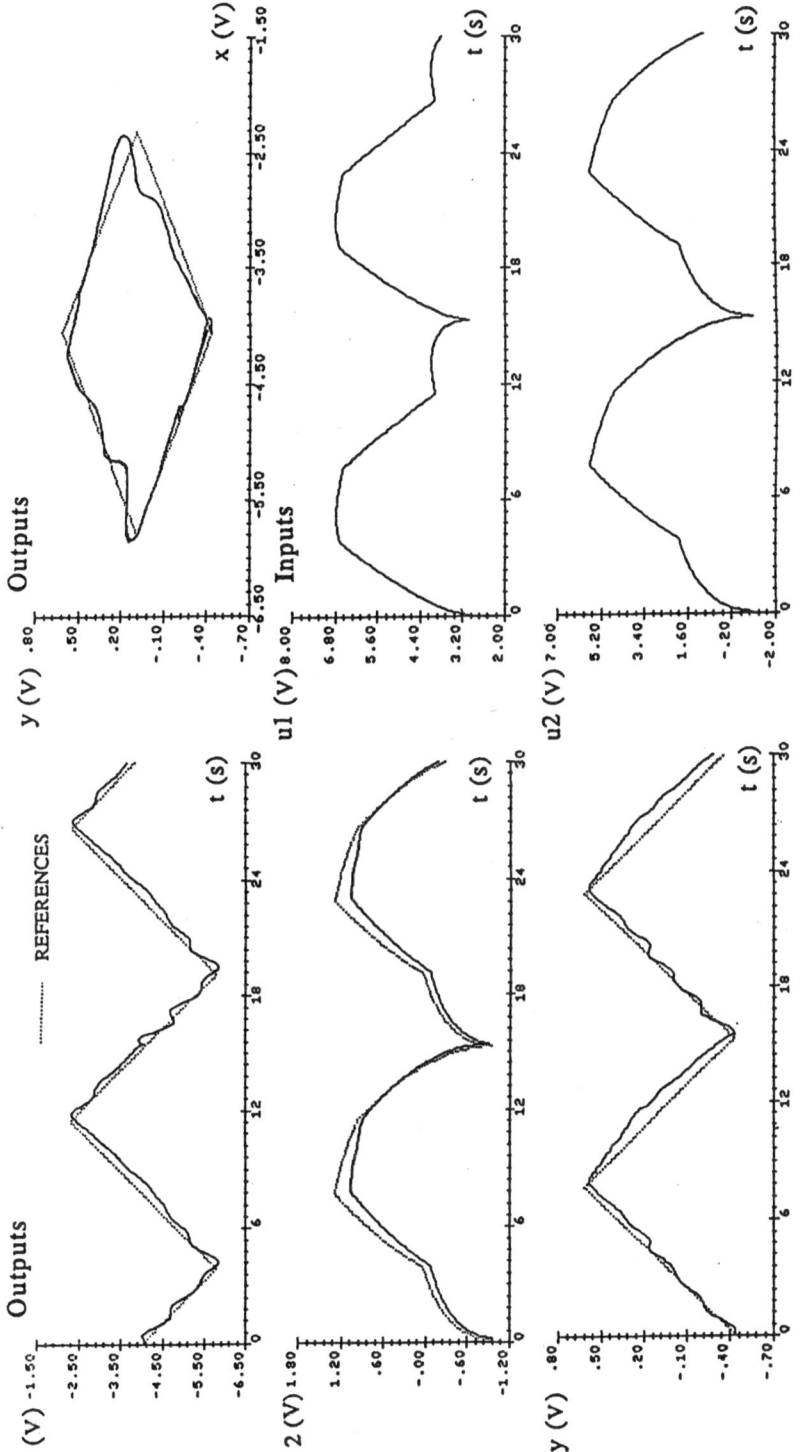

Fig.11 : Open Loop Control with Static Gain

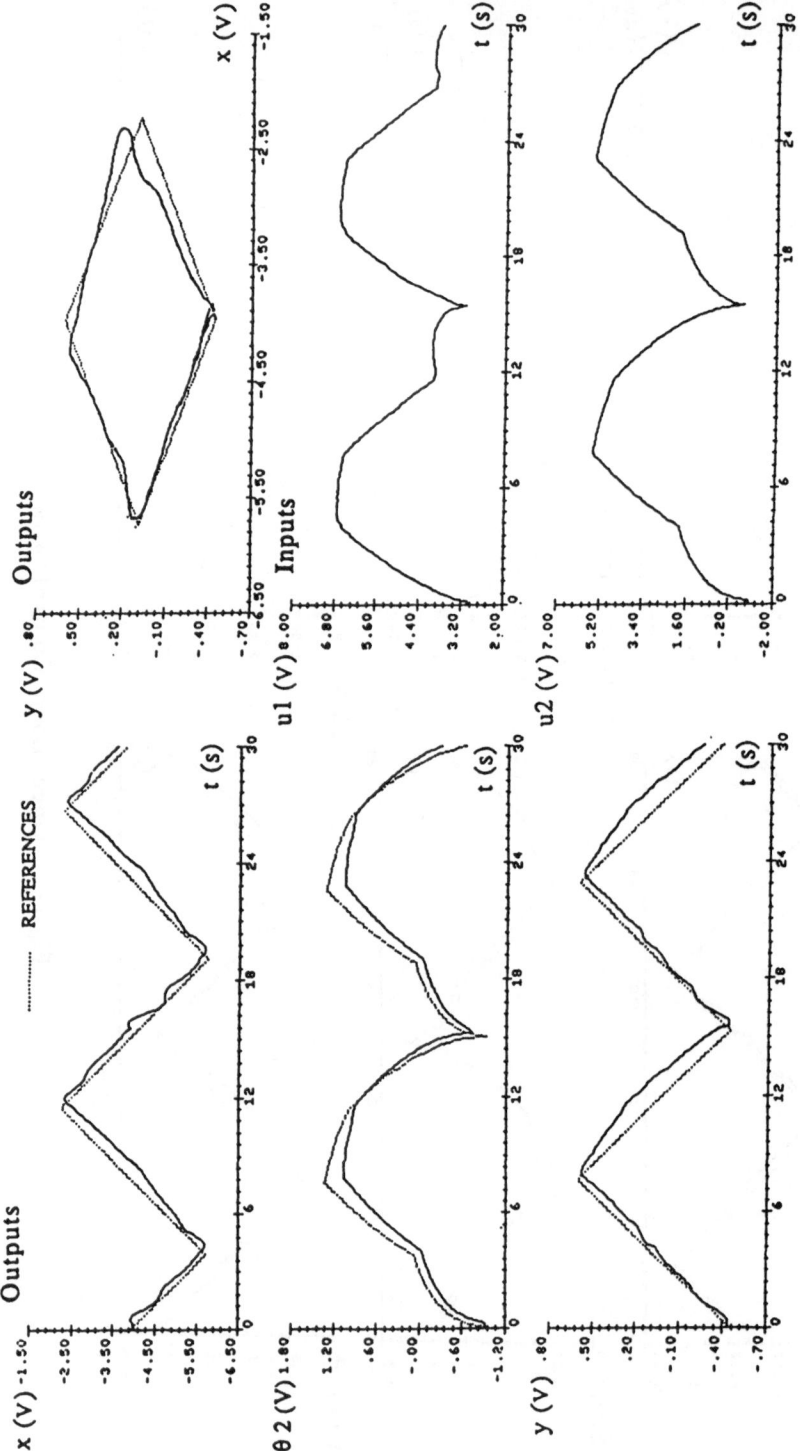

Fig.12 : Closed Loop Control with Static Gain. Pole=0.2

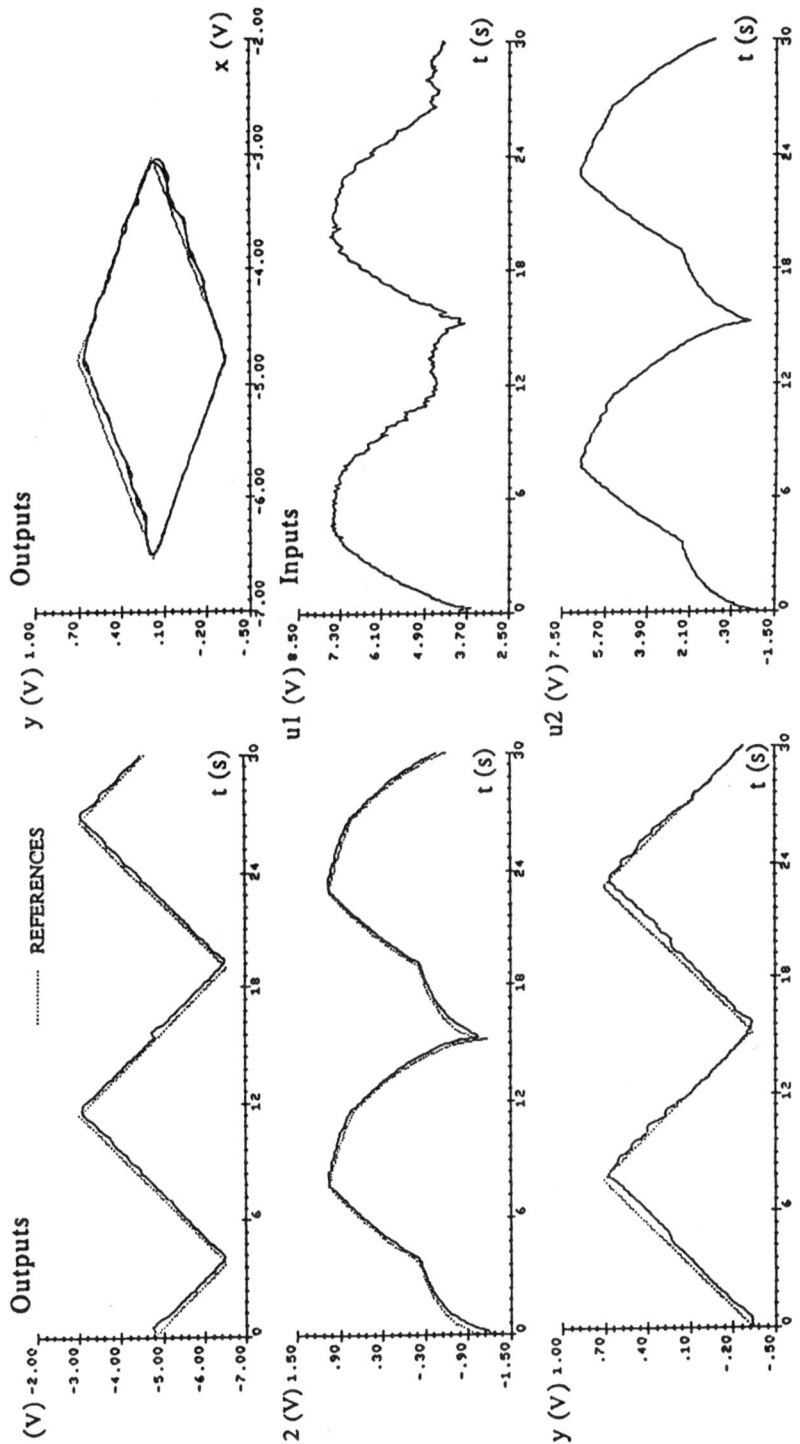

Fig.13 : G.P.C without Precompensator.hp=5; hc=2; Q=0.3.I; na=nb=3

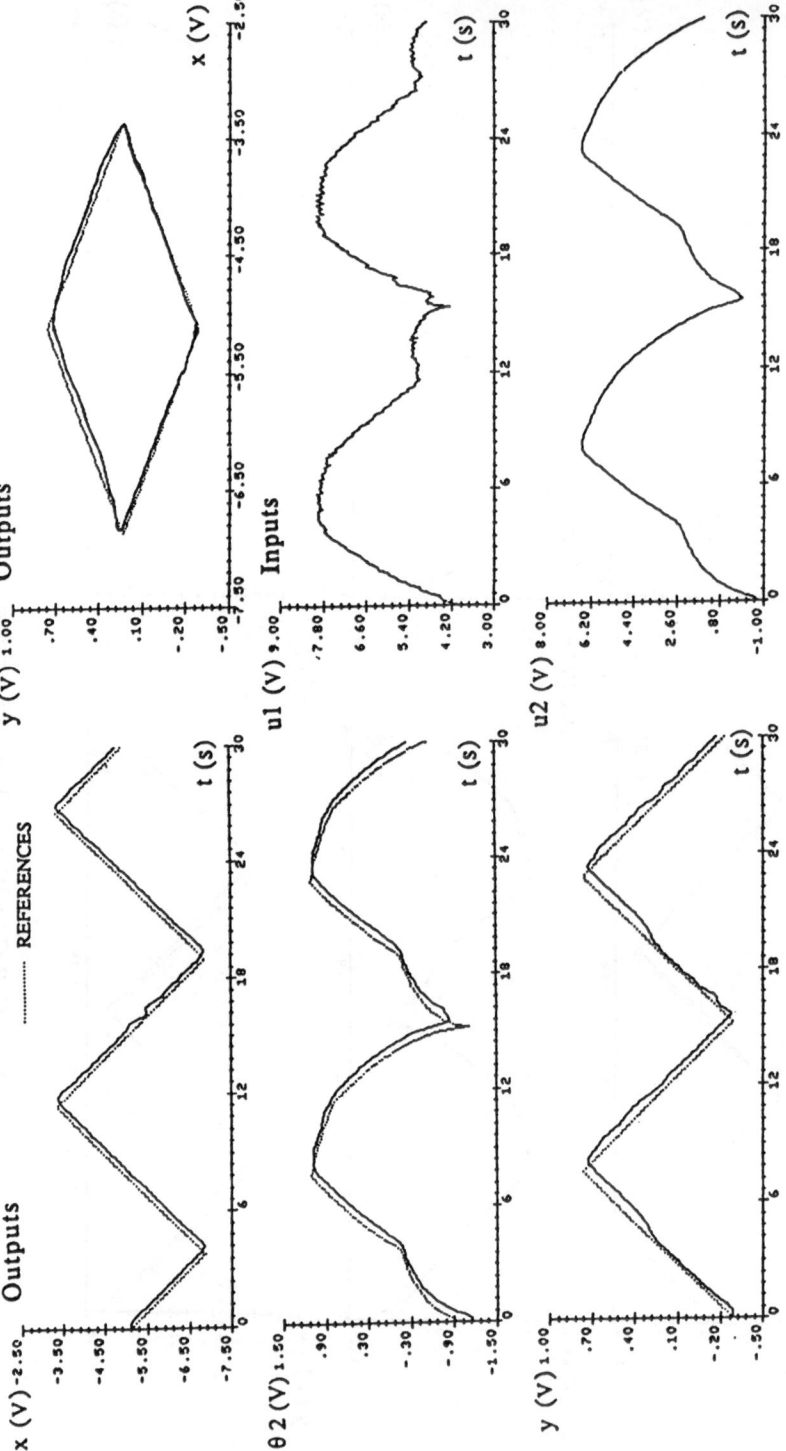

Fig.14 : G.P.C with Precompensator. hp=5; hc=2; Q=0.005.I; na=nb=3

# CONTROL OF ROBOT MANIPULATORS WITH JOINTS FLEXIBILITY

**A. BENALLEGUE and N. K. M'SIRDI**

*Laboratoire de Robotique de Paris (UA CNRS 1305). UPMC Tour 66 - 2$^e$ ét.*
*4,place Jussieu, 75252 PARIS Cedex 05. FAX : (1) 43 54 71 70*
*and GRECO (SARTA) CNRS Bp 46 St Martin d'Hères 38402 France*

**Abstract :** *In this paper we consider two stage controllers for motion control of robot manipulators with joint flexibilities. The dynamic equations of a manipulator with flexible joints are reparametrized in a model, with two blocs, suitable for control laws design. A passivity based adaptive compensator is proposed to control the reduced complexity model (the first bloc) with Integral (I) adaptation algorithms. The second stage uses a computed torque with a robust stabilization loop. The stability and robustness properties of this control scheme are analysed and the performances of the controllers are evaluated in simulation study and compared to other schemes. The proposed scheme allows the use of the controllers proposed for rigid manipulators.*

**Key Words :** Robot Manipulators, Flexible Joints, Fast Motion Tracking, Adaptive Motion Control, Robust Control, Two Stages controller.

## 1. INTRODUCTION

The problem of precise control of fast desired motions for robotic manipulators has been extensively studied in the literature. Actually robotic applications require effective control laws that achieve accurate tracking despite the variations of inertia and the gravitational loads of the manipulator during operation. Adaptive controllers taking in acount the non-linear dynamic characteristics, provide viable schemes for rigid robot control. In the adaptive control recent works, the non linear dynamic equation of the manipulator is parametrized in a model with constant parameters [1] and a large number of nonlinear control strategies have been proposed for rigid manipulators.

A classification of these adaptive controllers can be made by considering the control objective. The first control strategy provide adaptive versions of the inverse dynamics and is based on feedback linearization [1-4]. The desired objective is to obtain a closed loop system which is linear and decoupled [1-3]. For the second, the main objective is the preservation of the passivity property of the rigid manipulator in closed loop [5-7]. Figure 1 presents the adaptive controllers derived from this two main objectives. Robust compensation techniques have been also considered for rigid manipulators [5] [8-9]. All these controllers assume a rigid model for the manipulator and then are limited in their applicability to real robots.

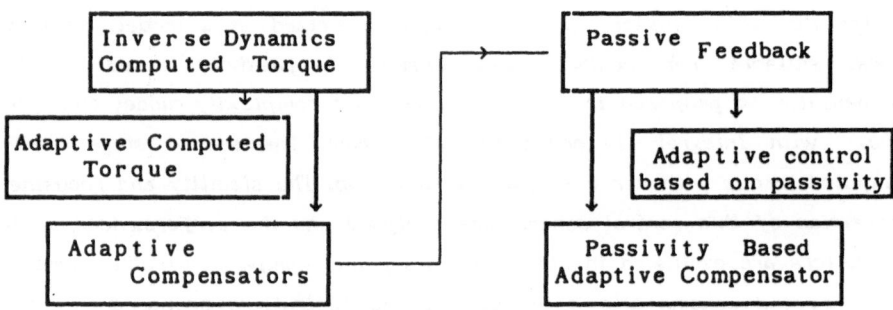

Figure 1a: Controllers for rigid manipulators

In applications some physical phenomena like backlash and joint flexibilties can no more be neglected when fast and precise motions are desired. In robotics, flexibility is attractive for compliance in case of collisions with external environment and contact. It has been shown, for elastic joints, that robots dynamic presents a resonance/anti-resonance characteristic [10-12] which have a great influence in the behaviour. This is the case for robot manipulators that use motors with harmonic drives, elastic transmission belts or hydraulic lines in actuators. Neglecting this phenomenon in control design impose restrictive bandwith and may cause stability problems if we use control algorithms designed assuming a rigid model.

The introduction of flexibilities in the system modeling increases the model complexity and intricates the control design problem. The control strategies must be redesigned tacking in account high frequency dynamics (flexibilities) and parameters uncertainty which become more important in case of flexible joints.

Several design methods have been proposed in literature for robots with flexibilities, using singular perturbation techniques, integral manifold design, variable structure systems and linearizing method [12–20]. It is worth to point out that despite implementation complexity, theoretically, it is possible to obtain as good performances as for the rigid link manipulators case. However, complexity remains important and practical implementation is difficult and depends on parameter and model knowledge. In this paper, the joint flexibilities are included in the model as torsional torques with constant coefficient. We consider the non linear model established in [12] and reparametrize this one as in [18]. This reparametrization is made in order to separate the link variables (positions and velocities) in a first bloc and other ones due to forces (torques) at the joints in a second bloc.

The obtained model, with two blocs, is suitable for control laws design. A passivity based adaptive compensator is proposed to control the reduced complexity model (the first bloc) with Integral (I) adaptation algorithms. It consists of a computed torque plus P–D feedback and an adaptation law to overcome precise knowledge of the manipulator parameters and joint flexibility. The second stage uses a robust stabilization loop in order to drive the tracking errors to zero. The proposed controller is conceptually different from the ones in [18][22] and belong to the class of passive feedback controllers. It uses a reference velocity model. The stability and robustness properties of this control scheme are analysed and the performances of the controllers are evaluated in simulation study and compared to other schemes. The proposed scheme allows the use of controllers proposed for rigid robots and simplify the stability analysis and the control parameters choice.

The layout of the paper is as follows. In section 2 problem formulation and properties of the robot manipulators with flexible joints are presented. In section 3 the proposed controllers for the two stages, in case of known parameters, are presented and their stability analyzed. Section 4 considers the adaptive version of this control technique. Finally, conclusion and some discussions on practical implementation, parameters choice and simulations are made in section 5.

## 2. FLEXIBLE JOINT MODEL AND PROPERTIES

Consider an n rigid link manipulator with flexible joints actuated by DC motors. The elasticity of each joint is modeled, for simplicity, as a linear torsional spring with elasticity constant k. We use the dynamic equations derived by spong in [12]. The simplified dynamics of the robot manipulator is described by the following 2n differential equations:

$$M(q)\ddot{q} + C(q,\dot{q})\dot{q} + G(q) + K(q - N^{-1}q_m) = 0 \qquad (1)$$

$$J \ddot{q}_m + B \dot{q}_m - N^{-1}K (q - N^{-1}q_m) = \tau_m \qquad (2)$$

where $q_m$ and $q$ are angular positions of motors and links respectively. M(q) is the inertia (nxn) matrix which is Symetric Positive Definite (SDP). $C(q,\dot{q})\dot{q}$ represent centrifugal and Coriolis terms. G(q) is the gravity vector and $\tau$ is the input torque vector. J and B are positive diagonal (nxn) SPD matrices representing motor inertia and viscous friction coefficients. N and K are diagonal positive (nxn) matrices representing the gear ratio and the spring constant; For simplicity of equations we shall consider N = I for the gear ratio and K=kI (with I the identity matrix).

If we let $z = q_m - q$, the model of (1) and (2) may be written:

$$M(q)\ddot{q} + C(q,\dot{q})\dot{q} + G(q) = Kz \qquad (3)$$

$$J \ddot{z} + B_z(q,\dot{q},\dot{z}) + G_z(q,z) = \tau \qquad (4)$$

with $\qquad B_z(q,\dot{q},\dot{z}) = B \dot{z} + B \dot{q} - J M(q)^{-1}C(q,\dot{q})\dot{q}$

and $\qquad G_z(q,z) = J M(q)^{-1}K z - J M(q)^{-1}G(q) + K z \qquad (5)$

This reparametrization emphasizes the model decomposition in two stages (3) and (4) (see Figure 1b). These stages are coupled and use the same parameters. The outputs of the first stage (eq 3) (q and $\dot{q}$) are the position and velocities of the links. The second stage (eq 4) has as outputs the differences beetween angular positions and velocities of the actuators and the links (z and $\dot{z}$). The parameters of this stage are non linear functions of those of the first and the second one. This parametrization presents the advantage of separation of rigid mode and the remaining dynamic which will be refered to as flexible dynamic.

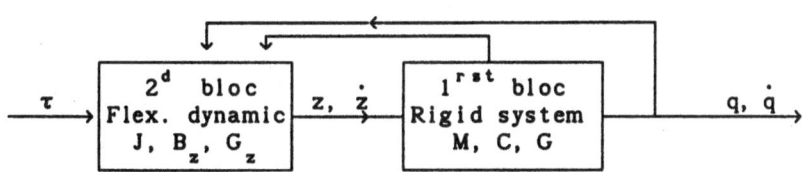

**Figure 1 b:** System Reparametrization

The model of equations (3) and (4) will be used here for the controller design. It is worthwhile to note that this parametrization preserves the good properties of the rigid robots at least for the first stage and this properties can be used for the control law synthesis.

**Property 1:** The inertia matrices $M(q)$ and $J$ are (SPD) and their inverses $M^{-1}(q)$ and $J^{-1}$ are uniformly bounded functions of link angular position q.

**Property 2:** With an appropriate definition of $C(q,\dot{q})$ the matrix, $\mathcal{A}(q,\dot{q})$ defined as follows is skew symmetric [2]:

$$\mathcal{A}(q,\dot{q}) = \dot{M}(q) - 2C(q,\dot{q}) \qquad (6a)$$

**Property 3:** We can rewrite the model as:

$$M(q)\ddot{q}_r + C(q,\dot{q})\dot{q}_r + G(q) = \Psi(q,\dot{q},\dot{q}_r,\ddot{q}_r)\theta \qquad (6b)$$

where $\Psi(q,\dot{q},\dot{q}_r,\ddot{q}_r)$ is an (nxr) matrix of known functions (regressor) and $\theta$ is an (rx1) vector of constant parameters.

**Property 4:** For any bounded $\theta$, there exist constants $C_m$, $C_c$, $C_g$ such that we have for any (q, $\dot{q}$):

$$\|M(q,\theta)\| \leq C_m \qquad (6c)$$

$$\|C(q,\dot{q},\theta)\dot{q}\| \leq C_c\|\dot{q}\|^2 \qquad (6d)$$

$$\|G(q,\theta)\| \leq C_g \qquad (6e)$$

These properties come from the physical nature of the robot and will be used for stability analysis. The control objective is that the manipulator track a set of given joint positions and velocities with an imposed dynamic for error rejection. The desired trajectories will be noted $q^d(t)$, $\dot{q}^d(t)$ and $\ddot{q}^d(t)$.

**Remark:** In case of slow motions, $\dot{z} = 0$ and $\ddot{z} = 0$, we obtain from equation (4), $Kz = \tau - B\dot{q} - J\ddot{q}$. Then equation (3) gives the rigid robot model of eq (7).

$$(M(q) + J)\ddot{q} + B\dot{q} + C(q,\dot{q})\dot{q} + G(q) = \tau \qquad (7)$$

The dynamic model of equations (3) and (4) and its properties will be used in the following sections for control law design.

## 3. ROBUST MOTION CONTROL WITH FIXED PARAMETERS

We consider, in this section for the controller design, the joint flexibility and the parameters known precisely. Once the model have been decomposed in two blocs (with n DOF), the control is computed in two steps. The first bloc of equation (3) has z(t) as input. Let us note $z^d$ (defined in equation 10) the suitable input of this bloc such that the outputs q and $\dot{q}$ follow the desired trajectories. Then for the second bloc (equation 4), we must compute the input torque $\tau$ such that the output z will track $z^d$. Seemingly, this design way have been used for the first time in [18] where the first step controller is PD feedback with a feedforward term using ($q^d$, $\dot{q}^d$, $\ddot{q}^d$) and the second step is a stabilization loop.

Examination of equations (3) and (4) shows that any of the globally stable controllers proposed for rigid manipulators can be considered for each bloc separately. Stability of the global control law (with the two stages) remain an open issue. An adaptive computed torque controller (for the first step) in conjunction with a robust stabilization loop for the second bloc, have been considered in [21]. The control law proposed here is based on passivity of the first stage in closed loop and robust compensation for the second one. The first stage is defined by the link variables which represent angular positions and velocities of the robot links. The variables, representing the gap between links and motors angular positions and velocities, define the second stage. The applicability of the proposed controller impose the following two asumptions which will be used in the sequel.

A1) The desired trajectory $q^d(t)$ is $C^4$ and has its derivatives $\dot{q}^d$, $\ddot{q}^d$, $q^{d(3)}$, $q^{d(4)}$ uniformly bounded.

A2) The variables q, $\dot{q}$, $\ddot{q}$ and z, $\dot{z}$ ( or $q_m$, $\dot{q}_m$ ) are measurable.

### 3.1. Passive Control Law For the Rigid Subsystem

The tracking error variable for the second bloc is noted $e_z$:

$$e_z = z - z^d \tag{8}$$

With this definition and equation (3), we can write:

$$M(q)\ddot{q} + C(q,\dot{q})\dot{q} + G(q) = K(z^d + e_z) \tag{9}$$

For the first bloc input, if K is bounded, we propose a passive feedback control law as in [6][7][9]:

$$z^d = K^{-1}\left(M(q)\ddot{q}_r + C(q,\dot{q})\dot{q}_r + G(q)\right) - F_v s - F_p e \qquad (10)$$

where $F_v$, $F_p$ are positive definite diagonal matrices and trajectory error is:

$$e = q - q^d, \qquad \dot{e} = \dot{q} - \dot{q}^d \qquad (11)$$

The reference velocity model is defined by:

$$\dot{q}_r = \dot{q}^d - \Lambda e \qquad (12)$$
$$s = \dot{e} + \Lambda e = \dot{q} - \dot{q} \qquad (13)$$

From equations (9) to (13), we obtain the closed loop equation for the first subsystem:

$$M \dot{s} + C s = - K F_v s - K F_p e + K e_z \qquad (14)$$

Then, as expressed by equation (14), the tracking objective, for this bloc, will be attained if $e_z$ goes to zero and $z^d$ can be realised. Determination of $z^d$ remain for the second bloc. From equation (14), we see that if the error signal $e_z$ tend to zero, we obtain a passive feedback system [3][7][9].

## 3.2. Robust Control Law For the Flexible Subsystem

Assumptions A1 and property 4 will be used to prove the boundedness of $z^d$ ,$\dot{z}^d$ and $\ddot{z}^d$. Application of an observer for reconstruction of $q_m$ and $\dot{q}_m$ from q, $\dot{q}$ and the model can be considered and is an open problem. We must compute now the input $\tau$ which ensure that z will track $z^d$ as follows by a linearizing feedback controller :

$$\tau = J \,\omega + B_z(q,\dot{q},\dot{z}) + G_z(q,z) \qquad (15)$$

Equations (4) and (15) lead to :     $J (\ddot{z} - \omega) = 0$

Then         $\ddot{z} = \omega$ \qquad (16)

and the control variable $\omega$ is chosen as follows :

$$\omega = \ddot{z}^d - \alpha \, \dot{e}_z - \alpha \, e_z \qquad (17)$$

Note that this control law uses the first and the second derivatives of $z^d$ and signals z and $\dot{z}$. The trajectories $\dot{z}^d$ and $\ddot{z}^d$ are obtained by derivation of the equation (10) using:

$$\ddot{q}_r = \ddot{q}^d - \Lambda \dot{e} , \quad \text{and} \quad \dot{z}^d = K^{-1}\left(M(q)\ddot{q}_r + C(q,\dot{q})\dot{q}_r + G(q)\right) - F_v s - F_p e$$

After derivation of $z^d$, $\ddot{q}$ appear in expression of $\dot{z}^d$, we replace it from equation (3). Proceeding in the same way for $\ddot{z}^d$ we obtain the expressions:

$$\dot{z}^d = h_1(q^d, \dot{q}^d, \ddot{q}^d, q^{d(3)}, q, \dot{q}, e_z, \theta, t) \tag{18}$$

$$\ddot{z}^d = h_2(q^d, \dot{q}^d, \ddot{q}^d, q^{d(3)}, q^{d(4)}, q, \dot{q}, e_z, \dot{e}_z, \theta, t)$$

From properties 1, 3 and 4 it results that $h_1()$ and $h_2()$ are smooth functions of their arguments. Then, with assumption A1, for the boundedness of $h_1(.)$ and $h_2(.)$, it remain to prove that $q$, $\dot{q}$, $e_z$ and $\dot{e}_z$ are bounded. This will be established in section 3.3 with the stability of the system in closed loop. In case of modelization error, boundedness of (18) is garanted by property 4 for bounded $\theta$.

With the equations (15) to (18) we obtain the closed loop :

$$\ddot{e}_z + \alpha_v \dot{e}_z + \alpha_p e_z = 0 \tag{19}$$

The closed loop dynamic, for rejection of the error $e_z$, is determined by the choice of $\alpha_p$ and $\alpha_v$ (eg for a second order dynamic $\alpha_v = 2\xi_r \omega_r$ and $\alpha_p = \omega_r^2$). From these equations, it is obvious that the dynamic of the system (19) must be faster than the one of the first bloc in closed loop, in order to insure the tracking objective. In other words, the robot can be considered as composed of two coupled subsystems, first bloc capturing the rigid modes and a second one traducing the flexible dynamics. The first bloc is excited by the output (error) of the second bloc (see figure 1), then the output error of the second bloc must be damped first. The closed loop dynamic imposed for the second bloc must be faster than for the first bloc in order to insure a good tracking.

### 3.3. Stability Analysis of the global closed loop

The fast subsystem of equation (19) in closed loop can be written in state space form:

$$\dot{x} = A x \tag{21}$$

with
$$x = \begin{pmatrix} e_z & \dot{e}_z \end{pmatrix}^T \quad \text{and} \quad A = \begin{bmatrix} 0 & I \\ -\alpha_p & -\alpha_v \end{bmatrix}$$

With the matrices $\alpha_p$ and $\alpha_v$ appropriately chosen, equation (21) represent an exponentially stable system. Then we can find an SPD matrix $P$, solution of the Lyapunov equation with $Q$ chosen SPD:

$$A^T P + P A = -Q \tag{23}$$

The matrix $Q$ determines the stability of the fast system alone. Then to analyse the stability of the global system in closed loop, we chose as Lyapunov function candidate:

$$V(t) = \frac{1}{2} s^T M \, s + \frac{1}{2} e^T K F_p e + x^T P \, x \tag{24}$$

The time derivation of $V(t)$ and the use of equations (14), (23) and (13) lead us to the expression (recall that $\Lambda$, $K$ and $F_p, F_v$ are diagonal matrices):

$$\dot{V}(t) = - s^T K F_v s - e^T \Lambda K F_p e + s^T K e_z - x^T Q \, x \tag{25}$$

Define the vector : $\quad y^T = \left( s \quad e \quad e_z \quad \dot{e}_z \right) \tag{26}$

Equation (25) can be written in the following form :

$$\dot{V}(t) = - y^T H \, y \tag{27}$$

with
$$H = \left[ \begin{array}{cc|cc} KF_v & 0 & -K/2 & 0 \\ 0 & \Lambda KF_p & 0 & 0 \\ \hline -K/2 & 0 & & \\ 0 & 0 & & Q \end{array} \right] \tag{28}$$

To proof the asymptotic stability of the closed loop system, it remain to show that $\dot{V}(t)$ is negative definite or the matrix $H$ is SPD. To this end, we need the following Lemma on a condition for positive definiteness of a symmetric matrix. This result is presented and used in [18].

**Lemma 1** : Given a symmetric matrix $W = \left[ \begin{array}{cc} U_1 & V \\ V^T & U_2 \end{array} \right] \tag{29}$

where $U_1 \in \mathbb{R}^{n \times n}$ and $U_2 \in \mathbb{R}^{m \times m}$ are symmetric positive definite matrices. Then $W$ is SPD if :

$$\lambda_{min}(U_1) \, \lambda_{min}(U_2) \geq \|V\|^2 \tag{30}$$

∎

Then for global asymptotic stability of our closed loop system ($\dot{V}(t) < 0$) we obtain the following condition on the feedback gains $\Lambda$, $F_p$, $F_v$, $\alpha_p$ and $\alpha_v$ :

$$\lambda_{min}\left( \left[ \begin{array}{cc} KF_v & 0 \\ 0 & \Lambda KF_p \end{array} \right] \right) \lambda_{min}(Q) \geq \left\| \left[ \begin{array}{cc} \frac{K}{2} & 0 \\ 0 & 0 \end{array} \right] \right\|^2 \tag{31}$$

If the condition (31) is verified then $\dot{V}(t) < 0$ and we can conclude that $e = q - q^d \longrightarrow 0$, $s \longrightarrow 0$, $e_z \longrightarrow 0$ and $\dot{e}_z \longrightarrow 0$ when $t \longrightarrow \infty$. Recall that $s = \dot{e} + \Lambda e$ with $\Lambda$ SPD, then $\dot{e} = \dot{q} - \dot{q}^d \longrightarrow 0$. Finally, with assumptions A1 and A2 and these results, boundedness of $z^d$, $\dot{z}^d$ and $\ddot{z}^d$ is assured and the proposed controller is realizable. Note that Q define the closed loop dynamic of the flexible dynamic and depending on the choice for the control of this second bloc, we have conditions on the gains $F_p$, $F_v$ and $\Lambda$ for the first bloc. For example Q=40I leads to min($\Lambda F_p$,$F_v$)>0.0125, we then take $\Lambda$=20,$F_p$=0.1 and $F_v$=0.02 (see fig 2).

## 4. ADAPTIVE AND ROBUST MOTION CONTROL

In the previous control design, we have considered the parameters known precisely. In this section, we consider the case of parameters unknown or varying during task operation. Then the parameters, used in the first stage for the passive feedback controller, will be adapted on line. The second stage has its parameters function of the actuators ones and those of the first bloc. Thus linearization become hardly achievable due to parameters uncertainty and variations, so a robust compensation loop seems more advisable. For this application, we shall state some additional assumptions on modelization errors. In the proposed control law, the system parameters K, M, C and G, will be replaced by their on line estimates $\hat{K}$, $\hat{M}$, $\hat{C}$ and $\hat{G}$ and for the second step J, $B_z$ and $G_z$ are replaced by $\hat{J}$, $\hat{B}_z$ and $\hat{G}_z$ which are function of the on line estimates and the a priori ones.

### 4.1. First step: Control Law For the Rigid Subsystem

For the first stage, the passive feedback control law is given in what follows with all signals defined by equations (11) to (13):

$$\hat{z}^d = \hat{K}^{-1}\left[\hat{M}(q)\ddot{q}_r + \hat{C}(q,\dot{q})\dot{q}_r + \hat{G}(q)\right] - F_v s - F_p e \tag{32}$$

Using the linear in parameters model of property 3, we can write:

$$\hat{z}^d = \Psi(q,\dot{q},\dot{q}_r,\ddot{q}_r)\,\hat{\theta} - F_p e - F_v s \tag{33}$$

$\hat{\theta}$ is the on line parameter vector estimate for the slow subsystem. The input error is noted $\hat{e}_z$, in the adaptive control case.

$$\hat{e}_z = z - \hat{z}^d \tag{34}$$

The system, the model and the control law equations (32) and (34) lead to the closed loop error equation as follows:

$$K^{-1}M\,\dot{s} + K^{-1}C\,s = \tilde{M}\,\ddot{q}_r + \tilde{C}\,\dot{q}_r + \tilde{G} - F_v s - F_p e + \hat{e}_z \tag{35a}$$

with $\quad \tilde{M} = \hat{K}^{-1} \hat{M}(q) - K^{-1} M(q)$

$$\tilde{C} = \hat{K}^{-1} \hat{C}(q,\dot{q}) - K^{-1} C(q,\dot{q}) \tag{35b}$$

$$\tilde{G} = \hat{K}^{-1} \hat{G}(q) - K^{-1} G(q)$$

Or equivalently using property 3:

$$K^{-1}M \dot{s} + K^{-1}C s = \Psi(q,\dot{q},\dot{q}_r,u)\tilde{\theta} - F_v s - F_p e + \hat{e}_z \tag{36}$$

with $\tilde{\theta} = \hat{\theta} - \theta$ defining parameter estimation error for the slow subsystem. The Parameter Adaptation Algorithm (PAA) used here is the following.

$$\dot{\hat{\theta}} = - \Gamma \Psi^T(q,\dot{q},\dot{q}_r,u). s \tag{37}$$

The adaptation gain is $\Gamma > 0$. This PAA is of Integral type, Proportional Integral PAA can be used here for enhancement of the transient behaviour [7][21]. For implementation, we use projection to keep the estimates inside positive intervalls. Only I PAA are considered here for stability analysis.

### 4.2. Second step: Control Law For the Fast Subsystem

The linearization and robust compensation [4][23] used for the second stage (4) is given by:

$$\tau = \hat{J} (\omega + \delta\omega) + \hat{B}_z(q,\dot{q},\dot{z}) + \hat{G}_z(q,z) \tag{38}$$

with $\quad \omega = \ddot{z}^d - \alpha_v \dot{\hat{e}}_z - \alpha_p \hat{e}_z$

The same comments as for equation (18) hold for the desired trajectories. Deriving (33) and using (37) we have:

$$\dot{\hat{z}}^d = \dot{\Psi} \hat{\theta} - \Psi\Gamma\Psi s - F_v \dot{s} - F_p \dot{e}$$

$$\ddot{\hat{z}}^d = \ddot{\Psi} \hat{\theta} - 2 \dot{\Psi}\Gamma\Psi s - \Psi\Gamma (\dot{\Psi}s + \Psi\dot{s}) - F_v \ddot{s} - F_p \ddot{e}$$

Note that $q^{(3)}$, used in the last equation can be estimated by filtering the acceleration. Using equation (4) the fast subsystem can be rewritten:

$$\ddot{z} = J^{-1} \hat{J} (\omega + \delta\omega) + J^{-1}\left( \tilde{B}_z(q,\dot{q},\dot{z}) + \tilde{G}_z(q,z)\right) \tag{39}$$

with $\quad \tilde{B}_z(q,\dot{q},\dot{z}) = \hat{B}_z(q,\dot{q},\dot{z}) - B_z(q,\dot{q},\dot{z})$

and $\quad \tilde{G}_z(q,z) = \hat{G}_z(q,z) - G_z(q,z)$

or finally, using the expression of the input $\omega$, we obtain the error equation:

$$\dot{x} = Ax + B\left(\delta\omega + \eta(\omega,q,\dot{q},z,\dot{z})\right) \qquad (40)$$

with
$$A = \begin{bmatrix} 0 & I \\ -\alpha_p & -\alpha_v \end{bmatrix} \qquad B = \begin{bmatrix} 0 \\ I \end{bmatrix} \qquad x = \begin{bmatrix} e_z \\ \dot{e}_z \end{bmatrix}$$

and
$$\eta(\omega,q,\dot{q},z,\dot{z}) = E\,\delta\omega + E\,\omega + J^{-1}\Delta H \qquad (41)$$

with
$$E = J^{-1}\hat{J} - I \qquad \text{and} \qquad \Delta H = \tilde{B}_z(q,\dot{q},\dot{z}) + \tilde{G}_z(q,z)$$

where the matrices $\alpha_p$ and $\alpha_v$ have been defined in (20).

For the rejection of the error $\hat{e}_z$ and the stabilization of the overall system, the signal $\delta\omega$ remain to be determined. We use for this bloc a robust controller [23][8]. The knowledge of the term $\eta(\omega,q,\dot{q},z,\dot{z})$ is not obvious but uncertainty on the parameters may be bounded a priori. So some additional assumptions on the composition of the term $\eta(\omega,q,\dot{q},z,\dot{z})$ are needed. These are the following:

**Assumption 1** : The matrix J is SPD and there exist some constants $\bar{J}$ and $\underline{J}$ such that :
$$\underline{J} \le \|J^{-1}\| \le \bar{J} < \infty \qquad (42)$$

**Assumption 2** : The choice $\hat{J}$ is such that for some positive constant $\gamma$ ($\gamma < 1$) we have :
$$\|E\| \le \|J^{-1}\hat{J} - I\| \le \gamma \qquad (43)$$

**Assumption 3** : There exist some positive constants $\delta_1$, $\delta_2$, $\delta_3$ and $\delta_4$ such that for all $(q,\dot{q},z,\dot{z}) \in \mathbb{R}^{4n}$ we have :
$$\|\Delta H\| = \|\tilde{B}_z(q,\dot{q},\dot{z}) + \tilde{G}_z(q,z)\| \le \delta_1\|\dot{q}\| + \delta_2\|\dot{z}\| + \delta_3\|z\| + \delta_4 = \Phi(q,\dot{q},z,\dot{z}) \qquad (44)$$

■ Assumption 2 determine the choice of the matrix $\hat{J}$ as estimation of J. In practice $\hat{J}$ is chosen function of the actuator parameters and a priori knowledge on the system in order to make $\gamma$ of assumption 2 as small as possible. The niminal estimationsand the parameters projection intervalls are used to determine the constants in assumption 3. For determination of the input $\delta\omega$ of the stabilization loop we suppose there exist a function $\rho(\hat{e}_z,t)$ bounded in time verifying:

$$\|\delta\omega\| < \rho(\hat{e}_z,t) \qquad (45)$$
$$\|\eta\| < \rho(\hat{e}_z,t) \qquad (46)$$

The function $\rho(\hat{e}_z,t)$ can be obtained using (45), (46), assumptions 1, 2, 3 and the expression of $\eta$ given by (40). This leads to:

$$\|\eta\| \leq \| E \delta\omega + E (\ddot{z}^d - \alpha_v \dot{\hat{e}}_z - \alpha_p \hat{e}_z) + J^{-1}\Delta H \| \tag{47}$$

or $$\|\eta\| \leq \gamma \rho(\hat{e}_z,t) + \gamma \|\omega\| + \bar{J} \Phi(q,\dot{q},z,\dot{z}) = \rho(\hat{e}_z,t) \tag{48}$$

and finally:

$$\rho(\hat{e}_z,t) = \frac{1}{1-\gamma} \left( \gamma \|\omega\| + \bar{J} \Phi(q,\dot{q},z,\dot{z}) \right) \tag{49}$$

The input $\delta\omega$ must satisfy the condition (45), it can be chosen in different ways [4][8][18] in order to obviate some chattering problems. We consider here for simplicity of presentation the choice:

$$\delta\omega = \begin{cases} - \rho(\hat{e}_z,t) \dfrac{W}{\|W\|} & \text{if } \|W\| \neq 0 \\[2mm] 0 & \text{if } \|W\| \equiv 0 \end{cases} \tag{50}$$

where W is given by the relation $W = B^T P x$, and matrix P is defined in the Lyapunov equation (23). In simulations we have used the same choice as in [4] [8], the stability analysis can be extended easily in case where we use a treshold in the lower values of $\|W\|$ to compute $\delta\omega$.

## 4.3. Stability Analysis of the Adaptive and Robust Controller

The overall system in closed loop is defined by equations (36) and (40). The stability analysis will be conducted based on the Lyapunov theory. The stability properties are summarized by the following theorem.

**Theorem :** The system of equations (3) and (4) is asymptotically stable when controlled by the feedback law defined by equations (32), (38) and (50) and the control gains verify the condition (31).

**Proof :** Consider the Lyapunov function candidate :

$$V(t) = \frac{1}{2} s^T M s + \frac{1}{2} e^T K F_p e + x^T P x + \frac{1}{2} \tilde{\theta}^T K \Gamma^{-1}\tilde{\theta} \tag{51}$$

with the (SPD) matrix P, solution of the Lyapunov equation given by (23), where Q is SPD. The matrix $\Gamma$ is diagonal positive definite. After derivation of the function V(t), we obtain :

$$\dot{V}(t) = s^T M \dot{s} + s^T \dot{M} s + e^T K F_p \dot{e} + \dot{x}^T P x + x^T P \dot{x} + \tilde{\theta}^T K \Gamma^{-1}\dot{\tilde{\theta}} \tag{52}$$

Using equations (36), (40) and (23) in (52), we have:

$$\dot{V}(t) = -s^T K F_v s - e^T \Lambda K F_p e + s^T K \hat{e}_z + s^T K \Psi \tilde{\theta} + \tilde{\theta}^T K \Gamma^{-1}\dot{\tilde{\theta}} - x^T Q x + 2W^T(\delta\omega+\eta) \tag{53}$$

Noting that $\dot{\tilde{\theta}} = \dot{\hat{\theta}}$ and K=kI, substituting (37) in (53) we have:

$$\dot{V}(t) = -s^T KF_v s - e^T \Lambda KF_p e + s^T K \hat{e}_z - x^T Q \, x + 2 \, W^T(\delta\omega + \eta) \tag{54}$$

Consider the last term of equation (54), it may be evaluated as follows:

if $W \equiv 0$ then $\delta\omega = 0$ and : $\qquad W^T(\delta\omega + \eta) \equiv 0 \tag{55}$

if $W \neq 0$ then from (50) and (46) we obtain:

$$W^T(\delta\omega + \eta) = W^T\left(-\rho(\hat{e}_z, t)\,\frac{W}{\|W\|} + \eta\right) = -\rho(\hat{e}_z, t)\,\frac{W^T W}{\|W\|} + W^T \eta \tag{56}$$

Equation (56) leads to :

$$W^T\left(-\rho(\hat{e}_z, t)\,\frac{W}{\|W\|} + \eta\right) \le -\rho(\hat{e}_z, t)\,\|W\| + \|W\|\|\eta\| \tag{57}$$

$$= \|W\|\left(-\rho(\hat{e}_z, t) + \|\eta\|\right) \le 0$$

Relation (57) is satisfied because $\|\eta\| \le \rho$ from (46).

Redefine the vector y as in (26) and the vector x as in (21), we can then rewrite equation (54) as follows:

$$\dot{V}(t) = -\, y^T H \, y + 2W^T(\delta\omega + \eta) \tag{58}$$

where the matrix H is defined by equation (27).

The second term in RHS of (58) has been shown negative. Then, to have an stable system, we obtain the same conditions as for the known parameter case. Consequently, if the matrix H verify the condition (31), we can conclude that s, e, x and $\tilde{\theta}$ are bounded and s $\rightarrow$ 0, e $\rightarrow$ 0 and x $\rightarrow$ 0 when t $\rightarrow$ ∞.

## 4.4. Simulation Results

Various adaptive controllers ([18],[21]) and the proposed adaptive compensator) are compared in application for a two-degree freedom IBM SCARA manipulator. Simulations are conducted with SIMNON software package The model of the considered manipulator is presented in appendix 1. The desired closed loop caracteristics are: $\alpha_p = 80$; $\alpha_v = 1600$; $F_p = 0.1$; $F_v = 0.002$; $\Lambda = 20$ with a sampling time $T = 10^{-2} \sec$ and k=1000 Nm/rd.

The desired trajectories have been chosen to be $q_1^d = a_1 + b_1(\sin(t) + \sin(2t))$ and $q_2^d = a_2 + b_2(\cos(4t) + \cos(6t))$, $\dot{q}^d$ and $\ddot{q}^d$ are mathematical derivatives of $q^d$ with $a_1 = 1$, $a_2 = 1$, $b_1 = 2$ and $b_2 = 1$. The figures present the positions and references, velocities and references, estimated parameters.

**Comments on the simulation results**

■ The transient behaviour is better for passive feedback controllers.

■ The estimated parameters converge near the true values for linearizing controllers with persistent excitation.

■ The input voltages are smooth and remain within the admissible zone for passive feedback controllers and adaptive computed torque.

■ Passive feedback controllers are more robust. They have better transient behavior than other controllers.

■ The convergence of the estimated parameters to the true values seems necessary for linearizing controllers (with persistant excitation) and not for the passive feedback controllers. The objective is the linearity of the closed loop in computed torque control. It is achieved if the parameters are well estimated. This is not the case for passive feedback controllers.


## 5. CONCLUSION

In this paper, the model of manipulator with flexible joints is reparametrized in two cascaded blocs with simplifyed coupled models. The considered reparametrization is more suitable for control laws design and avoids the complexity of analysis. This makes comprehensive the influence of the parameters used by the controllers. This method is applicable when the mechanical stiffness of the joints is not very large and no assumptions are made on the stiffness. For the first stage a passive feedback controller is used and the second stage uses a linearizing control with a robust feedback compensation loop. This is achievable easier than the design using a global linearizing control and no approximation are made as in singular perturbation based controllers.

The condition for stability of the closed loop system is derived and confirm the obtained simulation results. This condition is the same as for the case of known parameters. The estimation of all the parameters of the system is not necessary and the controllers developped for rigid robots can be extended to the case of flexible joints. Simulations results for 2 DOF robot with flexible joints (see table T1) emphasize the good performances of the two stages controllers and shows that the obtained results are comparable to the rigid case. This is in contrast to the adaptive globally linearizing controllers where approximations are needed and the stiffness constant is assumed large. The use of passive feedback controllers enhance the transient behaviour, the robustness and make the convergence to the true parameters not necessary, in adaptive case. Extensions to use PI adaptation laws and to other types of controllers for the second stage are possible and seem promizing.

**Acknowlegement** : The autors would like to thank the participants of the workshop for the helpfull comments and discussions.

## 6.REFERENCES

[1] J.J. Craig, Pi Hsu and S. S. Sastry (1986) "Adaptive control of Mechanical Manipulators". IEEE Int. Conf. Robotics and Automation, San Francisco, CA.
[2] R. Ortega and M. W. Spong (1988) "Adaptive Motion Control of Rigid Robots: A Tutorial", 27th IEEE Conf. on Decision and Control, Austin, Texas.
[3] B. Brogliato, I. D. Landau and R. Lozano (1990) "Adaptive Motion Control of Robot Manaiplutors: A Unified Approach Based on Passivity", Submitted for publication in Automatica. Internal report LAG, Grenoble.
[4] M. W. Spong and M. Vidyasagar (1988) "Robot Dynamics and Control", John Wiley and Sons, Inc., New York.
[5] C. Samson (1983). Robust non linear control of robotic manipulators. Proc 22nd IEEE C.D.C. San Antonio, TX, Dec 1983.
[6] N. Sadegh and R. Horowitz (1987) "Stability Analysis of an Adaptive Controller for Robotic Manipulators", IEEE I. C. on Robotics and Automation.
[7] I. D. Landau and R. Horowitz (1989) "Applications of the Passive Systems Approach to the Stability Analysis of Adaptive Controllers for Robot Manipulators", Int. J. of Adaptive Control and Signal Processing, V3.
[8] L. Cai and A. A. Goldenberg (1988) "Robust Control of Unconstrained Manoeuver and Collision for a Robot manipulator with Bounded Parameter Uncertainty", Proc. IEEE Conf. on Robotics and Automation, 24-29 April 1988, Philadelphia, Pensylvania.
[9] N. K. M'Sirdi and A. Benallegue (1990)"Adaptive Motion Control for Mechanical Manipulators: A Comparative Study", ACSP 90 (IASTED), 10-12 October 1990, New York USA (and Report Submitted to Int. Journal of ACSP).
[10] L.M. Sweet and M.C. Good (1985),"Redefinition of Robot Motion Control Problem" IEEE Control Systems Magazine, August, 1985, pp. 18-25.
[11] C. Canudas De wit and O. Lys (1988),"Robust Control and Parameter Estimation of Robots with Flexible Joints", Proc. IEEE Conf. on Robotics and Automation, 24-29 April 1988, Philadelphia, Pensylvania.
[12] M. W. Spong (1987),"Modeling and Control of Elastic Joint Robots", ASME Journal of Dynamic Systems, Measurement, and Control Vol. 109, pp. 310-319.
[13] J. H. Chow and P. V. Kokotovic (1978),"Two-Time-Scale Feedback Design of a class of Nonlinear Systems", IEEE Trans. on Automatic and Control, Vol AC-23, No. 3, June 1978.
[14] J.J. Slotine and S. Hong (1986),"Two-time Scale Control of Manipulators with Flexible Joints", American Control Conference, 1986, Seattle.
[15] R. Marino and M.W. Spong (1986),"Nonlinear Control Techniques for Flexible Joint Manipulators: A Single Link case Study", Proc. IEEE Conf. on Robotics and Automation
[16] F. Ghorbel, J. Y. Hung and M. W. Spong (1989),"Adaptive Control of Flexible Joint Manipulators", Proc. IEEE Conf. on Robotics and Automation.
[17] K. Khorasani (1989), "Robust Adaptive Stabilisation of Flexible Joint Manipulators", Proc. IEEE Conf. on Robotics and Automation.
[18] Kun-Pei Chen and Li-chen Fu (1989),"Nonlinear Adaptive Motion Control for a Manipulator with Flexible Joints", Proc. IEEE Conf. on Robotics and

Automation.

[19] Alessandro De Luca (1988),"Dynamic Control of Robots with Joint Elasticity", Proc. IEEE Conf. on Robotics and Automation, 24-29 April 1988, Philadelphia, Pensylvania.

[20] A. Ficola, R. Marino and S. Nicosia (1983),"A Singular Perturbation Approach to the Control of Elastic Robots", Proc. 21st Alterton Conf. on Communication, Control and Computing, Univ. of ILLINOIS.

[21] N. K. M'Sirdi and A. Benallegue (1990)"A two stages Controller for Robot Manipulators with Flexible Joints", Submitted to ECC 1991.

[22] F. Mrad and S. Ahmad (1990). Adaptive Control of Flexible Joint Robots with Stability in the sense of Lyapunov. Proc. 29th CDC, Honolulu Hawaii 1990.

[23] G. Lietmann (1981). On the Efficacy of Nonlinear Control in Uncertain Linear Systems. J. Dyn; Sys Meas. and Control, V 103, PP95-102.

**APPENDIX 1. Application for a 2 DOF Robot with Flexible Joints**

The model of two-degree freedom IBM SCARA robot is given by :

$$M(q)\ddot{q} + C(q,\dot{q})\dot{q} + G(q) + K(q - q_M) = 0$$
$$J\ddot{q}_M + B\dot{q}_M - K(q - q_M) = \tau$$

$$M(q) = \begin{pmatrix} Jo_1 + Jo_2 + m_2 L_1 L_2 c_2 & Jo_2 + \frac{1}{2} m_2 L_1 L_2 c_2 \\ Jo_2 + \frac{1}{2} m_2 L_1 L_2 c_2 & Jo_2 \end{pmatrix}$$

$$C(q,\dot{q}) = \frac{1}{2} m_2 L_1 L_2 s_2 \begin{pmatrix} -\dot{q}_2 & -(\dot{q}_2 + \dot{q}_1) \\ \dot{q}_1 & 0 \end{pmatrix} \qquad B = \begin{pmatrix} N_1^2 f_1 \\ N_2^2 f_2 \end{pmatrix} \qquad G(q) = \begin{pmatrix} 0 \\ 0 \end{pmatrix}$$

$$J = \begin{pmatrix} N_1^2 J_1 & 0 \\ 0 & N_2^2 J_2 \end{pmatrix} \qquad K = \begin{pmatrix} K_1 & 0 \\ 0 & K_2 \end{pmatrix}$$

$$Jo_1 = \frac{1}{3} m_1 L_1^2 \quad Kg.m^2 \qquad \text{and} \qquad Jo_2 = \frac{1}{3} m_2 L_2^2 \quad Kg.m^2$$
$$c_2 = \cos(q_2) \qquad \text{and} \qquad s_2 = \sin(q_2).$$

The parameter vector that will be estimated is given by $\theta = k^{-1}(m_1 \; m_2)^T$. For the remaining parameters we use a priori fixed estimates.

| Parameters | values |
|---|---|
| Mass of first link m1 | 15.91 KG |
| Mass of second link m2 | 11.36 KG |
| Inertia J1=J2 of motors 1 & 2 | 0.0001 KG.m$^2$ |
| Friction f1=f2 | 0.0007 N/rd$^2$/s |
| Gear ratio N1 | 120 |
| Gear ratio N2 | 50 |
| Motors constant Km | 0.18 N.m/A |
| Stiffness Constant K1=K2 | 1000 N.m/rd |

Table T1

**Figure 2:Bloc 1: Passive Feedback, Bloc 2: Inverse Dynamic. Known parameters.**
$\alpha_p$ =80; $\alpha_v$ =1600; $F_p$ =0.1; $F_v$ =0.002; $\Lambda$=20, sampling time T=10$^{-2}$sec, k=1000 Nm/rd.

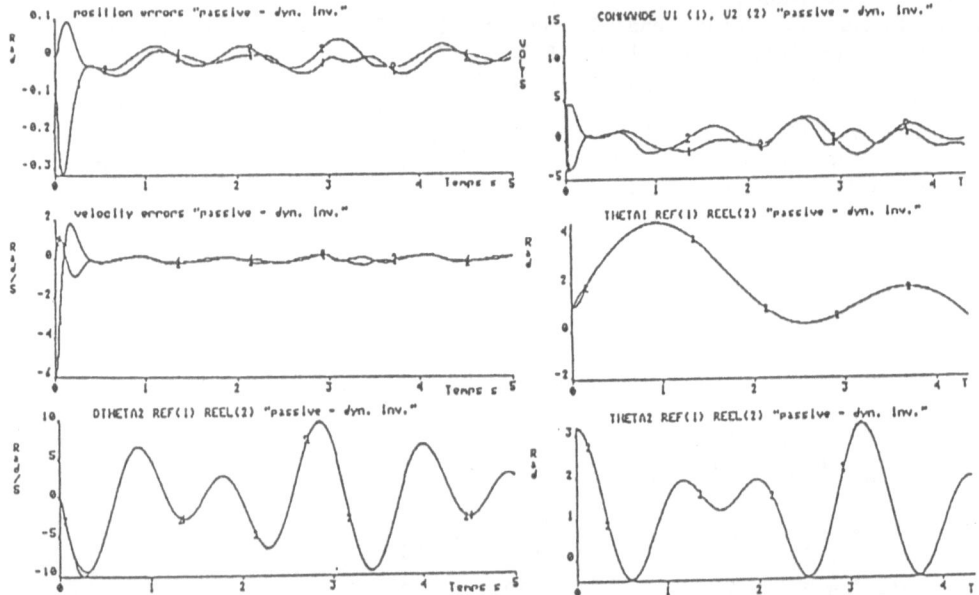

**Figure 3:Bloc 1: Passive Feedback, Bloc 2: Robust control. Unknown parameters.**
$\alpha_p$ =80; $\alpha_v$ =1600; $F_p$ =0.1; $F_v$ =0.002; $\Lambda$=30, sampling time T=10$^{-2}$sec, k=1000 Nm/rd.

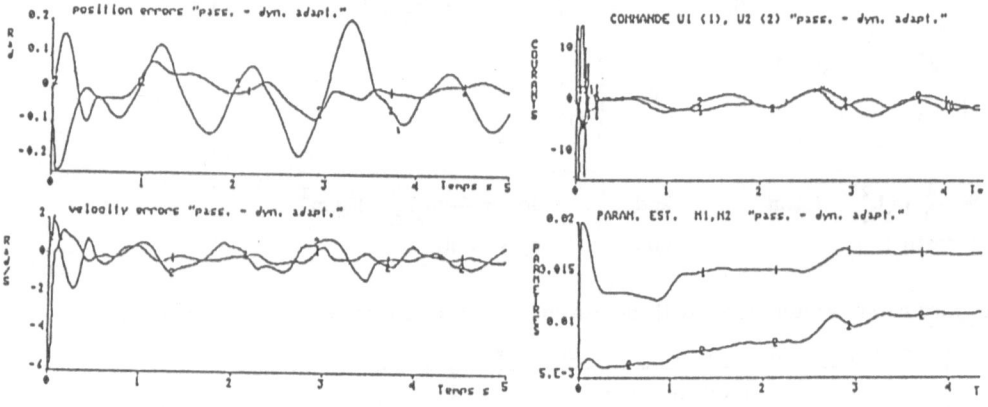

**Figure 4:**Bloc 1: Passive Feedback, Bloc 2: Passive Feedback + Robust control. Unknown parameters. $F_p^1$=0.1; $F_v^1$=0.002; $F_p^2$=0; $F_v^2$=20; $\Lambda_1$=30, $\Lambda_2$=60;

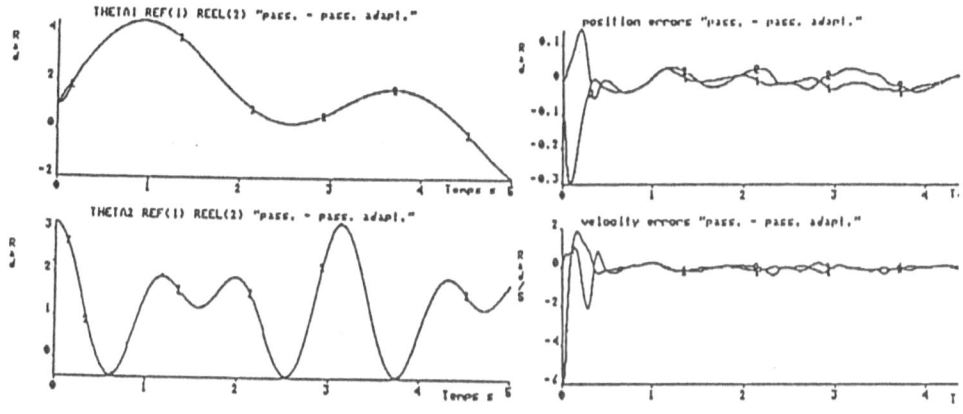

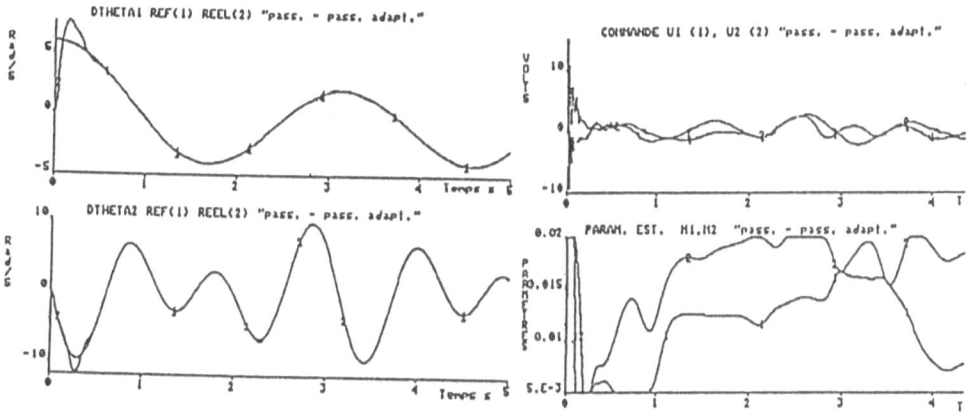

270

**Figure 5:**Bloc 1: Passive Feedback, Bloc 2: Inverse Dynamic. With and Without compensation δʊ. Without adaptation. Same as figure 2 with 30% error on parameters

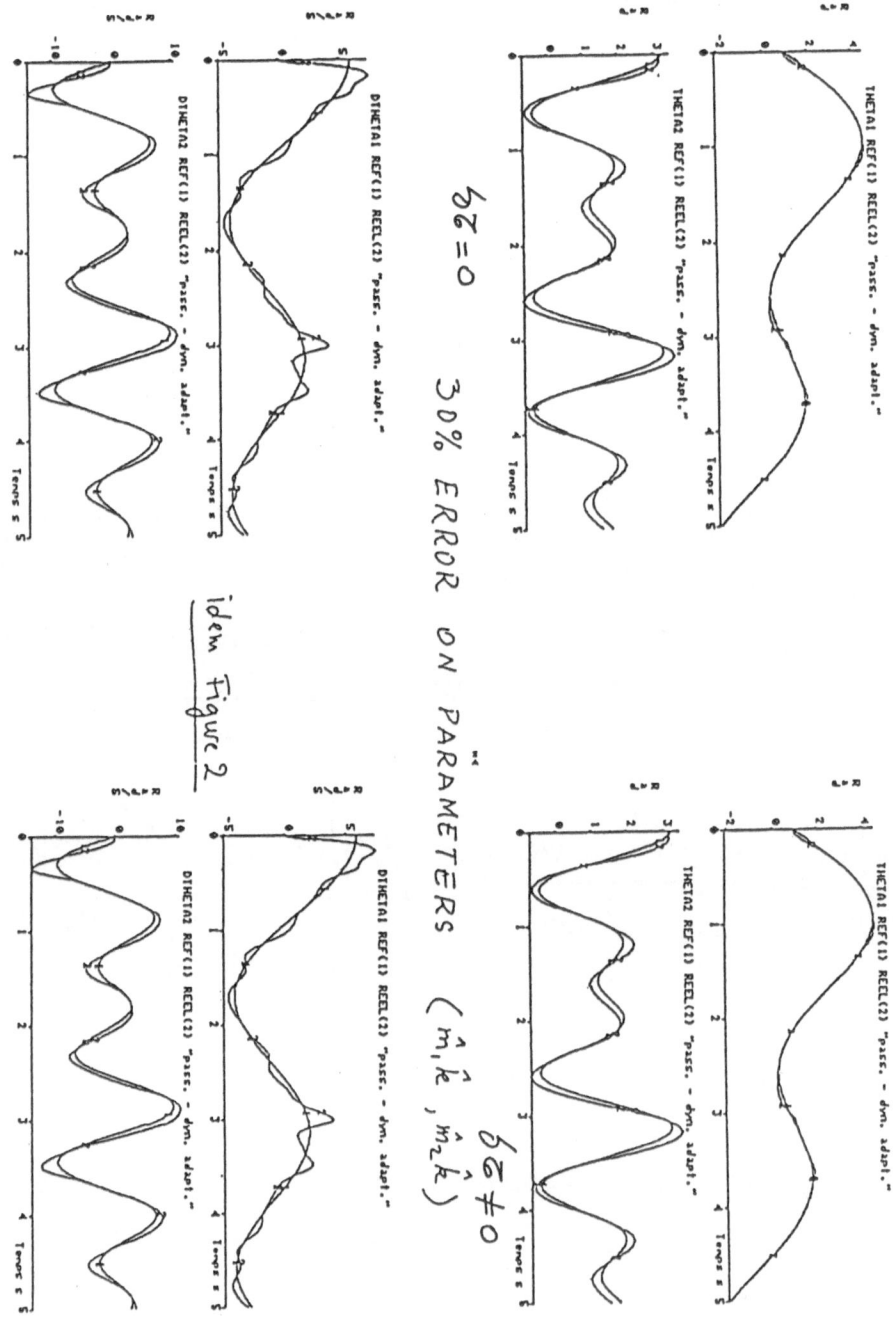

**Figure** 6:Bloc1: Passive Feedback, Bloc 2: Passive Feedback. With and Without compensation δυ. Without adaptation Γ=0. Same as figure 3 with 30% error on parameters

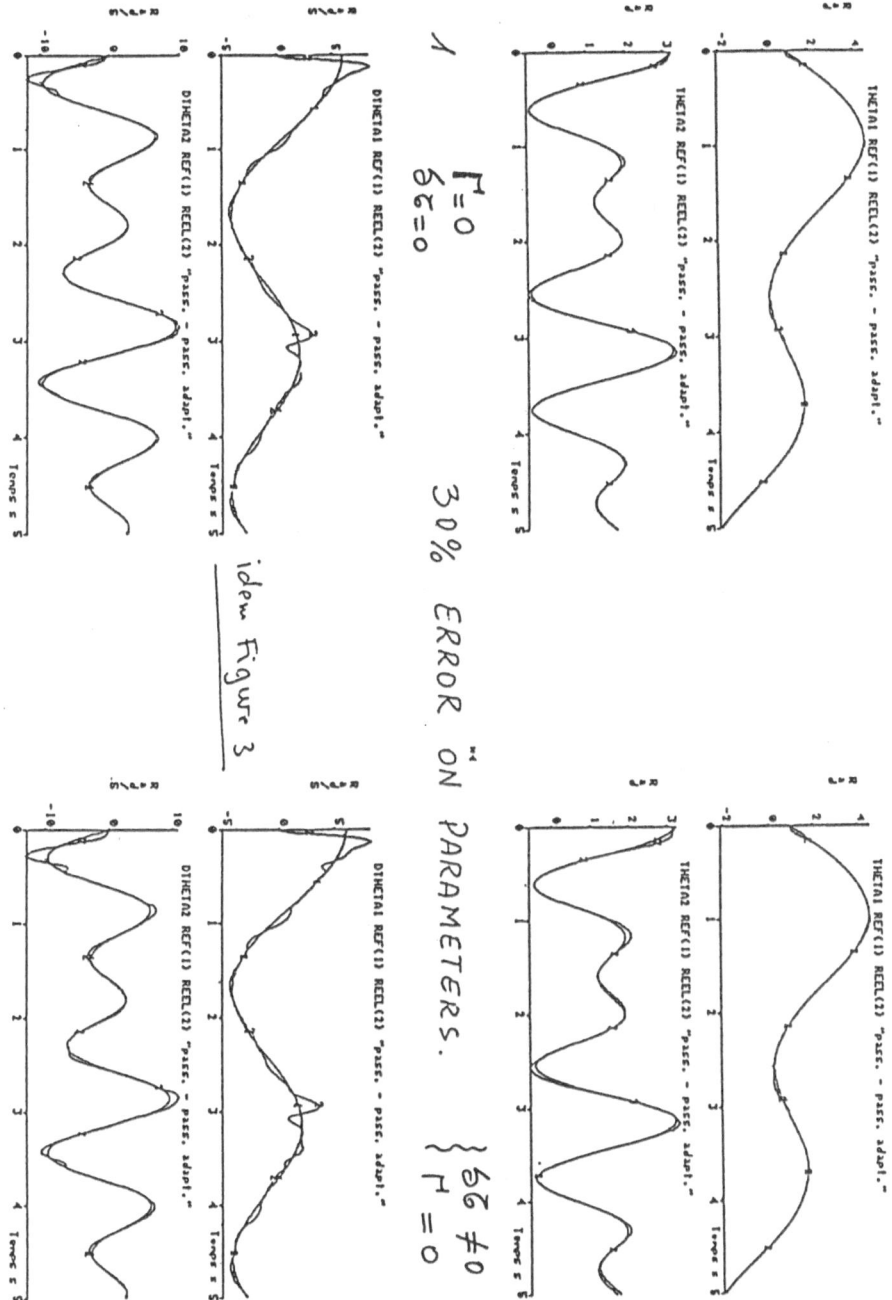

272

Figure 7: Blocl Passive Feedback, Bloc 2: Passive Feedback / Inverse Dynamic. With and Without compensation δυ. With adaptation Γ≠0.

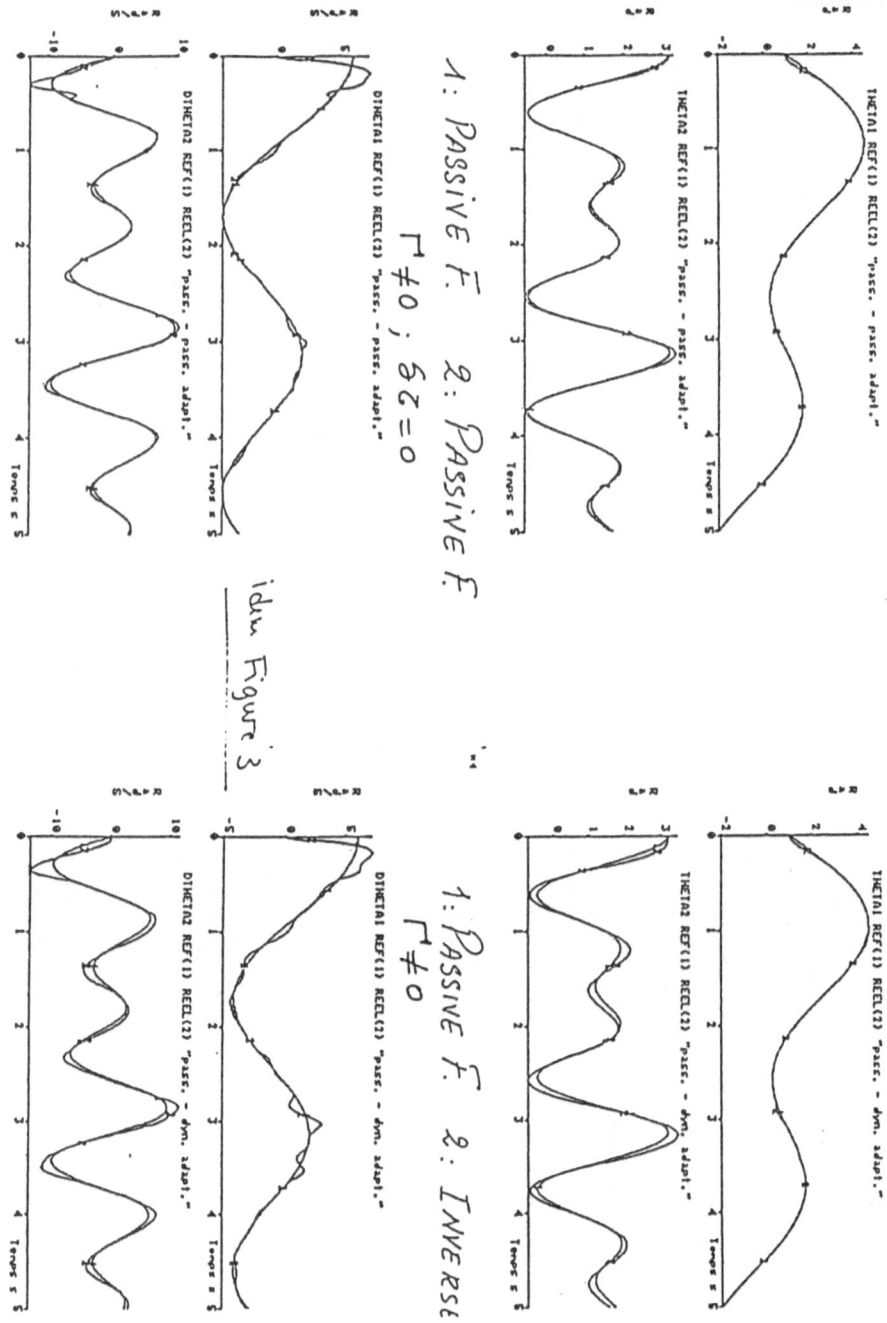

# OBSERVERS IN THE CONTROL OF RIGID ROBOTS*

S.Nicosia, A. Tornambè, P. Valigi

Dipartimento di Ingegneria Elettronica,
Seconda Università di Roma "Tor Vergata",
Via O. Raimondo 00173, Roma, Italy

**Abstract.** In this paper we propose a simple observer, with linear and decoupled structure, for the estimation of the generalized velocities of a rigid robotic manipulator. High-gain output injections are used in the attenuation of the effects of the non-linearities characterizing the dynamic behaviour of the robot upon the dynamic behaviour of the estimation errors. The Liapunov stability theory has been used to prove the practical stability of the error dynamics, in case of open-loop observers, and the asymptotic stability of the closed loop robotic system, in case of observer-based control laws.

## 1. INTRODUCTION

The control problem of industrial manipulators is a central issue in the robotics area. The models describing the robot dynamics are highly nonlinear and coupled and various control algorithms have been proposed, based on the assumption of complete knowledge of such a model [12]. In general, for the implementation of such control algorithms, the assumption of the availability of the whole state vector is unrealistic: some sensors are very noisy, e.g. the tachometers, and some state variables have no physical meaning, e.g. the approximate generalized coordinates describing the flexure of a beam.

In the last few years, a large effort has been devoted to the observer design for rigid [1,3,4,10,11] and elastic robots [6,7,9], also considering the possibility of implementing feedback control laws based on such state estimates [2,8]. Experimental validation have been carried out to study the accuracy of some of the proposed observer structures [10,11].

In this paper, with regard to the procedure proposed in [10,11], we consider the problem of designing state observers for rigid robots. In particular it is proposed an observer scheme, which does not requires an accurate knoledge of the robot motion equations. Using the Liapunov stability theory, the practical stability of the corresponding error dynamics can be inferred, under some mild assumptions. In addition, for such an observer scheme, we consider the implementation of feedback control laws based on

---

*This work was supported by CNR funds, under contract n. 89.00531.67

such estimates. Using again the Liapunov stability theory, the asymptotic stability of the resulting closed loop robotic system can be inferred.

The outline of the paper is as follows. Section 2 addresses the estimation problem for the joint velocities. The proposed algorithm is described in Section 3: the problem of open-loop observers is considered in Section 3.1, the problem of observer-based feedback control laws is considered in Section 3.2.

## 2. PROBLEM DEFINITION

We are considering rigid robotic manipulators having $N$ degrees of freedom, in which the $i$-th joint is controlled by a torque $u_i$, $i = 1, \cdots, N$, exerted by a motor. As usual in practical applications, we assume the availability of one encoder for each joint so that all the generalized coordinates are measurable. Since the available velocity measurements are, in general, corrupted by noise, we assume that the measurements supplied by the tachometers are not available for the control purposes. A possible approach to overcome this problem is the use of state observers for the velocity estimation.

To derive an effective structure for the estimation algorithm, the dynamical model of the robot to be controlled is needed. The motion equations of the robot are obtained by using the Lagrangian approach. Let $q_i$, $i = 1, \cdots, N$, be the $i$-th joint angle, i.e. the relative displacement between the adjacent links $i$ and $i - 1$. Collecting the generalized coordinates in a vector $q := (q_1, \cdots, q_N)^T$ and, using the Lagrangian notation, we obtain the motion equations in the following form:

$$\frac{d}{dt} \frac{\partial T}{\partial \dot{q}} - \frac{\partial T}{\partial q} + \frac{\partial U}{\partial q} = u, \tag{1}$$

where $u := (u_1, \cdots, u_N)^T$ denotes the vector of the external torques applied to the joints. The kinetic energy $T$ and the potential energy $U$ are given by, respectively:

$$T(q, \dot{q}) = \frac{1}{2} \dot{q}^T B(q) \dot{q},$$

$$U(q) = U_g(q) + U_f(q),$$

with $B(q)$ being the positive definite inertia matrix, and $U_g$ and $U_f$ denoting the gravitation and the dissipation energy, respectively. The dynamic equations (1) can be rewritten in the following well known form:

$$B(q)\ddot{q} + c(q, \dot{q}) + e(q) + d(q, \dot{q}) = u, \tag{2}$$

where $c(q, \dot{q})$ is the vector of Coriolis and centripetal terms, $e(q)$ is the vector of gravity terms, and $d(q, \dot{q})$ is the vector of friction terms. A simple and widely used control law is the independent joint PD control

$$u = \eta(q, \dot{q}, t) = K_v(\dot{q}_d(t) - \dot{q}) + K_p(q_d(t) - q),$$

where $q_d(t)$ denotes the trajectory to be tracked; $K_p$ and $K_v$ are diagonal gain matrices. A more sophisticated control law, which takes into account the nonlinear and coupled structure of the robot, can be derived by using nonlinearity compensation and pole placement techniques,

$$u = \eta(q, \dot{q}, t) = c(q, \dot{q}) + e(q) + d(q, \dot{q}) + B(q)(K_v(\dot{q}_d(t) - \dot{q}) + K_p(q_d(t) - q) + \ddot{q}_d(t)),$$

obtaining a closed loop robotic system described by

$$(\ddot{q}_d(t) - \ddot{q}) + K_v(\dot{q}_d(t) - \dot{q}) + K_p(q_d(t) - q) = 0.$$

Throughout the paper, without referring ourself to a particular control strategy, we will assume the existence of a feedback control law,

$$u = \eta(q, \dot{q}, t), \tag{3}$$

such that the robot end effector tracks the desired trajectory and the closed loop robotic system is asymptotically stable; notice that, this assumption is not restrictive.

Under the assumption that the continuous observation vector $y(t) \in R^N$ is constituted by the measurement of the generalized coordinates $y(t) = q(t)$, denote with $U_t$ and $Y_t$ the sequences of the available measurements at time instant $t$ for the input and output variables

$$U_t = \{u(\tau), \tau \in [0, t]\},$$
$$Y_t = \{y(\tau), \tau \in [0, t]\}.$$

The estimation problem we will deal with can be stated as follows:

**Observation problem**
Given the pair $(U_t, Y_t)$, compute an asymptotic estimate $(\hat{q}(t), \hat{\dot{q}}(t))$ of the robot state vector $(q(t), \dot{q}(t))$ based on $(U_t, Y_t)$. $\quad \triangledown$

For control purposes, we wish to obtain a solution to the above problem ensuring that the observer based control law,

$$u = \eta(\hat{q}, \hat{\dot{q}}, t), \tag{4}$$

makes the closed loop robotic system asymptotically stable and, furthermore, the robot end effector tracks the desired trajectory.

## 3. THE PROPOSED PROCEDURE

As already discussed in the above section, let us assume that measurements are available for the generalized coordinates $q$. In this section, we will show how an observer can be derived to obtain dynamic estimates for the generalized velocities $\dot{q}$.

The following state and output variables can be defined:

$$\left.\begin{aligned} x_1 &:= q \\ x_2 &:= \dot{q} \\ y &:= q \end{aligned}\right\} . \tag{5}$$

Motion equations (2) can be rewritten in the following state space form:

$$\left.\begin{aligned} \dot{x}_1 &= x_2 \\ \dot{x}_2 &= f(x) + g(x_1)u \\ y &= x_1 \end{aligned}\right\} , \tag{6}$$

where

$$\begin{aligned} f(x) &:= B(x_1)^{-1}\left[-c(x_1, x_2) - e(x_1) - d(x_1, x_2)\right], \\ g(x_1) &:= B(x_1)^{-1}, \\ x &:= (x_1^T, x_2^T)^T. \end{aligned}$$

In order to illustrate the structure of the proposed observer, let us consider the regulation problem. In this case, the feedback control law (3) can be rewritten as

$$u = \eta(x_1, x_2, x_{d,1}), \tag{7}$$

where $x_d := (x_{d,1}^T, 0^T)^T$ is a given regulation point. The more general case of the tracking problem can be handled in a similar manner, without loss of generality. The observation problem will be faced in two different steps. In the first one we will assume that control law (7) can be implemented by a direct measure of the state variables and we will consider the estimation problem for the resulting closed loop robotic system (6),(7). In the second step, we will consider the case of observer-based control laws.

### 3.1 Open Loop Observer
In this subsection, we will assume the availability of the whole state vector for the implementation of the control law (7). The resulting closed loop system is

$$\left.\begin{aligned} \dot{x}_1 &= x_2 \\ \dot{x}_2 &= \phi(x_1, x_2, x_{d,1}) \\ y &= x_1 \end{aligned}\right\} , \tag{8}$$

where $\phi(x_1, x_2, x_{d,1}) := f(x) + g(x_1)\eta(x_1, x_2, x_{d,1})$.
Consider the following linear filter

$$\left.\begin{aligned} \dot{\hat{x}}_1 &= \hat{x}_2 + \frac{1}{\epsilon}H_p(y - \hat{x}_1) \\ \dot{\hat{x}}_2 &= \frac{1}{\epsilon^2}H_v(y - \hat{x}_1) \end{aligned}\right\} , \tag{9}$$

with $\hat{x}_1$, $\hat{x}_2$ denoting respectively the dynamic estimates of the generalized coordinates and velocities, and $\hat{x} := (\hat{x}_1^T, \hat{x}_2^T)^T$. The constant matrices $H_p := \text{diag}(h_{p,1}, \cdots, h_{p,N})$, and $H_v = \text{diag}(h_{v,1}, \cdots, h_{v,N})$ are choosen so that the spectra of the characteristic polynomials $p_i(\lambda) = \lambda^2 + \lambda h_{p,i} + h_{v,i}$, $i = 1, \cdots, N$ are in the open left half plane, and $\epsilon$ is choosen as a small positive parameter. Let us now show that system (9) actually constitutes an observer for the nonlinear system (8).

The dynamics corresponding to the error between (8) and (9) take the form

$$\left. \begin{aligned} \dot{\tilde{x}}_1 &= \tilde{x}_2 - \frac{1}{\epsilon} H_p \tilde{x}_1 \\ \dot{\tilde{x}}_2 &= -\frac{1}{\epsilon^2} H_v \tilde{x}_1 + \phi(x_1, x_2, x_{d,1}) \end{aligned} \right\}, \tag{10}$$

where $\tilde{x}_i := x_i - \hat{x}_i$, $i = 1, 2$, and $\tilde{x} := (\tilde{x}_1^T, \tilde{x}_2^T)^T$.

In order to put the error dynamics into a suitable form, let us introduce the following transformation:

$$\left. \begin{aligned} \tilde{z}_1 &:= \tilde{x}_1 \\ \tilde{z}_2 &:= \epsilon \tilde{x}_2 \\ \tilde{z} &:= (\tilde{z}_1^T, \tilde{z}_2^T)^T \end{aligned} \right\}. \tag{11}$$

In the new coordinates the error dynamics can be rewritten as

$$\epsilon \dot{\tilde{z}} = A\tilde{z} + \epsilon^2 b \phi(x_1, x_2, x_{d,1}), \tag{12}$$

where

$$A = \begin{pmatrix} -H_p & I \\ -H_v & 0 \end{pmatrix}, \quad b = \begin{pmatrix} 0 \\ I \end{pmatrix}.$$

A linear system can be formally associated with the error dynamics (12) by defining a fast time scale $\tau = t/\epsilon$ and setting $\epsilon = 0$ into the resulting system:

$$\frac{d\tilde{z}}{d\tau} = A\tilde{z}. \tag{13}$$

A Liapunov function for the linear system (13) can be used to infer some stability properties of the error dynamics (12). Since matrices $H_p$ and $H_v$ have been choosen so that the spectra of $p_i(\lambda)$, $i = 1, \ldots, N$, are in the open left half plane, there exists a positive definite matrix $P$ solution of the following matrix equation $A^T P + P A = -I$. Let us define the following bound for the nonlinear terms apperaring in the error dynamics (12):

$$c_T := \sup_{t \in [0,T]} \gamma(t),$$

where $\gamma(t) := \|Pb\phi(x_1(t), x_2(t), x_{d,1})\|$.

**Theorem 1.**

The following stability properties hold for the error dynamics (12):

(i) for any $\mu > 0$, under the assumption that $c_T < \infty$, with $T$ being a finite time instant, and that $\epsilon \leq \dfrac{\mu}{2\,c_T}$, if $\|\tilde{x}(0)\| \leq \mu$, then $\|\tilde{x}(t)\| \leq \mu \,\forall t \in [0, T]$,

(ii) for any $\mu > 0$, if $c_\infty < \infty$ and $\epsilon \leq \dfrac{\mu}{2\,c_\infty}$, then, for any initial condition $\tilde{x}(0)$, there exists a finite time instant $t_\epsilon$ such that $\|\tilde{x}(t)\| < \mu \,\forall t \geq t_\epsilon$,

(iii) if $\lim_{t \to \infty} \gamma(t) = 0$ and $\epsilon \leq \dfrac{\mu}{2\,c_\infty}$, then $\lim_{t \to \infty} \|\tilde{x}(t)\| = 0$.

**Proof.**

Let us use the following Liapunov function for (13)

$$W(\tilde{z}) = \tilde{z}^T P \tilde{z}, \tag{14}$$

as a Liapunov function candidate for the error dynamics (12). Computing the derivative of $W(\tilde{z})$ along the solutions of (12), we obtain

$$\frac{dW}{dt} = \frac{1}{\epsilon}\left[\tilde{z}^T(A^T P + PA)\tilde{z} + 2\,\epsilon^2\,\phi(x_1, x_2, x_{d,1})^T b^T P \tilde{z}\right]$$

$$= -\frac{1}{\epsilon}\tilde{z}^T \tilde{z} + 2\,\epsilon\,\phi(x_1, x_2, x_{d,1})^T b^T P \tilde{z}. \tag{15}$$

From (15) we obtain the inequality:

$$\frac{dW}{dt} \leq -\frac{1}{\epsilon}\|\tilde{z}\|^2 + 2\epsilon\|Pb\phi(x_1, x_2, x_{d,1})\|\|\tilde{z}\|. \tag{16}$$

Under the assumption that the solution of (8) exists for any $t \in [0, T]$, the inequality (16) can be rewritten as follows:

$$\frac{dW}{dt} \leq -\frac{1}{\epsilon}\|\tilde{z}\|^2 + \frac{1}{\epsilon}K_z(\epsilon, c_T)\|\tilde{z}\|, \; \forall t \in [0, T], \tag{17}$$

where $K_z(\epsilon, c_T) := 2\epsilon^2 c_T$.

It is noted from (17) that: if $\|\tilde{z}\| > K_z(\epsilon, c_T)$, then $\dfrac{dW}{dt} < 0, \,\forall t \in [0, T]$. According to the value of the supremum $c_T$, different cases can arise:

(a) Under the assumption of case (i), i.e. $c_T < \infty$ and $\epsilon \leq \dfrac{\mu}{2\,c_T}$, if $\|\tilde{z}(0)\| \leq K_z(\epsilon, c_T) \leq \epsilon\,\mu$, then $\|\tilde{z}(t)\| \leq \epsilon\,\mu, \,\forall t \in [0, T]$.

(b) Under the stronger assumption of case (ii), i.e. $c_\infty < \infty$ and $\epsilon \leq \dfrac{\mu}{2\,c_\infty}$, if $\|\tilde{z}(0)\| \leq K_z(\epsilon, c) \leq \epsilon\,\mu$ then $\|\tilde{z}(t)\| \leq \epsilon\,\mu, \,\forall t \in [0, \infty)$, otherwise, if $\|\tilde{z}(0)\| \geq K_z(\epsilon, c)$ then there exists a finite $t_\epsilon > 0$ such that $\|\tilde{z}(t)\| \leq \epsilon\,\mu, \,\forall t \in [t_\epsilon, \infty)$.

(c) Taking into account the definition of the coordinate transformation (11), it is easy to see that, for $\epsilon < 1$, $\|\tilde{x}\| \leq \frac{1}{\epsilon}\|\tilde{z}\|$; therefore, if $\|\tilde{z}\| \leq \epsilon p\,\mu$, then $\|\tilde{x}\| \leq \mu$, thus proving (i) and (ii).

(d) Under the assumptions of case (ii), i.e. $\lim_{t\to\infty}\gamma(t) = 0$ and $\epsilon \leq \dfrac{\mu}{2\,c_\infty}$, it is possible to show that $\|\tilde{z}\| \to 0$ as $t \to \infty$ [5, pag. 271], thus proving (iii).    ▽

## 3.2 Observer-based Control Law.

In this subsection, we will consider the case in which the control law (7) is implemented by means of observer state estimates:

$$u = \eta(\hat{x}_1, \hat{x}_2, x_{d,1}), \tag{18}$$

where measured state variables $x$ are replaced by the corresponding estimates $\hat{x}$ supplied by the proposed observer given in (9).

The extended system describing the closed loop dynamics constituted by (6),(9),(18) can be rewritten as

$$\left.\begin{aligned}
\dot{x}_1 &= x_2 \\
\dot{x}_2 &= f(x) + g(x_1)u \\
\dot{\hat{x}}_1 &= \hat{x}_2 + \frac{1}{\epsilon}H_p(y - \hat{x}_1) \\
\dot{\hat{x}}_2 &= \frac{1}{\epsilon^2}H_v(y - \hat{x}_1) \\
u &= \eta(\hat{x}_1, \hat{x}_2, x_{d,1})
\end{aligned}\right\}. \tag{19}$$

In order to put system (19) into a suitable form, we will again use transformation (11). The extented system becomes

$$\left.\begin{aligned}
\dot{x}_1 &= x_2 \\
\dot{x}_2 &= f(x) + g(x_1)u \\
\epsilon\dot{\tilde{z}}_1 &= -H_p\tilde{z}_1 + \tilde{z}_2 \\
\epsilon\dot{\tilde{z}}_2 &= -H_v\tilde{z}_1 + \epsilon^2\left[f(x) + g(x_1)u\right] \\
u &= \eta(x_1 - \tilde{z}_1, x_2 - \frac{1}{\epsilon}\tilde{z}_2, x_{d,1})
\end{aligned}\right\}, \tag{20}$$

where $\tilde{z}_1 = \tilde{x}_1 = (x_1 - \hat{x}_1)$, $\tilde{z}_2 = \epsilon\tilde{x}_2 = \epsilon(x_2 - \hat{x}_2)$.

To study the stability of the operating point $(x, \tilde{z}) = (x_d, 0)$ of system (20), some preliminary assumptions on system (6),(7) are required. Throughout the rest of this subsection, we will make use of the following notation: $U_{x_d}$ denotes the compact neighborhood of $x = x_d$ such that $\|x - \bar{x}\| \leq \rho_x$, $\rho_x > 0$, $U_0$ denotes the compact neighborhood of $\tilde{z} = 0$ such that $\|\tilde{z}\| \leq \rho_z$, $\rho_z > 0$, $\xi := (\xi_1^T, \xi_2^T)^T \in R^{2N}$ and $\zeta := (\zeta_1^T, \zeta_2^T)^T \in R^{2N}$ denote two auxiliary vectors.

## Assumption 1.

There exists a Liapunov function $V(x - x_d)$ for system (6),(7) such that:

$$\frac{\partial V(\xi)}{\partial \xi_1}\bigg|_{\xi=x-x_d} x_2 + \frac{\partial V(\xi)}{\partial \xi_2}\bigg|_{\xi=x-x_d} \phi(x_1, x_2, x_{d,1}) \leq -a_1\|x - \bar{x}\|^2, \tag{21}$$

$$\forall x \in U_{x_d}, \quad a_1 > 0,$$

$$\left\|\frac{\partial V(\xi)}{\partial \xi_2}\bigg|_{\xi=x-x_d}\right\| \leq b\|x - \bar{x}\|, \quad \forall x \in U_{x_d} \quad b > 0. \quad \nabla \tag{22}$$

Notice that the assumption 1 implies the asymptotic stability of the equilibrium point $x = x_d$ of the system (6),(7).

## Assumption 2.

The matrix $g(\xi_1)$ is bounded

$$\|g(\xi_1)\| \leq c, \quad \forall \xi \in U_{x_d}, \quad c > 0. \quad \nabla \tag{23}$$

## Assumption 3.

The vector function $\phi(x_1, x_2, x_{d,1})$ satisfies the inequality:

$$\|\phi(x_1, x_2, x_{d,1})\| \leq h_1\|x - \bar{x}\| + h_2\|x - \bar{x}\|^2, \quad \forall x \in U_{x_d}, \quad h_1, h_2 > 0. \quad \nabla \tag{24}$$

## Assumption 4.

The input control law (7) satisfies the inequality

$$\|\eta(\xi_1, \xi_2, x_{d,1}) - \eta(\zeta_1, \zeta_2, x_{d,1})\| \leq l_1\|\xi - \zeta\| + l_2\|\xi - \zeta\|^2$$
$$\forall \xi \in R^{2N}, \forall \zeta \in R^{2N} : \|\xi - \zeta\| \leq \rho_z, l_1, l_2 > 0, \tag{25}$$

with $l_1$ and $l_2$ independent of $x_{d,1}$. $\quad \nabla$

Before proceeding and recalling the definition of the Liapunov function for system (13), it is noted that the following relations hold

$$\left\|\frac{\partial W(\tilde{z})}{\partial \tilde{z}_2}\right\| \leq 2\lambda\|\tilde{z}\|, \quad \forall \tilde{z} \in R^{2N}, \tag{26}$$

$$\frac{\partial W(\tilde{z})}{\partial \tilde{z}_1}(-H_p\tilde{z}_1 + \tilde{z}_2) + \frac{\partial W(\tilde{z})}{\partial \tilde{z}_2}(-H_v\tilde{z}_1) = -\|\tilde{z}\|^2, \quad \forall \tilde{z} \in R^{2N}, \tag{27}$$

with $\lambda = \max\{\lambda(P)\}$.

The validity of the following inequalities I1 - I3 can easily be proved:

**Inequality I1.**
From the inequalities (22),(23) and (25) of assumptions 1,2 and 4 one obtains:

$$\frac{\partial V(\xi)}{\partial \xi_2}\bigg|_{\xi=x-x_d} g(x_1)\left(\eta(x_1 - \tilde{z}_1, x_2 - \frac{1}{\epsilon}\tilde{z}_2, x_{d,1}) - \eta(x_1, x_2, x_{d,1})\right) \leq$$

$$\leq \frac{\beta_1}{\epsilon}\|x - \tilde{x}\|\,\|\tilde{z}\| + \frac{\beta_2}{\epsilon^2}\|x - \tilde{x}\|\,\|\tilde{z}\|^2, \quad \forall x \in U_{x_d}, \forall \tilde{z} \in U_0, \forall \epsilon < 1, \qquad (28)$$

where $\beta_1 = bcl_1 > 0$, $\beta_2 = bcl_2 > 0$.

**Inequality I2.**
From the inequality (24) of assumption 3 and from inequality (26) one obtains:

$$\frac{\partial W(\tilde{z})}{\partial \tilde{z}_2}\phi(x_1, x_2, x_{d,1}) \leq \gamma_1\|x - \tilde{x}\|\,\|\tilde{z}\| + \gamma_2\|x - \tilde{x}\|^2\|\tilde{z}\|, \forall x \in U_{x_d}, \forall \tilde{z} \in U_0, \quad (29)$$

where $\gamma_1 = 2\lambda h_1 > 0$, $\gamma_2 = 2\lambda h_2 > 0$.

**Inequality I3.**
From inequalities (23) and (25) of assumptions 2 and 4, and from inequality (26) one obtains:

$$\frac{\partial W(\tilde{z})}{\partial \tilde{z}_2}g(x_1)\left(\eta(x_1 - \tilde{z}_1, x_2 - \frac{1}{\epsilon}\tilde{z}_2, x_{d,1}) - \eta(x_1, x_2, x_{d,1})\right) \leq$$

$$\leq \frac{\delta_1}{\epsilon}\|\tilde{z}\|^2 + \frac{\delta_2}{\epsilon^2}\|\tilde{z}\|^3 \,\forall x \in U_{x_d}, \forall \tilde{z} \in U_0, \forall \epsilon < 1, \qquad (30)$$

where $\delta_1 = 2\lambda c\, l_1 > 0$, $\delta_2 = 2\lambda c\, l_2 > 0$.

**Theorem 2**
Under the assumptions 1 up to 4, for any $\omega < \dfrac{a_1}{\gamma_2} =: \omega^*$, if

$$\epsilon < \frac{4(a_1 - \gamma_2\omega)}{(\beta_1 + \beta_2\omega + \gamma_1)^2 + 4(\delta_1 + \delta_2\omega)(a_1 - \gamma_2\omega)} =: \epsilon^*(\omega),$$

then the equilibrium point $(x, \tilde{x}) = (x_d, 0)$ is asymptotically stable and an estimate of the region of attraction is:

$$\Omega_x = \{(x, \tilde{x}) \in U_{x_d} \cup U_0 : \|\tilde{x}\| \leq \omega\}.$$

**Proof.**
Consider the following Liapunov function candidate for the extended system (20)

$$v(x - x_d, \tilde{z}) = V(x - x_d) + d\,W(\tilde{z}), \qquad (31)$$

with $d > 0$. Computing the time derivative of $v$ along the solution of (20) we obtain

$$\frac{dv(x - x_d, \tilde{z})}{dt} =$$

$$\left.\frac{\partial V(\xi)}{\partial \xi_1}\right|_{\xi=x-x_d} x_2 + \left.\frac{\partial V(\xi)}{\partial \xi_2}\right|_{\xi=x-x_d} \phi(x_1, x_2, x_{d,1})$$

$$+ \left.\frac{\partial V(\xi)}{\partial \xi_2}\right|_{\xi=x-x_d} g(x_1)\left(\eta(x_1 - \tilde{z}_1, x_2 - \frac{1}{\epsilon}\tilde{z}_2, x_{d,1}) - \eta(x_1, x_2, x_{d,1})\right)$$

$$+ \frac{d}{\epsilon}\left[\frac{\partial W(\tilde{z})}{\partial \tilde{z}_1}(-H_p\tilde{z}_1 + \tilde{z}_2) + \frac{\partial W(\tilde{z})}{\partial \tilde{z}_2}(-H_v\tilde{z}_1)\right]$$

$$+ \frac{d}{\epsilon}\left[\frac{\partial W(\tilde{z})}{\partial \tilde{z}_2}\epsilon^2\phi(x_1, x_2, x_{d,1})\right] \tag{32}$$

$$+ \frac{d}{\epsilon}\left[\frac{\partial W(\tilde{z})}{\partial \tilde{z}_2}g(x_1)\epsilon^2\left(\eta(x_1 - \tilde{z}_1, x_2 - \frac{1}{\epsilon}\tilde{z}_2, x_{d,1}) - \eta(x_1, x_2, x_{d,1})\right)\right].$$

Using the inequalities I1 - I3 and equation (27), from equation (32) we obtain the inequality:

$$\dot{v} \leq -a_1\|x - \bar{x}\|^2 + \frac{\beta_1}{\epsilon}\|x - \bar{x}\|\,\|\tilde{z}\| + \frac{\beta_2}{\epsilon^2}\|x - \bar{x}\|\,\|\tilde{z}\|^2$$

$$- \frac{d}{\epsilon}\|\tilde{z}\|^2 + d\epsilon\gamma_1\|x - \bar{x}\|\|\tilde{z}\| + d\epsilon\gamma_2\|x - \bar{x}\|^2\|\tilde{z}\| +$$

$$+ d\delta_1\|\tilde{z}\|^2 + \frac{d\delta_2}{\epsilon}\|\tilde{z}\|^3, \ \forall x \in U_{x_d}, \forall \tilde{z} \in U_0. \tag{33}$$

Inequality (33) can be rewritten as

$$\dot{v} \leq -(\|x - \bar{x}\| \quad \|\tilde{z}\|)\,M\left(\begin{matrix}\|x - \bar{x}\| \\ \|\tilde{z}\|\end{matrix}\right), \tag{34}$$

with

$$M = \begin{pmatrix} a_1 - d\epsilon\gamma_2\|\tilde{z}\| & -\frac{1}{2}\left(\frac{\beta_1}{\epsilon} + \frac{\beta_2}{\epsilon^2}\|\tilde{z}\| + d\epsilon\gamma_1\right) \\ -\frac{1}{2}\left(\frac{\beta_1}{\epsilon} + \frac{\beta_2}{\epsilon^2}\|\tilde{z}\| + d\epsilon\gamma_1\right) & \frac{d}{\epsilon} - d\delta_1 - \frac{d\delta_2}{\epsilon}\|\tilde{z}\| \end{pmatrix}.$$

Thus, for $v$ to be a Liapunov function, and, therefore, for $(x, \tilde{z}) = (x_d, 0)$ to be an asymptotically stable equilibrium for (20), matrix $M$ must be positive definite, i.e.

$$(a_1 - d\epsilon\gamma_2\|\tilde{z}\|) > 0,$$

$$(a_1 - d\epsilon\gamma_2\|\tilde{z}\|)(\frac{d}{\epsilon} - d\delta_1 - \frac{d\delta_2}{\epsilon}\|\tilde{z}\|) - \frac{1}{4}\left(\frac{\beta_1}{\epsilon} + \frac{\beta_2}{\epsilon^2}\|\tilde{z}\| + d\epsilon\gamma_1\right)^2 > 0. \tag{35}$$

If we define d $= \dfrac{1}{\epsilon^2}$ and consider the region $\|\tilde{z}\| \leq \omega\epsilon$, then (34) can be rewritten as follows

$$\left.\begin{array}{l} a_1 - \gamma_2\omega > 0 \\[2mm] \dfrac{1}{\epsilon^2}\left[(a_1 - \gamma_2\omega)(\dfrac{1}{\epsilon} - \delta_1 - \delta_2\omega) - \dfrac{1}{4}(\beta_1 + \beta_2\omega + \gamma_1)^2\right] > 0 \end{array}\right\}. \qquad (36)$$

If we choose

$$\omega < \frac{a_1}{\gamma_2} = \omega^*, \qquad (37)$$

then the first inequality in (36) is satisfied. From the second one, after some algebra, one obtains:

$$\epsilon < \frac{4(a_1 - \gamma_2\omega)}{(\beta_1 + \beta_2\omega + \gamma_1)^2 + 4(\delta_1 + \delta_2\omega)(a_1 - \gamma_2\omega)} = \epsilon^*(\omega),$$
$$\omega < \omega^*. \qquad (38)$$

We have shown that the extended system (20) can be made asymptotically stable taking $\epsilon < \epsilon^*(\omega)$, with $\omega < \omega^*$, provided that the initial conditions $(x(0), \tilde{z}(0))$ belong to the region $\Omega_z := \{(x, \tilde{z}) \in U_{x_d} \cup U_0 : \|\tilde{z}\| \leq \omega\epsilon\}$; then, it follows that $\Omega_z$ constitutes an estimate of the attraction region for the system (20). Notice that, for $\epsilon < 1$, $\|\tilde{x}\| \leq \dfrac{1}{\epsilon}\|\tilde{z}\|$, therefore the attraction region estimate in the $(x, \tilde{x})$ variables becomes the set $\Omega_x = \{(x, \tilde{x}) \in U_{x_d} \cup U_0 : \|\tilde{x}\| \leq \omega\}$, with $\omega < \omega^*$, as was to be proved. $\quad\triangledown$

In order to choose the parameter $\omega$, and therefore $\epsilon$, (or viceversa) one can consider the further goal of maximizing the attraction region, in the $\tilde{z}$ or $\tilde{x}$ variables. Two different cases can be considered:

(i) In order to maximize the attraction region estimate $\Omega_z$, the function $\rho(\omega) := \omega\epsilon^*(\omega)$ has to be maximized in the domain $\omega \in (0, \omega^*)$, which is a continuous function of $\omega$. Notice that $\rho(0) = \rho(\omega^*) = 0$ and $\rho(\omega) \geq 0$, $\forall\omega \in (0, \omega^*)$; therefore it must exists a point $\bar{\omega} \in (0, \omega^*)$ such that $\rho(\bar{\omega}) \geq \rho(\omega)$, $\forall\omega \in (0, \omega^*)$. It follows that, choosing $\omega = \bar{\omega}$, and $\bar{\epsilon} = \epsilon^*(\bar{\omega})$, it is possible to maximize the attraction region estimate $\Omega_z$.

(ii) The attraction region estimate in the $\tilde{x}$ variables is given by $\Omega_x = \{(x, \tilde{x}) \in U_{x_d} \cup U_0 : \|\tilde{x}\| \leq \omega\}$. In order to maximize this convergence region, a fixed positive parameter $\bar{\omega}$ should be choosen, with the only condition that such a $\bar{\omega}$ must be *strictly* less than $\omega^*$; the value of the parameter $\epsilon$ can be derived by equation (38) as $\bar{\epsilon} = \epsilon^*(\bar{\omega})$.

## 4. CONCLUSIONS

In this paper we have proposed a simple asymptotic observer for the estimation of the joint velocities of a rigid robotic system. A high-gain output injection has been used in the attenuation of the effects of the coupled and nonlinear terms characterizing the dynamic behaviour of the robot upon the estimate errors, which can be made arbitrarily small by a proper choice of the gain parameters. The Liapunov theory has been used in the stability analysis of the error dynamics showing that: (i) if the open-loop observer is considered, then the practical stability of the corresponding error dynamics is ensured, (ii) if the observer is used for the implementation of a feedback control law, the resulting closed loop system can be made asymptotically stable.

## 5. REFERENCES.

1 Blauer, M., and Belonger, P.R., State and parameter estimation for robotic manipulator using force measurements, *IEEE Trans. Aut. Control*, AC-32, 1055-1066 (1987).

2 Canudas De Wit, C., Astrom, K.J. and Fixot, N., Robot control via nonlinear observer, *Robust Control of Linear Systems and Nonlinear Control*, Proc. Int. Symp. MTNS-89, Vol. 2, pp. 539-551 (1989).

3 Canudas De Wit, C., Fixot, N., Robot control via robust state estimate feedback, *New Trends in Systems Theory*, Genova, Italy (1990).

4 Canudas De Wit, C., Slotine, J.J., Sliding observers for robot manipulators, *1989 Nonlinear Control Systems Design Conference*, Capri, Italy (1989).

5 Hahn, W., *Stability of motion*, New-York, Spinger-Verlag (1967).

6 Nicosia, S., Tomei, P., and Tornambè, A., A nonlinear observer for elastic robots, *IEEE Trans. Robotics Automation*, RA-4, no. 1, 45-52 (1988).

7 Nicosia, S., Tomei, P., and Tornambè, A., Nonlinear control and observation algorithms for a single-link flexible robot arm, *Int. J. Control*, 49, 827-840 (1989).

8 Nicosia, S., Tomei, P., Robot control by using only joint position measurements, *IEEE Trans. on Automatic Control*, AC-35, no. 9, pp. 1058-1061 (1990).

9 Nicosia, S., and Tornambè, A., High-gain observers in the state and parameter estimation of robots having elastic joints, *Systems & Control Letters*, 13, 331-337 (1989).

10 Nicosia, S., and Tornambè, A. and Valigi, P., Experimental validation of asymptotic observers for robotic manipulators, *1990 IEEE Int. Conference on Robotics and Automation*, Cincinnati, USA, May 13-18, (1990).

11 Nicosia, S., and Tornambè, A. and Valigi, P., Experimental results in state estimation of industrial robots, *29-th IEEE Conference on Decision and Control*, Honolulu, USA, December 5-7 (1990).

12 Paul, R.P. *Robot Manipulators: Mathematics, Programming and Control*, MIT Press, Cambridge, MA (1981).

# Control of Robotic Systems through Singularities

Stefano Chiaverini   Lorenzo Sciavicco   Bruno Siciliano

Dipartimento di Informatica e Sistemistica
Università degli Studî di Napoli "Federico II"
via Claudio 21, 80125 Napoli, Italy

## Abstract

*The goal of this work is to provide an overview of major control techniques that manage the occurrence of singularities for robotic systems. The common feature of these methods is a modification of the inverse differential kinematic mapping which is ill-conditioned in the neighbourhood of a singularity. The following solutions are discussed; namely, the Jacobian transpose, the Jacobian pseudoinverse, and the damped least-squares Jacobian inverse.*

## Introduction

At a singular configuration the linear mapping which relates the joint-space velocity vector to the task-space velocity vector through the Jacobian matrix of a given manipulator becomes rank-deficient and the solution of the inverse kinematics problem is undefined; furthermore, very high values and discontinuities of the joint-space velocity result in the neighbourhood of the singularity due, respectively, to the unfeasible components of the assigned task-space velocity vector and to sudden crossings through the singular configuration itself.

When a preprogrammed reference task-space trajectory is to be tracked, it is normally possible to plan the trajectory so that singular configurations are avoided. Alternatively, it is possible to interpolate the joint-space solution close to the singularities. On the other hand, sensory control applications —where the reference trajectory is not known in advance— demand for singularity-robust inverse kinematics algorithms to guarantee the effectiveness of the robot control system all over the manipulator's workspace.

In order to overcome singularities, the basic numerical solution to the inverse kinematics problem which can be obtained by simply inverting the direct kinematic mapping is replaced by suitably-defined mappings relating the task-space to the joint-space; these are designed to ensure well-behaved joint-space solutions close to a singularity, while guaranteeing accurate trajectory tracking far from singularities.

The present work is aimed at surveying the most effective inverse differential kinematics techniques which allow control of robotic systems through singularities. We will focus our attention only to the singularity handling issues, while eventual kinematic redundancy will not be exploited.

A very simple inverse kinematics solution can be obtained by using the transpose of the Jacobian matrix to relate the joint-space velocity vector to the task-space location of the manipulator [1]; this corresponds to adopting an impedance control law for an ideal manipulator of simplified dynamics [2]. The approach leads to a closed-loop formulation which offers an iterative solution to the inverse kinematics problem; the solution is exact for a given constant task-space location while is approximate in the trajectory-tracking case. The solution is computationally inexpensive and performs very robust behaviour close and through singular configurations; as a drawback, optimal tuning of the algorithm is required [3] and degraded accuracy is experienced when tracking fast task-space trajectories.

A more accurate solution can be obtained by using a pseudoinverse of the Jacobian matrix to transform desired task-space trajectory into the corresponding joint-space motion [4]. This solution is defined even at singular configurations, but high joint velocities may result in the neighbourhood of singularities [5]. Compared to the transpose solution, the pseudoinverse solution is computationally demanding in view of real-time applications. An efficient procedure to compute the pseudoinverse of the Jacobian matrix can be devised by taking advantage of the kinematic analysis of the manipulator structure. In order to avoid excessive joint velocities close to singularities, the manipulator can be treated as singular in a suitably defined region around each singularity; inside the region the available extra degree(s) of freedom is used to achieve a continuous joint velocity solution [6].

Another approach to solve the inverse kinematics problem is the damped least-squares technique giving an approximate solution which is well-conditioned and defined everywhere in the manipulator's workspace [7,8]. The approach seems to be promising for real-time applications, as it implicitly removes the unfeasible motion components while it requires less computation than the previous method. A problem is to select a damping factor which results in a satisfactory trade-off between tracking accuracy and feasibility of the resulting joint-space solution. A varying damping factor gives better performance but is, in general, expensive to compute on-line [9]; simplified computation has recently been proposed [10]. Geometrical insight may prove very useful to perform the tuning of the damping factor [11].

## Differential kinematics

A velocity $\dot{q} \in R^n$ in joint-space coordinates is related to the corresponding velocity $\dot{x} \in R^m$ in task-space coordinates through the equation

$$\dot{x} = J(q)\dot{q}, \tag{1}$$

where $\mathbf{J}(\mathbf{q})$ is the $(m \times n)$ Jacobian matrix of the manipulator considered. In all cases of interest it is $n \geq m$; when $n > m$ the manipulator is said to be *redundant* and there exists an $(n - m)$-dimensional subspace of $\mathbf{R}^n$ in which any joint-space velocity gives a null velocity in the task space.

If for some configuration $\hat{\mathbf{q}}$ it happens that $\text{rank}(\mathbf{J}(\hat{\mathbf{q}})) = r < m$, the configuration is termed as *singular*. At a singular configuration the subspace of the joint velocity space which maps into the null velocity vector in the task space increases its dimension, as $\dim(\mathcal{N}(\mathbf{J}(\hat{\mathbf{q}}))) = n - r > n - m$. On the other hand, since $\dim(\mathcal{R}(\mathbf{J}(\dot{\mathbf{q}}))) = r < m$, only an $r$-dimensional subspace of task-space velocities can be spanned at a singularity; this subspace is the space of feasible motion for the manipulator. In general, at a singular configuration an assigned velocity vector in the task space may thus have both *feasible components*, lying in $\mathcal{R}(\mathbf{J}(\hat{\mathbf{q}}))$, and *degenerate components*, belonging to $\mathcal{R}^\perp(\mathbf{J}(\hat{\mathbf{q}}))$. It is clear that no joint velocity can provide a task-space velocity having components in a degenerate direction.

An effective tool to analyze the linear mapping from the joint velocity space into the task velocity space defined by (1) is offered by the singular value decomposition of the Jacobian matrix; this is given by

$$\mathbf{J} = \mathbf{U}\boldsymbol{\Sigma}\mathbf{V}^{\mathbf{T}} = \sum_{i=1}^{m} \sigma_i \mathbf{u}_i \mathbf{v}_i^{\mathbf{T}} \tag{2}$$

where $\mathbf{U}$ is the $m \times m$ matrix of the output singular vectors $\mathbf{u}_i$, $\mathbf{V}$ is the $n \times n$ matrix of the input singular vectors $\mathbf{v}_i$, and $\boldsymbol{\Sigma} = (\mathbf{S} \quad \mathbf{O})$ is the $m \times n$ matrix whose $(m \times m)$ diagonal submatrix $\mathbf{S}$ contains the singular values $\sigma_i$ of the matrix $\mathbf{J}$. If $r$ denotes the rank of $\mathbf{J}$, the following hold:

a) $\quad \sigma_1 \geq \sigma_2 \geq \ldots \geq \sigma_r > \sigma_{r+1} = \ldots = \sigma_m = 0$

b) $\quad \mathcal{R}(\mathbf{J}) = \text{span}\{\mathbf{u}_1, \ldots, \mathbf{u}_r\}$

c) $\quad \mathcal{N}(\mathbf{J}) = \text{span}\{\mathbf{v}_{r+1}, \ldots, \mathbf{v}_n\}$

Notice that the $m - r$ output singular vectors associated to the null singular values represent the degenerate directions in the given configuration. The singular value decomposition is continuous and well-behaved not only in singular values but also in the direction of the singular vectors; the $\mathbf{u}_i$ and $\mathbf{v}_i$ vectors will thus not change much in the neighbourhood of a singularity.

The solution of the inverse differential kinematics problem requires to find the joint velocity $\dot{\mathbf{q}}$ associated to an assigned task-space velocity $\dot{\mathbf{x}}$. In the remainder, even if the robotic system is redundant with respect to the given task, redundancy will not be exploited in solving the inverse kinematics.

## The Jacobian transpose method

In the previous section the differential kinematic mapping from the joint velocity space into the task velocity space was described in terms of the range space and null space of the Jacobian matrix. Therefore, the inverse mapping is characterized by the range space and null space of the transpose of the Jacobian matrix. In particular, the following well-known relations hold:

$$\mathcal{R}(\mathbf{J}) = \mathcal{N}^{\perp}(\mathbf{J}^T) \qquad \mathcal{N}(\mathbf{J}) = \mathcal{R}^{\perp}(\mathbf{J}^T) \tag{3}$$

which somewhat indicate the possibility of building an inverse mapping based on the Jacobian *transpose* matrix.

On the other hand, it is also known that the Jacobian transpose describes the static mapping from the task force space into the joint torque space, i.e.

$$\mathbf{t} = \mathbf{J}^T(\mathbf{q})\mathbf{f} \tag{4}$$

where $\mathbf{t} \in \mathbb{R}^n$ is the vector of joint torques and $\mathbf{f} \in \mathbb{R}^m$ is the vector of end-effector forces. Notice that eq. (4) can be derived by application of the principle of virtual work to eq. (1). As a consequence, the relations in (3) fully characterize the inherent duality existing between the kinematic mapping and the static mapping.

In view of solving the inverse kinematics problem, the mapping (4) can be conveniently adopted to devise an iterative solution algorithm [2]. In detail, let

$$\mathbf{e} = \mathbf{x}_d - \mathbf{x} \tag{5}$$

denote the difference between an assigned task-space position $\mathbf{x}_d$ and the task-space position $\mathbf{x}$ that can be reconstructed from the current joint space solution. This difference can be used to construct an elastic force vector $\mathbf{Ke}$ —with $\mathbf{K}$ positive definite and usually constant, diagonal— which has to be applied at the end effector of a 'virtual' manipulator, with the same kinematics as the actual manipulator but with null mass and unit viscous damping, in order to drive its end-effector position $\mathbf{x}$ to the given position $\mathbf{x}_d$.

Upon these premises, one may adopt an impedance control law for the virtual manipulator which in turn will provide the inverse kinematics solution for the actual manipulator. The sought law then results in

$$\dot{\mathbf{q}} = \mathbf{J}^T(\mathbf{q})\mathbf{Ke}. \tag{6}$$

A simple Lyapunov argument can be used to prove the stability of the closed-loop feedback virtual system and then the convergence of the inverse kinematics algorithm [1]; notice that $\mathbf{J}^T\mathbf{K}$ determines the convergence rate.

If the assigned end-effector position is time-varying, that is one is interested to inverting a task-space trajectory into a corresponding joint-space trajectory, two possibilities exist for the application of solution (6): Get a significant sampling of the given trajectory, let the algorithm solve for each point, and then interpolate between solutions in the joint space. Let the algorithm run on-line along the given trajectory, thus directly generating the joint-space trajectory. In both cases, inversion errors must be tolerated: In the former, those are due to the interpolation. In the latter, those descend from the 'dynamic' formulation of the problem, i.e. from the tracking properties of the algorithm.

As a matter of fact, the on-line solution is more appealing for control purposes. In that case, it can be recognized that the problem of algorithm tracking performance becomes crucial; the matrix $\mathbf{J}^T(\mathbf{q})\mathbf{K}$ and the given velocity $\dot{\mathbf{x}}_d$ are the relevant quantities affecting performance. In particular, in practical discrete-time implementation of the algorithm, an upper bound exists on the norm of $\mathbf{J}^T\mathbf{K}$ which varies with the current configuration.

An enhancement of the algorithm can be achieved by rendering the matrix $\mathbf{J}^T\mathbf{K}$ less sensitive to variations of joint configuration along the trajectory; this is accomplished by choosing a configuration-dependent $\mathbf{K}$ which compensates for variations of $\mathbf{J}$, for instance see [3].

At a kinematic singularity $\hat{\mathbf{q}}$, when $\mathbf{Ke} \in \mathcal{N}(\mathbf{J}^T(\hat{\mathbf{q}}))$ with $\mathbf{e} \neq 0$, it is $\dot{\mathbf{q}} = 0$ and the algorithm may in principle get 'stuck'. It can be easily shown, however, that such an equilibrium point is unstable as long as the time evolution of $\mathbf{x}_d$ drives $\mathbf{Ke}$ outside $\mathcal{N}(\mathbf{J}^T)$. This result is not surprising since, in force of the relations in (3), vectors belonging to the subspace of the degenerate components cannot be accomodated by both the virtual and the actual manipulator.

Further insight into the features of this solution can be gained by considering the singular value decomposition of the Jacobian transpose; from (2) it is

$$\mathbf{J}^T = \sum_{i=1}^{m} \sigma_i \mathbf{v}_i \mathbf{u}_i^T \tag{7}$$

which reveals a continuous, smooth behaviour of the solution close and through singular configurations.

## The Jacobian pseudoinverse method

The most natural approach to solve differential kinematics is based on the inversion of the mapping (1) using [4]

$$\dot{\mathbf{q}} = \mathbf{J}^T(\mathbf{q}) \left( \mathbf{J}(\mathbf{q})\mathbf{J}^T(\mathbf{q}) \right)^{-1} \dot{\mathbf{x}}. \tag{8}$$

Notice that when $n = m$ solution (8) simplifies to $\dot{\mathbf{q}} = \mathbf{J}^{-1}\dot{\mathbf{x}}$. Solution (8) presents two major limitations: It is not defined at a singular configuration. In the neighbourhood of a singularity, it gives an exact solution that may result in high joint velocities.

For this reason, an improved solution is obtained using the *pseudoinverse* of the Jacobian. The pseudoinverse $\mathbf{J}^\dagger$ of $\mathbf{J}$ is a unique matrix satisfying the conditions [12]

$$\mathbf{J}^\dagger\mathbf{J}\mathbf{a} = \mathbf{a} \qquad \forall \mathbf{a} \in \mathcal{N}^\perp(\mathbf{J}) \tag{9}$$

$$\mathbf{J}^\dagger\mathbf{b} = 0 \qquad \forall \mathbf{b} \in \mathcal{R}^\perp(\mathbf{J}) \tag{10}$$

$$\mathbf{J}^\dagger(\mathbf{a}+\mathbf{b}) = \mathbf{J}^\dagger\mathbf{a} + \mathbf{J}^\dagger\mathbf{b} \qquad \forall \mathbf{a} \in \mathcal{R}(\mathbf{J}) \quad \forall \mathbf{b} \in \mathcal{R}^\perp(\mathbf{J}). \tag{11}$$

Solution (8) is then modified into

$$\dot{\mathbf{q}} = \mathbf{J}^\dagger(\mathbf{q})\dot{\mathbf{x}} \tag{12}$$

and satisfies the condition

$$\min_{\dot{\mathbf{q}}} \|\dot{\mathbf{q}}\| \tag{13}$$

of all $\dot{q}$ that fulfill

$$\min_{\dot{q}} \|\dot{x} - J(q)\dot{q}\|. \tag{14}$$

Although solution (12) is defined even for singular configurations, high joint velocities will still result in the neighbourhood of singularities [5]. This happens because eq. (12) is equivalent to eq. (8) at non-singular configurations, while it discontinuously offers an approximate solution at singularities; the latter aspect is a source of additional problems since it implies discontinuous solutions in joint space. In detail, by using (12), first the set of all joint velocities ensuring best accuracy is determined via (14) and then the minimum norm joint velocity in that set is chosen via (13) as the solution. Since a minimum norm criterion is used after the accuracy requirement has been satisfied, it is difficult to guarantee feasibility of the solution. This issue becomes crucial when the manipulator attains near-singular configurations; in these cases, indeed, the Jacobian matrix is full-rank but ill-conditioned. As a consequence, an exact solution is possible but very large joint velocities are required if the assigned $\dot{x}$ has components along directions which become degenerate at the singularity.

In the framework of the singular value decomposition, with reference to eq. (2), the pseudoinverse solution (12) can be written in the form

$$\dot{q} = \sum_{i=1}^{m} \frac{1}{\sigma_i} v_i u_i^T \dot{x}. \tag{15}$$

When a singularity is approached, the smallest singular value tends to zero; this makes the solution very sensitive to the component of the commanded velocity in the $u_m^T$ direction which thus requires a large joint velocity to be performed. On the other hand, at the singularity the direction $u_m^T$ becomes degenerate and joint velocity discontinuity is experienced if a non-null task velocity is specified in that direction.

In order to exploit the potential of solution (12), a systematic and efficient procedure to compute the pseudoinverse of the Jacobian matrix is needed as well as the continuity of the joint-space solution must be ensured.

The first problem can be faced by suitably transforming the Jacobian matrix as

$$PJ = \begin{pmatrix} J_1 \\ J_2 \end{pmatrix} \tag{16}$$

where $P$ is an $(m \times m)$ transformation matrix such that $J_1$ is an $(r \times n)$ matrix of full rank, and $J_2$ is the $((m-r) \times n)$ matrix resulting from the above transformation. Once $J_1$ has been derived, it is straightforward to compute the pseudoinverse of the Jacobian as [13]

$$J^\dagger = J_1^T \left( J_1 J^T J J_1^T \right)^{-1} J_1 J^T. \tag{17}$$

The problem thus remains to find the transformation matrix as simply as possible. It can be recognized that $P$ is a projector onto a base of the task velocity space, whose first $r$ rows span the subspace of feasible motion and last $(m-r)$ rows represent directions of motion that linearly depend on the first.

Geometrical insight into the kinematic structure is helpful to derive, case by case, the required projector. For typical manipulators it is possible to identify classes of singular configurations which allow the determination of **P** in a systematic manner; to this end, it is convenient to analyze the singularity in a suitable link-fixed frame. This has been demonstrated, for instance, for a six-degree-of-freedom PUMA-like geometry with zero offsets [6]: For the three well-known types of singularities (wrist, elbow, and shoulder) it turns out that the projectors can be expressed in terms of the two rotation matrices from the base frame to the frames of links 5 and 2, respectively. In these cases, the computation of **P** is inexpensive as those matrices are already available for Jacobian computation.

Secondly, the problem of continuity of the joint-space solution can be tackled by treating the manipulator as singular in a suitably defined region around each singularity; this implies that only $r$ task space directions are accomplished. Thus, $(n-r)$ degrees of freedom are available in the region, of which $(n-m)$ are the usual redundant degrees of freedom —that are of no interest for the present work— and $(m-r)$ are the extra degrees of freedom generated by regarding the manipulator as if it were singular. The latter are exploited to ensure the continuity of the joint velocity solution.

In order to accomplish the above, the computation of the pseudo inverse is carried out as in (17), using a matrix $\mathbf{J}_1$ which has the same structure in the whole region as in the singularity. Let $\tilde{\mathbf{J}}_1$ denote such matrix; the modified pseudoinverse is computed as

$$\mathbf{J}^- = \tilde{\mathbf{J}}_1^T \left( \tilde{\mathbf{J}}_1 \mathbf{J}^T \mathbf{J} \tilde{\mathbf{J}}_1^T \right)^{-1} \tilde{\mathbf{J}}_1 \mathbf{J}^T. \tag{18}$$

Continuity can be accomplished by interpolating between the singularity and a *border point* $\mathbf{q}_b$. Inside the region, the solution (12) is modified into

$$\dot{\mathbf{q}} = \mathbf{J}^-(\mathbf{q})\dot{\mathbf{x}} + \left( \mathbf{I} - \mathbf{J}^-(\mathbf{q})\mathbf{J}(\mathbf{q}) \right) \dot{\mathbf{q}}_0 \tag{19}$$

where $(\mathbf{I} - \mathbf{J}^- \mathbf{J})$ is a projector onto $\mathcal{N}(\mathbf{J}_1)$ which approximates $\mathcal{N}(\mathbf{J})$ inside the region but coincides with it at the singularity.

In the case of a single singularity, a simple choice for $\dot{\mathbf{q}}_0$ is ·

$$\dot{\mathbf{q}}_0 = \alpha \mathbf{J}^\dagger(\mathbf{q}_b)\dot{\mathbf{x}} \tag{20}$$

where $\alpha$ is the interpolation factor which is zero at the singularity and unity at the border of the region. In the case of multiple singularities, one interpolation factor has to be used for each singularity to ensure continuity when the manipulator transits from one restricted region to another region associated with a singularity of different dimension; a continuous solution could be obtained using multiple border points, but this gives a complex solution with increased computation. In a PUMA-like structure, the wrist singularity is of primary concern since it is difficult to predict at the trajectory planning level. In this case, one may think of interpolating only for that singularity, tolerating instead discontinuities due to elbow and shoulder singularities; in fact, those singularities are naturally characterized in the task space and thus they can be avoided during the planning [6].

# The damped least-squares Jacobian inverse method

Another method which is especially directed to overcoming the problem of control through singularities makes use of the *damped least-squares inverse* of the Jacobian matrix for solving the inverse differential kinematics problem [7,8]. The method corresponds to using

$$\dot{q} = J^T(q) \left( J(q)J^T(q) + \lambda^2 I \right)^{-1} \dot{x} \tag{21}$$

in lieu of solution (12); in (21) $\lambda \in R$ is the damping factor. It is easily seen that when $\lambda$ is zero, solutions (8) and (21) become identical.

Solution (21) satisfies the condition

$$\min_{\dot{q}} \| \dot{x} - J(q)\dot{q} \|^2 + \lambda^2 \| \dot{q} \|^2 \tag{22}$$

which, differently from (13,14), accounts for both accuracy and feasibility in choosing the joint-space velocity required to produce the given task-space velocity. In this regard, it is essential to select a suitable value for the damping factor; small values of $\lambda$ give accurate solutions but low robustness to the occurrence of singular and near-singular configurations, high values of $\lambda$ result in low tracking accuracy even where a feasible and accurate solution would be possible.

Resorting to the singular value decomposition (2), the damped least-squares solution (21) can be rewritten as

$$\dot{q} = \sum_{i=1}^{m} \frac{\sigma_i}{\sigma_i^2 + \lambda^2} v_i u_i^T \dot{x}. \tag{23}$$

It is clear that the components for which $\sigma_i \gg \lambda$ are little influenced by the damping factor as it is

$$\frac{\sigma_i}{\sigma_i^2 + \lambda^2} \approx \frac{1}{\sigma_i}. \tag{24}$$

On the other hand, when a singularity is approached, the smallest singular value tends to zero while the associated component of the solution is driven to zero by the factor $\sigma_i/\lambda^2$; this progressively reduces the joint velocity required to achieve near-degenerate components of the commanded task velocity. In comparison with the previous pseudoinverse method, solutions (12) and (21) behave identically as long as the singular values are significantly larger than the damping factor.

The damping factor serves as an index for the minimum singular value to establish whether the current configuration can be treated as near-singular. Further, it determines the degree of approximation introduced with respect to the pure least-squares pseudoinverse solution. Therefore, an optimal choice for $\lambda$ requires consideration of the smallest non-null singular value experienced along the whole trajectory and of the minimum damping needed to ensure feasible joint velocities. The use of a configuration-varying damping factor is advisable to achieve good performance in the entire manipulator's workspace.

The natural choice is to adjust $\lambda$ as a function of some measure of closeness to a singularity at the current configuration; the closer the arm is to the singularity, indeed, the higher the need is for damping. One proposal has been brought up in [7] as to use the manipulability measure to determine an appropriate value for the damping factor at each configuration; in order to avoid unnecessary damping far from the singularity, a threshold value of manipulability is assigned above which a pure pseudoinverse is used. The manipulability measure alone, however, does not constitute an absolute measure of closeness to a singularity; this is otherwise provided by the minimum singular value of the Jacobian matrix. A solution that tunes the damping factor according to numerical estimates of the minimum singular value has been developed in [9]; a selective filtering of the directions defined by the singular vectors is also accomplished by isolating the contribution pertaining to the minimum singular value, which has the advantage of removing velocity transformation errors along the feasible directions. A recent study [10] has shown that the singular value decomposition is considerably simplified when applied to the manipulator Jacobian matrix, and this appears promising in view of an on-line computation of the minimum singular value.

An effective solution has been proposed in [11] which attempts to combine the advantages of both the above techniques; namely, the analytical simplicity of a closed-form adjusting law for the damping factor as in [7] and the correctness of shaping the damping factor according to an estimate of the minimum singular value as in [9]. The key of the solution is to exploit kinematical insight and then define a region inside which a very simple estimate of the minimum singular value is devised to be used in the damped least-squares inverse; for instance, in the case of the wrist singularity, the angle itself causing the singularity has been recognized as a good estimate of the minimum singular value [11].

## Conclusion

In the above sections we have reviewed —what we believe— the most relevant methods for solving the differential kinematics of a manipulator in the neighbourhood of singular configurations.

The Jacobian transpose method utilizes the differential kineto-static mapping to solve the inverse kinematics problem for a given task-space position. For this reason, it necessitates of a feedback correction term in a closed-loop setting. The advantage is that it provides a computationally cheap solution which is also robust to the occurrence of singularities, although the tracking performance is inherently lower than any exact inverse differential kinematics solution.

The Jacobian pseudoinverse method and the damped least-squares Jacobian inverse method provide a possibly exact inversion of the differential mapping; this implies, however, that, if a joint position solution is desired, a feedback correction is needed to eliminate numerical drift arising from open-loop integration of the joint velocity solution. On the other hand, they are both computationally more demanding than the Jacobian transpose method; this point must be considered in sensory-based real-time applications.

Two counteracting parameters are at issue when evaluating the performance of either of the above two methods; namely, accuracy versus robustness of the solution. In this respect, it can be recognized that the pseudoinverse originates from the desire of finding an accurate solution to the differential kinematic mapping, which though needs to be relaxed in order to get satisfactory behaviour in the neighbourhood of singularities. On the contrary, the damped least-squares inverse arises from the necessity of devising a singularity-robust inverse solution to the differential kinematic mapping, which though needs to be properly tuned in order to increase its accuracy.

From the discussion in this work, it has emerged that there exist effective refinements of both the pseudoinverse and the damped least-squares inverse solutions which allow to trade accuracy off robustness. To this end, it is often decisive to perform a case-by-case analysis by taking advantage of geometrical insight into the kinematic structure under investigation.

# Acknowledgements

This paper reports research work supported by *Ministero dell'Università e della Ricerca Scientifica e Tecnologica* under 40% funds.

# References

[1] A. Balestrino, G. De Maria, and L. Sciavicco, "Robust control of robotic manipulators," *9th IFAC World Congress*, Budapest, H, July 1984.

[2] L. Sciavicco and B. Siciliano, "A solution algorithm to the inverse kinematic problem for redundant manipulators," *IEEE J. Robotics and Automation*, RA-4, 403–410, 1988.

[3] P. Chiacchio and B. Siciliano, "Achieving singularity robustness: An inverse kinematic solution algorithm for robot control," *IEE Int. Work. Robot Control: Theory and Application*, Oxford, GB, pp. 149–156, Apr. 1988.

[4] D.E. Whitney, "The mathematics of coordinated control of prosthetic arms and manipulators," *Trans. ASME J. Dynamic Systems, Measurement, and Control*, 303–309, 1972.

[5] C.A. Klein and C.H. Huang, "Review of pseudoinverse control for use with kinematically redundant manipulators," *IEEE Trans. Systems, Man, and Cybernetics*, SMC-13, 245–250, 1983.

[6] S. Chiaverini and O. Egeland, "A solution to the singularity problem for six-joint manipulators," *1990 IEEE Int. Conf. Robotics and Automation*, Cincinnati, OH, pp. 644-649, May 1990.

[7] Y. Nakamura and H. Hanafusa, "Inverse kinematic solutions with singularity robustness for robot manipulator control," *Trans. ASME J. Dynamic Systems, Measurement, and Control*, 108, 163–171, 1986.

[8] C.W. Wampler, "Manipulator inverse kinematic solutions based on damped least-squares solutions," *IEEE Trans. Systems, Man, and Cybernetics*, SMC-16, 93–101, 1986.

[9] A.A. Maciejewski and C.A. Klein, "Numerical filtering for the operation of robotic manipulators through kinematically singular configurations," *J. Robotic Systems*, 5, 527–552, 1988.

[10] A.A. Maciejewski and C.A. Klein, "The singular value decomposition: Computation and application to robotics," *Int. J. Robotics Research*, 8(6), 63–79, 1989.

[11] O. Egeland, M. Ebdrup, and S. Chiaverini, "Sensory control in singular configurations — Application to visual servoing," *IEEE Int. Work. Intelligent Motion Control*, Istanbul, TR, pp. 401–405, Aug. 1990.

[12] L. Zadeh and C.A. Desoer, *Linear System Theory*, New York, Mac-Graw-Hill, 1963, p. 577.

[13] D.Q. Mayne, "On the calculation of pseudoinverses," *IEEE Trans. Automatic Control*, 204–205, Apr. 1969.

# Manipulator control in singular configurations—Motion in degenerate directions

Olav Egeland and Inge Spangelo

Division of Engineering Cybernetics
The Norwegian Institute of Technology
N-7034 Trondheim, Norway

## Abstract

A manipulator may get stuck in a singular configuration if the commanded motion is in a direction that is not feasible. This is due to the rank deficiency in the velocity mapping from joint to end-effector coordinates. It is possible to use nullspace motion in the singularity to achieve a change in the Jacobian matrix so that although the end-effector cannot have a velocity it can be given an acceleration in a degenerate direction using the nullspace motion. In this paper a simple technique is presented which makes it possible to exit a singularity when the commanded motion is in a degenerate direction. The method is based on the damped least-squares solution in inverse kinematics.

## 1   Introduction

Problems are encountered close to the singularities of a manipulator where the end effector loses degrees of freedom and the usual inverse kinematic solution becomes undefined. This is experienced in the form of very high joint velocities which results in large control deviations. When a preprogrammed reference trajectory is to be tracked it is normally possible to plan the trajectory so that singularities are avoided. Alternatively it is possible to interpolate in joint coordinates close to the singularities [5]. This is not possible in sensory control where the reference is not known in advance.

Nakamura and Hanafusa [3] and Wampler [6] used the damped least-squares technique to solve the differential inverse kinematic problem. This gives an approximate solution which is well-conditioned and defined everywhere. A problem with the method is to find a damping factor which gives an acceptable compromise between the accuracy and feasibility of the solution. It is possible to use a constant damping factor [6], or to adjust the damping as a function of manipulability [3] or the smallest singular value of the Jacobian [2]. The computation of manipulability or singular values is in general expensive, but for some industrial manipulators it is possible to find simple solutions [7].

In a singular configuration the end-effector cannot have a velocity in directions outside the range space of the manipulator Jacobian. These directions are termed degenerate directions. As a result of this the manipulator may get stuck in a singular configuration if the commanded motion is in a degenerate direction. This will be the case when singularity handling techniques like the damped least-squares solution is used. However, it will normally be possible to achieve an acceleration in the degenerate direction as pointed out by Nielsen, Canudas de Wit and Hagander [4] who analyzed a simple planar mechanism. This means that although a trajectory cannot be tracked accurately in a degenerate direction it is possible to follow a path in a degenerate direction without errors.

The problem of leaving a singularity in a degenerate direction is analyzed in this paper in the damped least-squares formulation. The problem is analyzed in terms of the singular value decomposition of the manipulator Jacobian. The relevant state-space model is presented and a method for leaving the singularity by utilizing the nullspace motion is presented. A simple simulation study with a wrist mechanism is included.

## 2    Kinematics

The $n$-dimensional vector of joint coordinates is denoted by $\mathbf{q}$. Differential task-space motion is described by $\delta \mathbf{x} = \dot{\mathbf{x}} \delta t$ where $\delta t$ is a small time increment and $\dot{\mathbf{x}}$ is an $n$-dimensional task-space velocity. The $n \times n$ task Jacobian $\mathbf{J}(\mathbf{q})$ is defined by

$$\dot{\mathbf{x}} = \mathbf{J}(\mathbf{q})\dot{\mathbf{q}} \tag{1}$$

and incremental motion is given by

$$\delta \mathbf{x} = \mathbf{J}(\mathbf{q})\delta \mathbf{q} \tag{2}$$

where $\delta \mathbf{q} = \dot{\mathbf{q}} \delta t$.

In configurations where the task Jacobian $\mathbf{J}$ has full rank, the end effector has $n$ degrees of freedom. When the Jacobian is rank deficient so that $\mathrm{rank}(\mathbf{J}) = r$, $r < n$, the end effector has only $r$ degrees of freedom and the manipulator is said to be in a singular configuration.

## 3    Damped least-squares in inverse kinematics

A motion increment $\delta \mathbf{x}$ in end-effector coordinates can be transformed to joint coordinates by solving

$$\mathbf{J}(\mathbf{q})\delta \mathbf{q} = \delta \mathbf{x}. \tag{3}$$

using Gaussian elimination. This solution is not possible in singular configurations, and the solution becomes ill-conditioned close to the singularities.

Nakamura and Hanafusa [3] and Wampler [6] independently proposed to use the damped least-squares solution of eq. (3) in the inverse kinematics algorithm. This method was further developed by Maciejewski and Klein [2]. The solution minimizes

$$\|\delta \mathbf{x} - \mathbf{J}\delta \mathbf{q}\|^2 + \lambda^2 \|\delta \mathbf{q}\|^2 \tag{4}$$

and is obtained by solving the equation

$$(\mathbf{J}^T\mathbf{J} + \lambda^2\mathbf{I})\delta \mathbf{q} = \mathbf{J}^T\delta \mathbf{x}. \tag{5}$$

When **J** is full rank and $\lambda^2$ is zero the solution of eqs. (3) and (5) is identical. $\lambda^2$ is the damping factor which must be specified. A small $\lambda^2$ gives an accurate solution, but high joint velocities close to singular configurations. Feasible joint velocities can be obtained by increasing $\lambda^2$, so a compromise between the accuracy and feasibility of the solution must be made.

The solution can be analyzed using the singular value decomposition

$$\mathbf{J} = \sum_{i=1}^{n} \sigma_i \mathbf{u}_i \mathbf{v}_i^T \tag{6}$$

of the Jacobian. $\mathbf{v}_i$ and $\mathbf{u}_i$ denote the input and output singular vectors, and $\sigma_i$ denotes the singular values which are ordered so that

$$\sigma_1 \geq \sigma_2 \geq \ldots \geq \sigma_r \geq \sigma_{r+1} = \ldots = \sigma_n = 0 \tag{7}$$

where $r$ is the rank of **J**. By scaling the task space with the length of the arm, the singular values will be between zero and unity.

The last $n - r$ singular values are zero. The nullspace of the Jacobian is

$$\mathcal{N}(\mathbf{J}) = \mathrm{span}\{\mathbf{v}_{r+1}, \ldots, \mathbf{v}_n\} \tag{8}$$

and the range space is

$$\mathcal{R}(\mathbf{J}) = \mathrm{span}\{\mathbf{u}_1, \ldots, \mathbf{u}_r\}. \tag{9}$$

The solution of the usual inverse kinematic problem eq. (2) can be written

$$\delta\mathbf{q} = \sum_{i=1}^{n} \frac{1}{\sigma_i} \mathbf{v}_i \mathbf{u}_i^T \delta\mathbf{x}. \tag{10}$$

When a singularity with rank $\mathbf{J} = r$ is approached, the last $n - r$ of the singular values will tend to zero, and $1/\sigma_i \to \pm\infty$ for $i \in \{r+1, \ldots, n\}$. This means that the solution $\delta\mathbf{q}$ from eq. (10) becomes very sensitive to components $\delta\mathbf{x}_i = \mathbf{u}_i^T \delta\mathbf{x}$ along the last $n - r$ output singular vectors.

The damped least-squares solution of eq. (5) can be written

$$\delta\mathbf{q} = \sum_{i=1}^{n} \frac{\sigma_i}{\sigma_i^2 + \lambda^2} \mathbf{v}_i \mathbf{u}_i^T \delta\mathbf{x}. \tag{11}$$

Note that the singular vectors are unchanged when damping is used. It is clear that the components with $\sigma_i \gg \lambda$ will be little influenced by $\lambda$ as

$$\frac{\sigma_i}{\sigma_i^2 + \lambda^2} \approx \frac{1}{\sigma_i} \text{when } \sigma_i \gg \lambda \tag{12}$$

When the singular value approaches zero so that $\lambda \gg \sigma_i$, the associated component of the solution will approach zero depending on $\lambda$. This is illustrated in Fig. 1

The error in joint coordinates due to damping is

$$\mathbf{e}_q = \sum_{i=1}^{n} \frac{\lambda^2}{\sigma_i(\sigma_i^2 + \lambda^2)} \mathbf{v}_i \mathbf{u}_i^T \delta\tilde{\mathbf{x}} \tag{13}$$

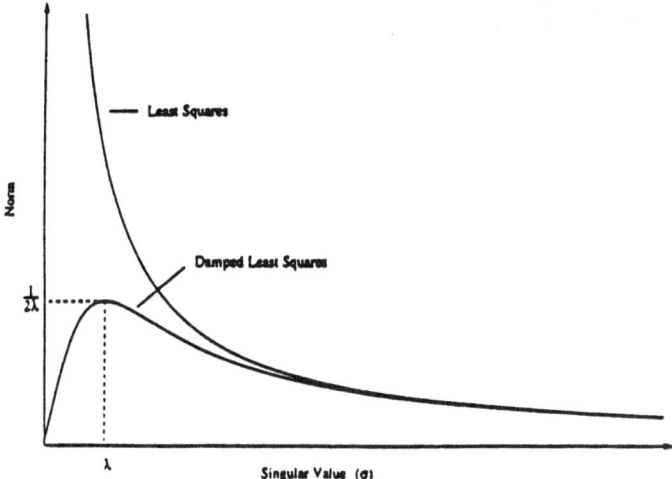

Figure 1: $\frac{\sigma}{\sigma^2+\lambda^2}$ as a function of $\sigma$ for different values of the damping factor $\lambda^2$

The error in $\mathbf{x}$ is $\mathbf{e}_x = \mathbf{J}\mathbf{e}_q$, by inserting eq. (6) in eq. (13) this can be written

$$\mathbf{e}_x = \sum_{i=1}^{n} \frac{\lambda^2}{\sigma_i^2 + \lambda^2} \mathbf{u}_i \mathbf{u}_i^T \delta\tilde{\mathbf{x}} \tag{14}$$

which can be approximated by

$$\mathbf{e}_x = \sum_{i=r+1}^{n} \frac{\lambda^2}{\sigma_i^2 + \lambda^2} \mathbf{u}_i \mathbf{u}_i^T \delta\tilde{\mathbf{x}} \tag{15}$$

It is possible to use a constant $\lambda$. Wampler [6] suggested $\lambda = 0.03$ which resulted in a conditioning number $\mathrm{cond}(\mathbf{J}^T\mathbf{J} + \lambda^2\mathbf{I}) \approx 1200$ for a well scaled problem.

If an estimate of the 'smallest singular value is available it is possible to have a more accurate solution in the transition area where $\sigma_n \approx \lambda$. An estimate of the smallest singular value can be found from geometric insight for simple manipulators, or numerically as in [2]. The damping factor can then be calculated as in [1] using

$$\lambda^2 = \begin{cases} 0 & \text{when } \hat{\sigma}_n > \epsilon \\ \epsilon^2 - \hat{\sigma}^2 & \text{otherwise} \end{cases} \tag{16}$$

which gives

$$\frac{\sigma_n}{\sigma^2 + \lambda^2} = \begin{cases} 1/\sigma_n & \text{when } \hat{\sigma}_n > \epsilon \\ \sigma_n/\epsilon^2 & \text{otherwise} \end{cases} \tag{17}$$

## 4 Control in singular configurations

A manipulator can get stuck in a singular configuration when the damped least-squares solution is used. This will happen if the commanded motion is in a degenerate direction in task space, that is $\delta x = \sum_{i=r+1}^{n} u_i^T \delta x$ and $r$ is the rank of J. For simplicity we assume $r = n - 1$.

In many cases it is obvious that although the manipulator cannot be given a differential motion in the degenerate direction, it is possible to move in the end effector in the degenerate direction after some initial nullspace motion in the joint space. This was reported for a simple planar mechanism in [4]. However we are not aware of any systematic method for controller design for this problem.

The Jacobian in the velocity mapping

$$\dot{x} = J\dot{q} \tag{18}$$

is rank deficient in a singularity, and a task velocity cannot be achieved in the $u_n$ direction. However, the acceleration mapping is

$$\ddot{x} = J\ddot{q} + \dot{J}\dot{q} \tag{19}$$

and it is clear that an acceleration in the $u_n$ direction can be achieved if $\dot{J}$ spans $u_n$. If this is to be of any practical importance,

$$\dot{J} = \frac{\partial J}{\partial q}\dot{q} \tag{20}$$

must span $u_n$ when the joint velocity is in the nullspace of the Jacobian, that is $\dot{q} = \dot{q}v_n$.

In terms of the singular value decomposition, $\dot{J}$ is written

$$\dot{J} = \sum_{i=1}^{n}(\dot{\sigma}_i u_i v_i^T + \sigma_i \dot{u}_i v_i^T + \sigma_i u_i \dot{v}_i^T) \tag{21}$$

The first and second term in the sum is of particular interest in connection with the damped least-squares method.

In an external singularity the position transformation is one-to-one although the velocity transformation is rank deficient. In this case any nullspace motion will make the manipulator leave the singularity. This means that a nullspace velocity $\dot{q} = \dot{q}v_n$ results in

$$\dot{\sigma}_n = \frac{\partial \sigma_n}{\partial q}\dot{q}v_n \neq 0 \tag{22}$$

and accordingly $\dot{J}$ spans $u_n$ when $\dot{q}$ has a component in the nullspace direction $v_n$. This means that an acceleration in $u_n$ can be achieved in the singularity if a nullspace velocity is commanded.

In an internal singularity the position and velocity transformations are not one-to-one, and the manipulator will remain singular if the joint velocity q is purely in the nullspace. The most common internal singularities are the wrist and shoulder singularities. The development below is correct for these types of singularities, and we conjecture that the results are valid for all internal singularities in robotic manipulators. A nullspace motion does not cause the manipulator to leave an internal singularity. This means that the first term in the sum in eq. (21) gives no component in the $u_n$ direction. The second term in the sum is more interesting in this case. Both the wrist and shoulder singularity have the following property:

$$(\sum_{i=1}^{n-1} \frac{\partial u_i}{\partial q} v_n)^T u_n \neq 0 \tag{23}$$

which means that a nullspace velocity will also in this case make $\dot{J}$ span $u_n$. The result for the internal singularity is somewhat different from that of the external singularity. Assume that a task velocity is commanded in the initial $u_n$ direction. The initial nullspace motion makes part of the commanded motion feasible and as a result the manipulator leaves the singularity.

The surprisingly simple result is: If the manipulator is stuck at a singularity due to a commanded motion that is not feasible and the damped least-squares solution is used, it is sufficient to make a small motion in the nullspace. The manipulator will then leave the singularity and the commanded motion becomes feasible.

# 5    A state-space representation of the system in singular configurations

Differential geometry was used in [4] to analyze the controllability of the manipulator in a singular configuration. As the velocity mapping (18) does not span all directions in the task space it was argued that the acceleration mapping (19) should be used instead. However if $\ddot{q}$ is used as the input to the system the problem of the rank deficient Jacobian is not eliminated and the analysis based on this approach failed to show that the system was controllable.

The velocity or acceleration mapping does not constitute a complete model of the manipulator kinematics. The fact that the Jacobian changes with $q$ must be included explicitly in the model. The appropriate model structure for the problem is

$$\dot{x} = J\dot{q} \tag{24}$$

$$\dot{J} = \frac{\partial J}{\partial q}\dot{q} \tag{25}$$

In order to get a model with n coordinates, the $n-1$ first coordinates in the singular value decomposition is taken from eq. (24), while the model for the last coordinate is found from eq. (25). This gives

$$\dot{x}_{u_i} = \sigma_i v_i^T \dot{q} , i = 1 \ldots n-1 \tag{26}$$

for the $n-1$ first coordinates. Here $\dot{x}_{u_i} = u_i^T \dot{x}$.

For external singularities the remaining equation is

$$\dot{\sigma}_n = \frac{\partial \sigma_n}{\partial q}\dot{q} \tag{27}$$

while for internal singularities it is

$$(\frac{d}{dt}\sum_{i=1}^{n-1} u_i)^T u_n = (\sum_{i=1}^{n-1} \frac{\partial u_i}{\partial q}\dot{q})^T u_n \tag{28}$$

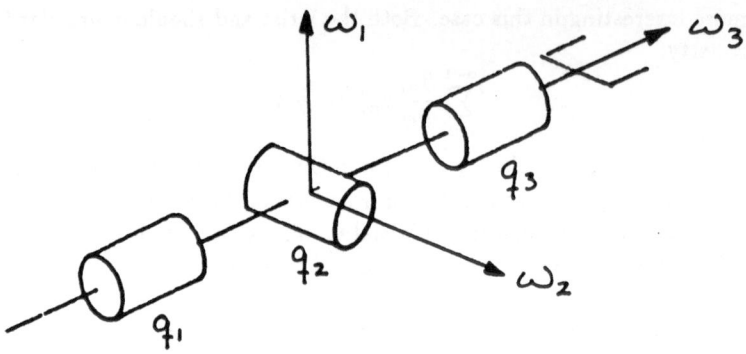

Figure 2: *Mechanism simulated in the case study*

Velocities in the first $n-1$ output singular directions $u_1, \ldots, u_{n-1}$ are then achieved through velocities in the input singular directions $v_1, \ldots, v_{n-1}$, while velocity in the remaining input singular direction $v_n$ can be used to exit the singularity.

# 6    A simple case study

A simple wrist mechanism was studied in a singular configuration. The joint angles are $q_1$, $q_2$ and $q_3$, and the angular velocity in base coordinates is

$$\omega = \begin{pmatrix} c_1 s_2 & -s_1 & 0 \\ s_1 s_2 & c_1 & 0 \\ c_2 & 0 & 1 \end{pmatrix} \dot{q} \tag{29}$$

where $\omega = \begin{pmatrix} \omega_1 & \omega_2 & \omega_3 \end{pmatrix}^T$ and $q = \begin{pmatrix} q_1 & q_2 & q_3 \end{pmatrix}^T$. The mechanism is singular with respect to $\omega$ when $s_2 = \sin q_2 = 0$. The mechanism is shown in Fig. 2.

The mechanism was simulated using the damped least-squares solution. The damping factor $\lambda$ was calculated from eq. (16) with $\epsilon = 0.01$. The initial position was $q_1 = q_2 = q_3 = 0$ which is a singularity, and the commanded velocity was

$$\omega = \begin{pmatrix} 0.002 \\ 0 \\ 0 \end{pmatrix}$$

In this position the smallest singular value is $\sigma_3 = 0$ and the corresponding singular vectors are

$$u_3 = \begin{pmatrix} 1 \\ 0 \\ 0 \end{pmatrix}$$

$$\mathbf{v}_3 = \begin{pmatrix} 1/\sqrt{2} \\ 0 \\ -1/\sqrt{2} \end{pmatrix}$$

The commanded motion is in purely in the $\mathbf{u}_3$ direction which is not feasible, and the mechanism will not move when the damped least-squares solution is used.

A nullspace motion increment

$$\delta\mathbf{q} = \sqrt{2}\alpha\mathbf{v}_3$$

was then introduced. This gave no output velocity $\omega$. The resulting change in the feasible output singular vector $\mathbf{u}_2$ was

$$\delta\mathbf{u}_2 = \alpha\mathbf{u}_3$$

This means that the feasible output singular vector $\mathbf{u}_2$ was given a component in the direction of the commanded motion. The damped least-squares solution projected the commanded motion on $\mathbf{u}_2$, and the mechanism started to move. A rotation in the $\mathbf{u}_2$ direction gave a rotation of the wrist angle $q_2$ and the mechanism became nonsingular. After an initial transient the angle $q_1$ reached $\pi/2$ and the rotational axis of joint 2 was aligned with the commanded rotation. The result is shown in Figs. 3, 4 and 5.

# 7 Conclusion

A method for leaving singularities in a degenerate direction has been presented and analyzed using singular value decomposition. The method is based on the damped least-squares solution for inverse kinematics. It was shown that a small nullspace motion makes the initially degenerate direction feasible, and an exact tracking of the commanded path is possible.

# 8 Acknowledgments

This work was supported by the Norwegian Space Center and NFT as.

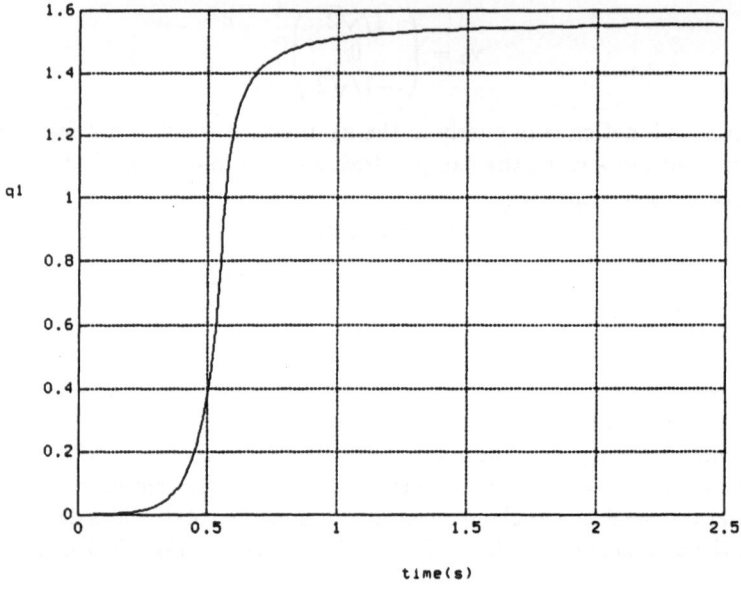

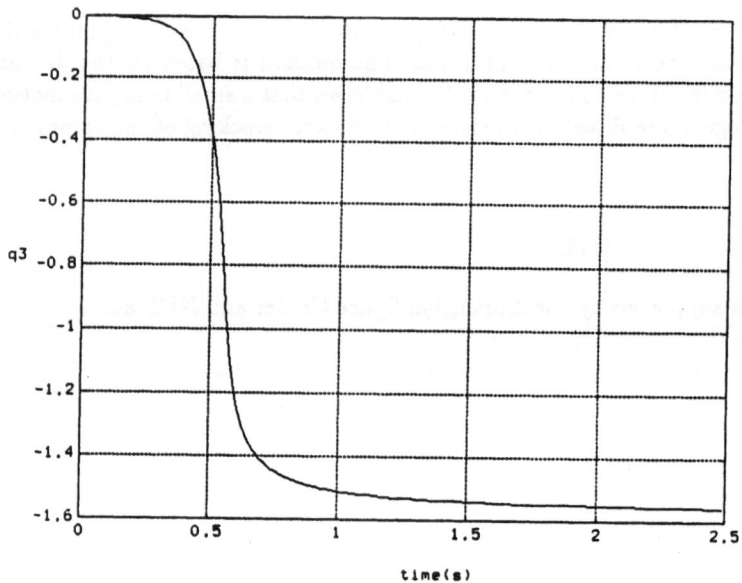

Figure 3: *Joint angles $q_1$ and $q_3$ for the wrist mechanism leaving the singularity in a degenerate direction. The stationary values $q_1 = -q_3 = \pi/2$ resulted in joint axis 2 being along the commanded rotation.*

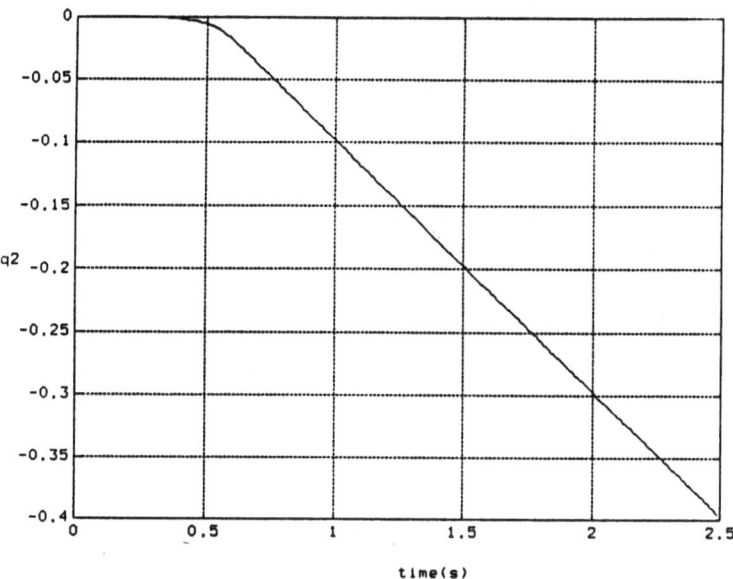

Figure 4: *Joint angle $q_2$ for the simulation of the wrist mechanism. The simulation started in a singular configuration with $q_2 = 0$. The initial nullspace motion rotated joint axis 2 so that a nonzero projection of the commanded motion was along axis 2. This resulted in a nonzero $q_2$ and the manipulator left the singularity.*

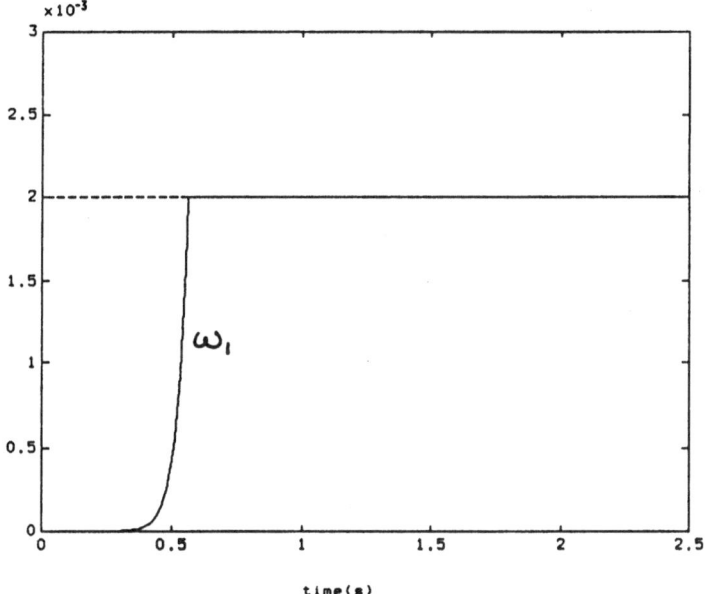

Figure 5: *Commanded and actual angular velocity in direction 1. An angular velocity in direction 1 is not feasible in the singularity, and the actual angular velocity was zero until the mechanism had left the singularity. Then the actual angular velocity increased to the commanded value.*

# References

[1] S. Chiaverini and O. Egeland, "A solution to the singularity problem in six-joint manipulators," in *Proc. 1990 IEEE Int. Conf. Robotics and Automation*, Cincinnati, Ohio, May 13–18, 1990.

[2] A. A. Maciejewski and C. A. Klein, "Numerical filtering for the operation of robotic manipulators through kinematically singular configurations," *J. Robotic Systems*, vol. 5, no. 6, pp. 527–552.

[3] Y. Nakamura and H. Hanafusa, "Inverse kinematic solutions with singularity robustness for robot manipulator control," *ASME J. Dynamic Syst., Meas., Contr.*, vol. 108, Sept. 1986, pp. 163–171.

[4] L. Nielsen, C. Canudas de Wit and P. Hagander, "Controllability issues of robots near singular configurations," in *Proc. 2nd Int. Workshop on Advances in Robot Kinematics*, Sept 10–12, 1990, Linz, Austria.

[5] R. H. Taylor, "Planning and execution of straight-line manipulator trajectories," *IBM Journ. of Research and Developments*, 1979, vol. 23, pp. 424–436.

[6] C. W. Wampler II, "Manipulator inverse kinematic solutions based on vector formulations and damped least-squares method," *IEEE Trans. Syst., Man, Cybern.*, vol. SMC-16, no. 1, Jan/Feb 1986, pp. 93–101.

[7] T. Yoshikawa, "Translational and rotational manipulability in robotic manipulators," *Proc. 1990 American Control Conference*, pp.228–233.

# Controllability Issues of Robots
# near Singular Configurations

Lars Nielsen*, Carlos Canudas de Wit**, Per Hagander*

*Abstract*  The problems of robot motion in the task space, such as Cartesian control and force control are widely studied. Here the relation between robot singularities and the concept of controllability is important. Simple examples contradict some statements usually made about the motion of robots in singular configurations. A robot may be controlled in an arbitrary direction from a singular configuration if the velocity profiles are shaped in a proper way. Nonlinear controllability introduced by differential geometry actually suggests lack of controllability at a singular point. Therefore the observations on possible motions made in this paper opens a number of questions. Further, the results may be of immediate and significant relevance to path programming and to path following.

## 1. Introduction

The dynamic behavior of the robot model in the task space around kinematic singularities is central to many control problems related to Cartesian control and force control. A complete investigation about robot model dynamics properties in this space, where the tasks to be performed would be specified, is lacking. In such a study, singularities and controllability are two central concepts. The singularities are the configurations for which the Jacobian loses rank, and degrees of freedom in the robot motion are lost in some sense. However, a closer investigation will show that the position may be controlled arbitrarily but that there may follow requirements on the velocities. One way of anticipating such a possibility is to view the Jacobian as a first term in a Taylor expansion of the kinematics. If the first term is zero then higher order terms will, if they exist, determine the kinematic behavior. Of course, such observations relate to the concept of nonlinear system controllability. Tools for studying controllability in nonlinear system were introduced by Hermann and Krener (1977) among others. See Isidori (1989) for a complete treatment. This theory, based on differential geometry, requires some smoothness conditions of the system to be analyzed. It turns out that the robot model formulated in the task space does not fulfill those conditions at the points corresponding to the Jacobian singularities. The controllability rank of the work-space robot model drops at the singularities but only there. This means that robot controllability, as defined by the differential geometry, would be lost at those points.

Kinematics, without relation to dynamics that is robust to singularities has been presented (Wampler, 1986; Nakamura and Hanafusa, 1986). They use approximation techniques based on the pseudo-inverse of the Jacobian or least-squares with Levenberg-Marquardt stabilization, i.e. basically general techniques to avoid problems with the linear part. We are instead aiming at nonlinear properties and most important at control possibilities. The presentation starts in Section 2 with kinematics and a closer look at

* Department of Automatic Control, Lund Institute of Technology, Box 118, 221 00 Lund, Sweden

** Laboratoire d'Automatique de Grenoble, Ensieg-B.P. 46, 38042 Saint-Martin-d'Heres, France

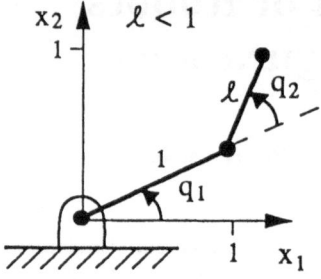
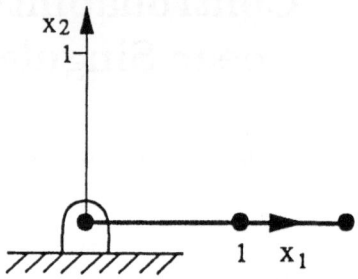

Figure 1. (a). A planar two link robot. (b). The two-link robot in a singular configuration. The robot can move along the $x_1$-axis, but there are restrictions on the possible velocities, $\dot{x}_1$. Nevertheless, it is reasonable to talk about two degrees of freedom, at least for *position* control.

a familiar kinematics example. Controllability based on differential geometry is treated in Section 3, and illustrated on a simple example and on a general rigid robot. The conclusions are given in Section 4.

## 2. Kinematics

The kinematics of a robot is defined as the Cartesian representation, $x \in R^n$ as a function of the joint angles, $q \in R^m$,

$$x = F(q) \tag{1}$$

The derivative of the kinematics, the Jacobian, is denoted

$$J(q) = \frac{dF(q)}{dq} \tag{2}$$

The Jacobian is thus the first order term in a Taylor expansion of the kinematics $F(q)$. The higher order terms usually do not vanish. The Jacobian also gives the relation between velocities in Cartesian space, $\dot{x}$, and joint rates, $\dot{q}$,

$$\dot{x} = J(q)\dot{q} \tag{3}$$

The singularities of the robot are defined as those configurations, $q$, for which $J(q)$ loses rank. The purpose of this paper is to further investigate the kinematic and dynamic properties of the robot close to singularities. We will later be interested in control possibilities, but we will start with a closer look at a familiar kinematics example (Asada and Slotine, 1986; Craig, 1989; Spong and Vidyasagar, 1989).

### A Familiar Kinematics Example

Consider the planar two-link, $m = n = 2$, manipulator in Figure 1 a. Without loss of generality one of the link lengths is normalized to 1. The kinematics is

$$\begin{cases} x_1 = \cos q_1 + \ell \cos(q_1 + q_2) \\ x_2 = \sin q_1 + \ell \sin(q_1 + q_2) \end{cases} \tag{4}$$

leading to the Jacobian

$$J(q) = \begin{pmatrix} -\sin q_1 - \ell\sin(q_1 + q_2) & -\ell\sin(q_1 + q_2) \\ \cos q_1 + \ell\cos(q_1 + q_2) & \ell\cos(q_1 + q_2) \end{pmatrix} \tag{5}$$

The Jacobian loses rank if $\det J(q) = -\ell \sin q_2 = 0 \Leftrightarrow q_2 = n\pi$. The singular configurations are thus those where the arm is fully stretched or completely folded back.

The inverse kinematics, i.e. $q$ as function of $z$, is derived using the notational conventions in (Craig, 1989). Start e.g. from the Cosine-theorem to first yield $q_2$ as $\cos q_2$ and $\sin q_2$ from $\cos q_2 = \frac{1}{2\ell}(z_1^2 + z_2^2 - 1 - \ell^2) = c(z)$ and $\sin q_2 = \pm\sqrt{1 - c(z)^2} = s(z)$. The final result is

$$q_1 = \text{Atan2}(z_2, z_1) - \text{Atan2}(\ell s(z), 1 + \ell c(z))$$

$$q_2 = \text{Atan2}(s(z), c(z)) \tag{6}$$

Consider the robot in the singular configuration where the robot is fully stretched along the $z_1$-axis, as seen in Figure 1 b. We will now study motions inward along the $z_1$-axis from this singular configuration. Such a motion is obtained if $z_2 = 0$ i.e. if the joint angles are related as:

$$\sin q_1 + \ell \sin(q_1 + q_2) = 0 \tag{7}$$

*An approximate analysis*  Before developing an explicit expressions for a motion along the $z_1$-axis, an approximate analysis based only on the leading terms in a Taylor expansion of the kinetics (4) is enlightening. The analysis clearly indicates how different time functions i.e. different trajectories, $(z_1(t), z_2(t))$ may describe possible motions.

Close to the singularity $(z_1, z_2) = (1 + \ell, 0)$ the angles $(q_1, q_2)$ are small. A Taylor expansion of (4) up to order two gives

$$\begin{cases} z_1 = 1 - q_1^2/2 + \ell\left[1 - (q_1 + q_2)^2/2\right] + O(q^4) \\ z_2 = q_1 + \ell(q_1 + q_2) + O(q^3) \end{cases} \tag{8}$$

Now requiring motion along the $z_1$-axis and neglecting higher order terms gives the constraint

$$q_1 = -\frac{\ell}{1 + \ell} q_2 \tag{9}$$

and

$$\begin{cases} z_1 = 1 + \ell - \frac{\ell}{1 + \ell} q_2^2/2 + O(q_2^4) \\ z_2 = 0 + O(q_2^3) \end{cases} \tag{10}$$

We thus see that it is possible to move approximately along the $z_1$-axis close to the singularity, since we have $\frac{\Delta z_2}{\Delta z_1} \to 0$, when $q_2 \to 0$.

The analysis so far has been without regard to velocities. Assume now motions along the $z_1$-axis away from the singularity

$$z_1 = 1 + \ell - p(t) \tag{11}$$

where $p(0) = 0$ and $p(t) \geq 0$. Note that

$$p(t) = \frac{\ell}{1 + \ell} q_2^2/2 + O(q_2^4) \tag{12}$$

We see that linear motion in $z$, i.e. $p(t) = t$ implies $q_2 \sim \sqrt{t} \Rightarrow \dot{q}_2 \sim \frac{1}{\sqrt{t}} \to \infty$ as $t \to 0$. However, quadratic motion, $p(t) = t^2$ implies $q_2 \sim t \Rightarrow \dot{q}_2 \sim 1$. The conclusion for this particular example is that constant speed, $\dot{z}_1$, leads to infinite joint rates, but that there exist softer starts that make the motion possible. To be more formal on the latter statement: The kinematics (4) and the motion

$$\begin{cases} q_1 = -\dfrac{\ell}{1+\ell} \cdot t \\ q_2 = t \end{cases} \tag{13}$$

gives $\frac{\Delta x_2}{\Delta x_1} \to 0$ as $t \to 0$, and all derivatives are bounded.

***Path following*** The approximate analysis gives a motion that locally starts out along the $x_1$-axis. We will now give a motion that really follows a path with $x_2 = 0$. We will use the inverse kinematics, eq. (6), with $x_2 = 0$. To give such an example, it is enough to study $q_i \in (-\pi/2, \pi/2)$ and hence to use arctan instead of Atan2. Introduce the time dependence $t = 2 \tan q_2/2$, which yields $\sin q_2(t) = \frac{t}{1+t^2/4}$ and $\cos q_2(t) = \frac{1-t^2/4}{1+t^2/4}$. Let the motion be such that it is defined over a time-interval and so that the singularity $t = 0$ is included, i.e. use e.g. $t \in [0, 1]$. Introduce this into the inverse kinematics, eq. (6), to obtain

$$q_1(t) = -\arctan \frac{\ell t}{1 + \ell + (1 - \ell)t^2/4}$$

$$q_2(t) = \arctan \frac{t}{1 - t^2/4} \tag{14}$$

Note that (14), for small $t$, simplifies to (13). It is straightforward to verify that (14) has all derivatives bounded and hence is a possible motion along the $x_1$-axis.

## Summing Up

The two-link robot is a standard object of study, and it is usually claimed that the only motions possible in the singular configuration are those perpendicular to the arm (Asada and Slotine, 1986, pp. 65-66; Craig, 1989, pp. 173-174; Spong and Vidyasagar, 1989, pp. 25-26). The motion (14) is a counterexample to this statement. The interpretation of a singularity for control purposes is thus nontrivial. The kinematics is a non-linear function and the fact that the first order term, the Jacobian, loses rank means that degrees of freedom are lost only in some sense. Usually there are higher order terms that determine the behavior. In particular, in the specific example treated here, we find it reasonable to talk about two degrees of freedom for position control, although one of these degrees of freedom cannot be extended outside the reachable space of the robot and has restrictions on the shape of the velocity profile.

## 3.    Controllability: The Differential Geometry Approach

We will now investigate the kinematic singularities with respect to robot controllability as defined by the differential geometry approach. Consider the following nonlinear system

$$\dot{x} = f(x) + \sum_{i=1}^{m} g_i(x)u_i \tag{15}$$

where $f(x)$, $g_i(x) \in \mathbb{R}^n$ are smooth functions belonging to $C^k$, i.e. with continuous partial derivatives of order $k$. The vector $x \in \mathbb{R}^n$ describes the state vector and $u_i \in \mathbb{R}$ are the system inputs. According to the concept of nonlinear controllability as defined by Herman

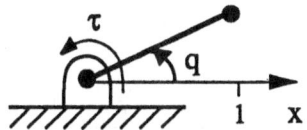

**Figure 2.** A second example. The variable $x$ is to be controlled.

and Krener (1977), the system (15) is said to be locally controllable if the vector fields $g_i, f$ and their consecutive Lie brackets $[f, g_i], [f, [f, g_i]], \ldots$, span the whole state space. The condition for a system of the form (15) to be controllable is thus:

$$\mathrm{rank}\Delta_c(x) = \mathrm{rank}\,(.., g_i, .., [f, g_i], .., [f, [f, g_i]], \ldots) = n \tag{16}$$

*A Second Example* These concepts will be studied first using a second simple example. Consider the system shown by Figure 2, where $q$ is the controlled joint angle and $\tau$ is the applied torque. The dynamic of the robot is simply described as

$$\ddot{q} = \tau \tag{17}$$

and the mappings between $q \mapsto x$, $(q, \dot{q}) \mapsto \dot{x}$ and $(q, \dot{q}, \ddot{q}) \mapsto \ddot{x}$, are given as

$$x = \cos q \tag{18a}$$

$$\dot{x} = -\dot{q}\sin q \tag{18b}$$

$$\ddot{x} = -\ddot{q}\sin q - \dot{q}^2 \cos q \tag{18c}$$

Since (18a) does not provide a bijective transformation between the joint coordinates and the Cartesian coordinates we constrain the joint motions to the set $S_q = \{q \in \mathbb{R} : 0 \le q \le \pi/2\}$ for which the transformation (18a) and its inverse are unique. Note that a kinematic singularity is included in $S_q$. The Jacobian, $\sin q$, is zero at $q = 0$, which corresponds to the position $x = 1$. The dynamics expressed in $x$ can now be obtained by combining (17) with (18a-c). This gives

$$\ddot{x} = -\frac{\dot{x}^2}{\sin^2 q}\cos q - (\sin q)\tau = \frac{-\dot{x}^2 x}{1 - x^2} - \sqrt{1 - x^2}\,\tau \tag{19}$$

Introducing the following state-space variables $x_1 = x$ and $x_2 = \dot{x}$, the system (19) can be rewritten as

$$\dot{x} = f(x) + g(x)u \tag{20}$$

with $\tau = u$ and $f(x)$ and $g(x)$ defined as follows:

$$f(x) = \begin{pmatrix} x_2 \\ -\dfrac{x_2^2 x_1}{1 - x_1^2} \end{pmatrix} = \begin{pmatrix} f_1 \\ f_2 \end{pmatrix} \qquad g(x) = \begin{pmatrix} 0 \\ -\sqrt{1 - x_1^2} \end{pmatrix} = \begin{pmatrix} 0 \\ g_2 \end{pmatrix} \tag{21}$$

Consider first position displacements excluding singular points, that is in $S'_{x_1} = \{x_1 \in \mathbb{R} : 0 \le x_1 < 1\}$ and define a corresponding space for the full state vector $(x_1, x_2)$ as $\Omega'_x = S'_{x_1} \times \mathbb{R}$. Vector fields $f(x)$ and $g(x)$ are thus analytic functions in $\Omega'_x$ and hence the system (20) becomes a member of the system class described by Equation (15). The controllability test can thus be performed in $\Omega'_x$ by inspecting condition (16). Computing the bracket $[f, g]$ gives:

$$[f,g] = \frac{\partial g}{\partial x} \cdot f - \frac{\partial f}{\partial x} \cdot g = \begin{pmatrix} -g_2 \\ \frac{\partial g_2}{\partial x_1} f_1 - \frac{\partial f_2}{\partial x_2} g_2 \end{pmatrix} = \begin{pmatrix} \sqrt{1-x_1^2} \\ -\frac{x_1 x_2}{\sqrt{1-x_1^2}} \end{pmatrix} \tag{22}$$

Subsequent brackets evaluate to zero for all $x$. Thus the condition for the matrix $\Delta_c(x)$ to be full rank is:

$$\det \begin{pmatrix} 0 & -g_2 \\ g_2 & \left(\frac{\partial g_2}{\partial x_1} f_1 - \frac{\partial f_2}{\partial x_2} g_2\right) \end{pmatrix} = g_2^2 = 1 - x_1^2 \neq 0$$

and the system (20) is thus controllable in $\Omega'_x$. This analysis shows that the controllability properties of the robot model in the joint space are preserved through the transformation (18) for motions in the set $\Omega'_x$ within which transformation (18 a-b) defines a diffeomorphism. Note that in a larger set including the singularity $(x_1, x_2) = (1,0)$, i.e., $\Omega_x = \Omega'_x \cup (1,0)^T$, the mapping (18 a-b) no more defines a diffeomorphism.

Intuition developed in the previous section by analyzing the kinematic relations suggests that motion might be possible to and from the singularity provided that certain constraints are imposed on the way that motion is performed. The following trajectory in the $x$-space, defined for $t \in [0, \sqrt{\pi/2\alpha}]$, confirms the result obtained from the kinematic relationships of the first example

$$x_1 = \cos \alpha t^2, \quad x_2 = -2\alpha t \sin \alpha t^2, \quad \dot{x}_2 = -4\alpha^2 t^2 \cos x t^2 - 2\alpha \sin x t^2 \tag{23}$$

which is one solution of the system (20) with the constant input torque $\tau(t) = 2\alpha$ and with the initial condition, $x(0) = (1,0)^T$. Actually (20) also has other solutions like $x(t) = x(0), \forall t$, also for $\tau \neq 0$, but they have no physical meaning. They are consequences of the lack of differential bijectivity of the relation (18) at the considered initial point. Note that the chosen torque yields "square motion" in the joint space, i.e., $q(t) = \alpha t^2$, just as expected from (17). The single link manipulator will thus move from $x_1(0) = 1$ to $x_1(\sqrt{\pi/2\alpha}) = 0$ with finite torque $\tau = 2\alpha$. This simple example thus also shows that motion with finite torque is possible starting in a singular configuration. The trajectory (23) also implies boundedness of the vector fields $f$ and $g$, $\forall t \in [0, \sqrt{\pi/2\alpha}]$, and in particular also at the singularity, i.e. by inserting the explicit time dependency (23) into (21) and (22) and thereafter taking the limit we get $\lim_{t \to 0} f(x(t)) = \lim_{t \to 0} g(x(t)) = \lim_{t \to 0} [f,g] = (0,0)^T$. The rank condition (16) can thus be evaluated to give $\text{rank}(g, [f,g], \cdots) = 0$. In conclusion, the controllability criterion (16) loses rank in the limit despite the obvious controllability of (17).

## General Robot Controllability Analysis

Consider the usual rigid body robot model, $H(q)\ddot{q} + C(q,\dot{q})\dot{q} + \tau_g(q) = \tau$, where $H(q)$ is the $n \times n$ inertia matrix, $C(q,\dot{q})\dot{q}$ represents are the Coriolis and centripetal forces, and where $\tau_g(q)$ and $\tau$ are the gravity vector and the applied forces respectively. The mapping between the joint coordinates $q \in \mathbb{R}^n$ and the work coordinates $x \in \mathbb{R}^n$ is given by (1), with first order derivative given by (3), and with second order derivative $\ddot{x} = J(q)\ddot{q} + \dot{J}(q)\dot{q}$. To rewrite the dynamic equations in terms of the work space coordinates we restrict our analysis to the set $S'_q$ and hence to the set $S'_x$ for which there exists a one-to-one mapping between $q$ and $x$. Thus the Jacobian can be rewritten in terms of $x$. Introduce the state vectors $x_1 = x$

and $x_2 = \dot{x}$, and the notation $\bar{J}(x) = J(F^{-1}(x_1))$, $\bar{H}(x) = H(F^{-1}(x_1))$, $\bar{\tau}_g(x) = \tau_g(F^{-1}(x_1))$, and define $\bar{C}$ so that $\bar{C}(x_1, x_2)x_2 = C(q_1, q_2)q_2$ for corresponding values of $q$ and $x$. We get the state space representation

$$\dot{x} = f(x) + \sum_{i=1}^{n} \bar{g}_i(x)u_i$$

with $f(x), \bar{g}_i(x) \in \mathbb{R}^n$, $\tau_i = u_i$, and with

$$f(x) = \begin{pmatrix} x_2 \\ -\bar{J}\bar{H}^{-1}[\bar{C}x_2 + \bar{\tau}_g] + \dot{\bar{J}}\bar{J}^{-1}x_2 \end{pmatrix} = \begin{pmatrix} f_1 \\ f_2 \end{pmatrix}$$

$$\bar{g}_i(x) = \begin{pmatrix} 0 \\ g_i(x_1) \end{pmatrix}$$

where $g_i(x_1)$ stands for the $i^{th}$ column vector of the matrix $\bar{G}(x) = \bar{J}\bar{H}^{-1}$. The Lie bracket $[f, \bar{g}_i]$ gives:

$$[f, \bar{g}_i] = \begin{pmatrix} 0 & 0 \\ \frac{\partial g_i}{\partial x_1} & 0 \end{pmatrix} \begin{pmatrix} f_1 \\ f_2 \end{pmatrix} - \begin{pmatrix} 0 & I \\ \frac{\partial f_2}{\partial x_1} & \frac{\partial f_2}{\partial x_2} \end{pmatrix} \begin{pmatrix} 0 \\ g_i \end{pmatrix} = \begin{pmatrix} g_i \\ \frac{\partial g_i}{\partial x_1}f_1 - \frac{\partial f_2}{\partial x_2}g_i \end{pmatrix} = \begin{pmatrix} g_i \\ z_i \end{pmatrix}$$

Now evaluate the rank of $\Delta_c$ defined as in (16) which according to the above calculations becomes:

$$\Delta_c = \begin{pmatrix} 0 & \bar{G} & \cdots \\ \bar{G} & Z & \cdots \end{pmatrix}$$

where $Z$ is the $n \times n$ matrix defined as $Z = [z_1, z_2 \ldots z_n]$. The rank of $\Delta_c$ can be proved to be full outside the kinematic singularities by inspecting

$$\det \begin{pmatrix} 0 & -\bar{G} \\ \bar{G} & Z \end{pmatrix} = \det(\bar{G}\bar{G}) = \det(\bar{J}\bar{H}^{-1}\bar{J}\bar{H}^{-1}) = \det(\bar{J})^2 \cdot \det(\bar{H}^{-1})^2$$

The rank controllability condition is thus satisfied outside the Jacobian singularities. This analysis also covers the case of the two-link robot in Section 2.

## 4. Conclusions

We have presented examples of trajectories that contradict well-known statements in the robotics literature concerning motion in certain directions out from Jacobian singularities. The question at this stage is if the system is controllable in the Cartesian task-space at the kinematic singularities. The answer depends on what definition we take for controllability. In the Lie-bracket sense the answer is *no*, when going to the limit using a feasible trajectory. However, recalling the original definition of controllability in linear systems: "To go from one state to another in arbitrary finite time with finite input", see e.g. (Brockett, 1970), the answer would obviously be *yes* for $\Omega_x$, since it is controllable in joint space, $\Omega_q$, and there is always a $q, \dot{q} \in \Omega_q$ that gives any desired $x, \dot{x} \in \Omega_x$.

Our closer look at properties of singularities could open up interesting possibilities for controller design. One may think of preshaping of velocity profiles, or of closed loop velocity filters activated only in directions where the Jacobian loses rank, or of other possibilities, instead of just giving up and saying that control in certain directions are impossible at a singularity. Another observation is that it is possible to obtain a bijection between the position coordinates $x$ and $q$, but not between their respective velocities. This, together with the structure of the robot dynamics, suggests that it is possible to follow a geometric path defined in Cartesian space. These issues are of immediate and significant relevance in path programming and path following, where a typical problem is to find and follow a time-optimal trajectory. The ability to follow a path, like e.g. in (14), then assures the existence of solutions and indicates control possibilities.

# 5.  References

ASADA, H., and J.-J. E. SLOTINE (1986): *Robot Analysis and Control*, John Wiley & Sons, Inc., USA.

BROCKETT, R.W. (1970): *Finite Dimensional Linear Systems*, John Wiley & Sons, Inc., USA.

CRAIG, J. J. (1989): *Introduction to Robotics Mechanics and Control*, Second Edition, Addison-Wesley Publishing Company, Inc., USA.

HERMANN, R., and A. J. KRENER (1977): "Nonlinear controllability and observability," *IEEE Transactions of Automatic Control*, **AC–22**, 728–740.

ISIDORI, A. (1989): *Nonlinear Control Systems*, Springer Verlag Editions.

NAKAMURA, Y., and H. HANAFUSA (1986): "Inverse kinematic solutions with singularity robustness for robot manipulator control," *Journal of Dynamic Systems, Measurement, and Control*, **108**, 163–171.

SPONG, M. W., and M. VIDYASAGAR (1989): *Robot Dynamics and Control*, John Wiley & Sons, Inc., USA.

WAMPLER, C. W. II (1986): "Manipulator inverse kinematic solutions based on vector formulations and damped least-squares methods," *IEEE Transactions on Systems, Man, and Cybernetics*, **SMC–16**, 93–101.

# Lecture Notes in Control and Information Sciences

Edited by M. Thoma and A. Wyner

# Lecture Notes in Control and Information Sciences

Edited by M. Thoma and A. Wyner

# Lecture Notes in Control and Information Sciences

Edited by M. Thoma and A. Wyner